Lecture Notes in Mathematics 1591

Editors:
A. Dold, Heidelberg
B. Eckmann, Zürich
F. Takens, Groningen

Subseries: Scuola Normale Superiore, Pisa

Advisor: E. Vesentini

T0222899

Marco Abate Giorgio Patrizio

Finsler Metrics – A Global Approach

with applications
to geometric function theory

Springer-Verlag
Berlin Heidelberg New York
London Paris Tokyo
Hong Kong Barcelona
Budapest

Authors

Marco Abate
Dipartimento di Matematica
Università di Ancona
Via Brecce Bianche
60131 Ancona, Italy

Giorgio Patrizio
Dipartimento di Matematica
Università di Roma Tor Vergata
00133 Roma, Italy

Mathematics Subject Classification (1991): 32H15, 53C60

ISBN 3-540-58465-X Springer-Verlag Berlin Heidelberg New York

CIP-Data applied for

This work is subject to copyright. All rights are reserved, whether the whole or part of the material is concerned, specifically the rights of translation, reprinting, re-use of illustrations, recitation, broadcasting, reproduction on microfilms or in any other way, and storage in data banks. Duplication of this publication or parts thereof is permitted only under the provisions of the German Copyright Law of September 9, 1965, in its current version, and permission for use must always be obtained from Springer-Verlag. Violations are liable for prosecution under the German Copyright Law.

© Springer-Verlag Berlin Heidelberg 1994
Printed in Germany

Typesetting: Camera ready by author
SPIN: 10130166 46/3140-543210 - Printed on acid-free paper

Contents

Preface

When, some years ago, we started working on a differential geometric study of the structure of strongly convex domains in $\mathbb{C}^n$, we did not expect to end up writing a book on global Finsler geometry. But, along the way, we found ourselves needing several basic results on real and complex Finsler metrics that we were unable to find in the literature (or at least not in the form necessary to us). So we felt compelled to provide proofs — and this is the final result of our work.

Our exposition is very much in the vein of the work of Cartan [C], Chern [Ch1, 2], Bao-Chern [BC] and Kobayashi [K]; in particular the latter gave us the preliminary idea for our approach to smooth complex Finsler metrics. We would also like to say that we would have been very happy to know earlier of [BC], which, although only marginally related to our work, would have been of great help in solving questions which we treated independently.

Our starting point was the study of the existence and global behavior of complex geodesics for intrinsic metrics in complex manifolds. Our goal was to look at this question from a differential geometric point of view, with the hope of possibly reproducing in a wider class of complex manifolds what Lempert [L] was able to prove for strongly convex domains in $\mathbb{C}^n$. The idea was to treat complex geodesics through a point as images of disks through the origin in the tangent space at the point via the exponential map of a complex Finsler metric; thus we were led to study the local and global theory of geodesics of a Finsler metric. As in Hermitian (and Riemannian) geometry, the local theory of geodesics means the study of the first variation of the length integral, and of the associated Euler-Lagrange equation. The global theory, on the other hand, involves the accurate control of the second variation and hence of the curvature, together with Jacobi fields, conjugate points, the Morse index form and the like. In particular, we needed a Finsler version of the Cartan-Hadamard theorem (originally proved by Auslander [Au1, 2]), and a way to apply it in a complex situation.

The main difficulty at this point was that the problems we were interested in involved complex Finsler metrics, and whereas there is a clear understanding of the relationship between the complex geometry and the underlying real geometry of a Hermitian manifold, nothing of the kind was available to us in Finsler geometry. We then started following the tradition of "linearizing" the questions by passing from the study of Finsler metrics on the tangent bundle (real or complex) to the study of the associated Hermitian structure on the tangent bundle of the tangent bundle. At this level it is also possible to describe the correct relationship between the complex and the corresponding real structure of objects like connections and curvatures.

But our approach is different from the traditional one for two main reasons. First of all, we everywhere stress global objects and global definitions (in fact, we are interested in global results), using local coordinates almost uniquely as a computational tool (in a way not too far from the first chapter of Bejancu [B]). But the main difference is another one. Possibly because of our motivations, working in this area we discovered that there might be a danger of carrying out the linearization program previously described too far. In fact, the formal setting naturally leads to very general definitions which make proofs of theorems easier, but do not give much geometrical insight: we had the feeling that working only at the tangent-tangent level was too restrictive, too formal, too far away from the actual geometry of the manifold. For this reason, our point of view now is to stick to notions which really provide informations about the geometry of geodesics on the manifold, and about the curvature of the manifold. This approach leads to "minimal" definitions, which are probably more complicated to state and surely more difficult to handle, but nevertheless more effective and really conveying the geometry of the manifold. For instance, there are many ways of generalizing the notion of Kählerianity to Finsler metrics, but not all of them have non-trivial examples and applications. We shall show how the notions we singled out can be effectively used by illustrating their applications in complex geometric function theory.

The first two chapters of this book are devoted to the exposition of our approach to real and complex Finsler geometry. In the first chapter, after setting the stage introducing the necessary general definitions and objects, we define in a global way the classical Cartan connection, and we discuss the variation formulas of the length integral, the exponential map, Jacobi fields, conjugate points and the Morse index form up to provide a proof of the Cartan-Hadamard and Bonnet theorems for Finsler metrics suitable for our needs in complex geometry. In the exposition we stress the similarities with the standard Riemannian treatment of the subject, as naturally suggested by our global approach.

In the second chapter we study the geometry of complex Finsler metrics. After having adapted the general definitions of chapter 1 to the complex setting, we introduce (following Kobayashi [K]) the Chern-Finsler connection, which is our main tool. We discuss at some length several Kähler conditions, and we introduce the notion of holomorphic curvature of a complex Finsler manifold, showing the equivalence of the differential geometric definition with a variational definition previously used in function theory.

Finally the third chapter contains the results and applications that motivated our work. From a differential geometric point of view, it is devoted to the study of the function theory on Kähler Finsler manifolds with constant nonpositive holomorphic curvature; from a complex analysis point of view, it is devoted to the study of manifolds where there is a Monge-Ampère foliation with exactly the same properties as the one discovered by Lempert in strongly convex domains. In particular we prove the existence of nice foliations and strictly plurisubharmonic exhaustions satisfying the Monge-Ampère equation on Kähler Finsler manifolds with constant nonpositive holomorphic curvature. Furthermore we prove that the only complex manifold admitting such a metric with zero holomorphic curvature is $\mathbb{C}^n$, and we describe a characterization of strongly convex circular domains in terms of differential geometric properties of the Kobayashi metric.

Of course, this book is not intended as a definitive treatise on the subject; on the contrary, it is just the description of an approach to Finsler metrics that we found reasonable and fruitful, but still leaving a lot of open problems. Just to mention a couple of them: the comparison between the complex Finsler geometry and the underlying real one carried out in section 2.6 seems to suggest that the Cartan connection contains terms which have no direct influence on the geometry of the manifold — and so maybe it is not the correct connection to use even in real Finsler geometry. Or: in the third chapter we give a fairly complete description of the complex structure of Kähler Finsler manifolds of constant nonpositive holomorphic curvature, which is satisfying from a geometric function theory point of view, but it still leaves completely open the problem of classifying the metrics with these properties (we remark that there are many more such manifolds and metrics than in the Hermitian case: there are at least all the strongly convex domains in $\mathbb{C}^n$ endowed with the Kobayashi metric, thanks to Lempert's work [L]) — and in fact it is even still far from being completed the classification of simply connected real Finsler manifolds with constant (horizontal flag) real curvature. Or: it follows from chapter 3 that the only part of Lempert's results actually depending on the strong convexity of the domain is the smoothness of the Kobayashi metric. It would be then interesting to construct directly a smooth weakly Kähler Finsler metric of constant holomorphic curvature -4 on any strongly convex domain; then this metric would automatically be the Kobayashi metric of the domain, and we would have recovered the full extent of Lempert's work.

So we hope that the possibly new perspectives on Finsler geometry introduced in this book will eventually lead to new results in this field; and in particular in geometric function theory of complex Finsler manifolds, where all this work started.

CHAPTER 1

Real Finsler geometry

1.0. Introduction

As already discussed in the preface, this book is mainly devoted to the study of complex Finsler geometry; but of course such a study cannot leave out of consideration real Finsler manifolds. So this first chapter is devoted to a discussion of real Finsler geometry, starting from the very basics and ending with a proof of the appropriate versions of the Cartan-Hadamard and Bonnet theorems for Finsler manifolds, obtained using global Riemannian-style techniques.

Let M be a real manifold endowed with a Finsler metric, that is with a positively homogeneous function $F: TM \to \mathbb{R}^+$ smooth outside the zero section of TM and strongly convex on each tangent space. Roughly speaking, our main idea is to replace the given Finsler metric on TM by a Riemannian metric on a suitable sub-bundle of $T(TM)$ — in a certain sense we linearize the Finsler metric going one step upstairs — and then use the standard tools of Riemannian geometry there. A canonically defined isometric embedding of TM (outside the zero section, actually) into this bundle will then allow us to transfer information back and forth, thus giving geometrical results about the original manifold. For instance, we shall be able to recover for Finsler manifolds more or less all the results described in the first chapter of [CE] for Riemannian manifolds. We also refer to [C], [Rd1], [M], [Ch1] and [B] for a description of the standard theory of real Finsler metrics, and to [Ch2] and [BC] for a recent approach akin in spirit to ours.

To be more precise, let M be a manifold, and let $\pi: TM \to M$ denote the tangent bundle of M; $\tilde{M}$ will stand for $TM \setminus \{\text{zero section}\}$. The *vertical bundle* is $\mathcal{V} = \ker d\pi$, a sub-bundle of $T\tilde{M}$. Take a Finsler metric $F: TM \to \mathbb{R}^+$ on M, and set $G = F^2$. Then it is easy to see (section 1.4) that using the Hessian of G it is possible to define a Riemannian metric on $\mathcal{V}$ in such a way that a canonically defined section $\iota: \tilde{M} \to \mathcal{V}$ of $\mathcal{V}$ (see section 1.1) turns out to be an isometric embedding of $\tilde{M}$ into $\mathcal{V}$.

But this is not yet the setting mentioned before. The point is that to such a Riemannian metric on the vertical bundle it is possible to associate two objects: a linear connection D on $\mathcal{V}$ with respect to which the given Riemannian metric is parallel; and a horizontal bundle, that is a sub-bundle $\mathcal{H}$ of $T\tilde{M}$ such that $T\tilde{M} = \mathcal{H} \oplus \mathcal{V}$. The general theory of horizontal bundles yields a bundle isomorphism $\Theta: \mathcal{V} \to \mathcal{H}$; using Θ we can define a Riemannian metric and a connection on $\mathcal{H}$ — and hence on $T\tilde{M}$. This is the *Cartan connection*, the exact analogue in the Finsler setting of the Levi-Civita connection (e.g., its torsion is almost zero — in a very definite sense; furthermore, the torsion is identically zero if and only if the Finsler metric actually is Riemannian, and in this case the Cartan connection coincides with the Levi-Civita connection). The bundle $\mathcal{H}$ with this structure (and with its section $\chi = \Theta \circ \iota$) provides the aforementioned setting, where one can use Riemannian tools to get Finsler statements.

The main examples of this assertion are provided by the first and second variations of the length integral, derived in section 1.5; we get formulas formally identical to the Riemannian ones, just replacing the curvature of the Levi-Civita connection by (a suitable contraction of the horizontal part of) the curvature of the Cartan connection. Then we shall be able to recover the Hopf-Rinow theorem for Finsler manifolds (this is not too surprising, since it holds in much more general settings; see [Ri]) and the theory of Jacobi fields and of the Morse index form, in a way exactly parallel to the one presented in standard Riemannian geometry texts. In particular, in section 1.7 we shall be able to prove the generalizations (originally due to Auslander [Au1, 2]) to Finsler manifolds of the classical Cartan-Hadamard and Bonnet theorems.

In detail, the content of this chapter is the following. In section 1.1 we discuss at some length the general theory of horizontal bundles, horizontal maps (i.e., maps like our Θ above) and non-linear connections on M. In section 1.2 we introduce the concept of vertical connection (i.e., of linear connection on the vertical bundle), and we show how to associate to certain vertical connections (we call them the *good* ones) a horizontal bundle, and hence a non-linear connection on M and a linear connection on $\tilde{M}$. In section 1.3 we define and discuss the torsion and the curvature of a good vertical connection. In section 1.4 we define Finsler metrics, and we show that to any Finsler metric F is canonically associated a good vertical connection, the Cartan connection mentioned before. Section 1.5 is devoted to prove the first and second variation of the length integral for Finsler metrics; section 1.6 to parallel transport, geodesics, the exponential map and the Hopf-Rinow theorem for Finsler metrics. Finally in section 1.7 we shall define Jacobi fields and the Morse index form in this setting, and we shall prove the Finsler versions of the Cartan-Hadamard and Bonnet theorems.

1.1. Non-linear connections

1.1.1. Preliminaries

In this subsection we fix our notations and collect a few formulas concerning change of coordinates. We choose symbols and notations so to be compatible with the complex case we shall discuss in chapters 2 and 3; this is the reason behind some apparently slightly unusual choices (u instead of v to denote tangent vectors, and the like).

Let M be a real manifold of dimension m; we shall denote by TM its tangent bundle, and by $\pi: TM \to M$ the canonical projection, as usual. The cotangent bundle will be denoted by T^*M.

If $(x^1, \ldots, x^m)$ are local coordinates on M about a point $p_0 \in M$, a vector $u \in T_pM$ (with p close to p_0) is represented by

$$u = u^a \left.\frac{\partial}{\partial x^a}\right|_p,$$

where we are using the Einstein convention, and lowercase roman letters run from 1 to m. In particular, local coordinates on TM are given by $(x^1, \ldots, x^m, u^1, \ldots, u^m)$, and so a local frame of $T(TM)$ is given by $\{\partial_1, \ldots, \partial_m, \dot{\partial}_1, \ldots, \dot{\partial}_m\}$, where

$$\partial_a = \frac{\partial}{\partial x^a} \qquad \text{and} \qquad \dot{\partial}_b = \frac{\partial}{\partial u^b}.$$

We shall denote by $o: M \to TM$ the zero section of TM, that is $o(p) = o_p$ is the origin of T_pM, and we set $\tilde{M} = TM \setminus o(M)$, the tangent bundle minus the zero section. $\tilde{M}$ is naturally equipped with the projection $\pi: \tilde{M} \to M$, the restriction of the canonical projection of TM. Correspondingly, $T\tilde{M} \subset T(TM)$ comes equipped with a natural projection $\tilde{\pi}: T\tilde{M} \to \tilde{M}$, the restriction of the natural projection $\tilde{\pi}: T(TM) \to TM$.

We shall use uppercase roman letters to denote different coordinate patches. A coordinate patch (U_A, φ_A) in M determines a coordinate patch $(\tilde{U}_A, \tilde{\varphi}_A)$ in TM (and $\tilde{M}$) setting $\tilde{U}_A = \pi^{-1}(U_A)$ and

$$\forall u \in \tilde{U}_A \qquad\qquad \tilde{\varphi}_A(u) = d\varphi_A(u).$$

If $\varphi_A = (x_A^1, \ldots, x_A^m)$, then $\{(\partial/\partial x_A^j)|_p\}$ is a basis of T_pM for any $p \in U_A$. Writing $u = u_A^a(\partial/\partial x_A^a)$, then

$$\tilde{\varphi}_A(u) = (x_A^1, \ldots, x_A^m, u_A^1, \ldots, u_A^m).$$

On $U_A \cap U_B$ we have

$$dx_B^i = \frac{\partial x_B^i}{\partial x_A^j}\, dx_A^j, \qquad \frac{\partial}{\partial x_B^i} = \frac{\partial x_A^j}{\partial x_B^i} \frac{\partial}{\partial x_A^j},$$

where $\partial x^i_B/\partial x^j_A = \partial(\varphi_B \circ \varphi_A^{-1})^i/\partial x^j_A$. By the way, we set

$$\mathcal{J}_{BA} = \left(\frac{\partial x^i_B}{\partial x^j_A}\right);$$

clearly,

$$\mathcal{J}_{AB} = \mathcal{J}_{BA}^{-1} \circ (\varphi_A \circ \varphi_B^{-1}).$$

Taking $u \in \widetilde{U}_A \cap \widetilde{U}_B$ and expressing it in local coordinates, we find

$$u^i_B = dx^i_B(u) = \frac{\partial x^i_B}{\partial x^j_A}\, dx^j_A\left(u^k_A \frac{\partial}{\partial x^k_A}\right) = \frac{\partial x^i_B}{\partial x^j_A}\, u^j_A,$$

that is

$$u_B = \mathcal{J}_{BA}\, u_A.$$

Therefore

$$(x_B, u_B) = \tilde{\varphi}_B \circ \tilde{\varphi}_A^{-1}(x_A, u_A) = (\varphi_B \circ \varphi_A^{-1}(x_A), \mathcal{J}_{BA} u_A). \tag{1.1.1}$$

Up to now everything was quite standard. But now something different: change of coordinates in $T(TM)$. Define $(\tilde{\tilde{U}}_A, \tilde{\tilde{\varphi}}_A)$ by setting

$$\tilde{\tilde{U}}_A = \tilde{\pi}^{-1}(\widetilde{U}_A) = (\pi \circ \tilde{\pi})^{-1}(U_A)$$

and $\tilde{\tilde{\varphi}}_A(X) = d\tilde{\varphi}_A(X)$ for any $X \in \tilde{\tilde{U}}_A$.

A vector $X \in T(TM)$ in local coordinates is expressed by

$$X_A = X^i_A(\partial_i)_A + \dot{X}^j_A(\dot{\partial}_j)_A = h(X_A) + v(X_A).$$

Taking derivatives of (1.1.1), we find the Jacobian matrix for $T(TM)$:

$$\tilde{\mathcal{J}}_{BA} = \left(\begin{array}{c|c} \dfrac{\partial x^i_B}{\partial x^j_A} & \dfrac{\partial x^i_B}{\partial u^k_A} \\ \hline \dfrac{\partial u^h_B}{\partial x^j_A} & \dfrac{\partial u^h_B}{\partial u^k_A} \end{array}\right) = \left(\begin{array}{c|c} (\mathcal{J}_{BA})^i_j & 0 \\ \hline \dfrac{\partial^2 x^h_B}{\partial x^j_A \partial x^l_A}\, u^l_A & (\mathcal{J}_{BA})^h_k \end{array}\right).$$

Setting

$$(H_{BA})^h_{kl} = \frac{\partial^2 x^h_B}{\partial x^k_A \partial x^l_A},$$

we find

$$\begin{aligned}(x_B, u_B, h(X_B), v(X_B)) &= \tilde{\tilde{\varphi}}_B \circ \tilde{\tilde{\varphi}}_A^{-1}(x_A, u_A, h(X_A), v(X_A)) \\ &= \Big(\varphi_B \circ \varphi_A^{-1}(x_A), \mathcal{J}_{BA}\, u_A, \mathcal{J}_{BA}\, h(X_A), (H_{BA})^{\bullet}_{kl}\, u^l_A\, h(X_A)^k + \mathcal{J}_{BA}\, v(X_A)\Big).\end{aligned} \tag{1.1.2}$$

Now $\{\partial_i, \dot\partial_j\}$ is a local frame for $T(TM)$; let $\{dx^i, du^j\}$ be the dual coframe (note that $dx^i|_v$ is not the same as $dx^i|_p$). First of all, (1.1.2) yields

$$\begin{aligned} dx^i_B &= \frac{\partial x^i_B}{\partial x^j_A}\, dx^j_A = (\mathcal{J}_{BA})^i_j\, dx^j_A \\ du^h_B &= \frac{\partial x^h_B}{\partial x^k_A}\, du^k_A + \frac{\partial^2 x^h_B}{\partial x^k_A \partial x^l_A}\, u^l_A\, dx^k_A = (\mathcal{J}_{BA})^h_k\, du^k_A + (H_{BA})^h_{kl}\, u^l_A\, dx^k_A. \end{aligned} \tag{1.1.3}$$

Recalling that $\{dx^i, du^j\}$ is the dual frame of $\{\partial_i, \dot\partial_j\}$, we get

$$\begin{aligned} (\partial_i)_B &= \frac{\partial x^j_A}{\partial x^i_B}\,(\partial_j)_A - \frac{\partial x^r_A}{\partial x^h_B}\,\frac{\partial^2 x^h_B}{\partial x^k_A \partial x^l_A}\, u^l_A\, \frac{\partial x^k_A}{\partial x^i_B}\,(\dot\partial_r)_A \\ &= (\mathcal{J}_{BA}^{-1})^j_i\,(\partial_j)_A - (\mathcal{J}_{BA}^{-1})^k_i (H_{BA})^h_{kl} (\mathcal{J}_{BA}^{-1})^r_h\, u^l_A\,(\dot\partial_r)_A, \\ (\dot\partial_h)_B &= \frac{\partial x^k_A}{\partial x^h_B}\,(\dot\partial_k)_A = (\mathcal{J}_{BA}^{-1})^k_h\,(\dot\partial_k)_A. \end{aligned} \tag{1.1.4}$$

1.1.2. Horizontal and vertical bundles

Now we may introduce our first main actor.

DEFINITION 1.1.1: The *vertical bundle* of a manifold M is the vector bundle $\tilde\pi\colon \mathcal{V} \to TM$ of rank $m = \dim M$ given by

$$\mathcal{V} = \ker d\pi \subset T(TM).$$

In local coordinates,

$$\varphi_A \circ \pi \circ \tilde\varphi_A^{-1}(x_A, u_A) = x_A,$$

and so

$$\begin{aligned} \tilde{\tilde\varphi}_A \circ d\pi \circ \tilde{\tilde\varphi}_A^{-1}(x_A, u_A, X_A) &= d\varphi_A \circ d\pi \circ (d\tilde\varphi_A)^{-1}(x_A, u_A, X_A) \\ &= d(\varphi_A \circ \pi \circ \tilde\varphi_A^{-1})(x_A, u_A, h(X_A), v(X_A)) \\ &= (x_A, h(X_A)). \end{aligned}$$

This means that $\{\dot\partial_h\}$ is a local frame for $\mathcal{V}$. We get charts restricting $\tilde{\tilde\varphi}_A$, and in particular (1.1.2) yields

$$(x_B, u_B, V_B) = \tilde{\tilde\varphi}_B \circ \tilde{\tilde\varphi}_A^{-1}(x_A, u_A, V_A) = \left(\varphi_B \circ \varphi_A^{-1}(x_A), \mathcal{J}_{BA}\, u_A, \mathcal{J}_{BA}\, V_A\right).$$

Let $j_p\colon T_pM \to TM$ be the inclusion and, for $u \in T_pM$, let $k_u\colon T_pM \to T_u(T_pM)$ denote the usual identification. Then we get a natural isomorphism

$$\iota_u = d(j_{\pi(u)})_u \circ k_u\colon T_{\pi(u)}M \to \mathcal{V}_u. \tag{1.1.5}$$

DEFINITION 1.1.2: The *radial vertical vector field* is the natural section $\iota: TM \to \mathcal{V}$ given by

$$\iota(u) = \iota_u(u);$$

clearly, $\iota(u) \in \mathcal{V}_u$.

In local coordinates,

$$\iota_u \left(\left. \frac{\partial}{\partial x^j} \right|_{\pi(u)} \right) = \dot{\partial}_j|_u;$$

in particular, if $u = u^a(\partial/\partial x^a)$ then

$$\iota(u) = \iota \left(u^a \frac{\partial}{\partial x^a} \right) = u^a \dot{\partial}_a|_u.$$

Note that the derivatives with respect to x (coordinates in M) become derivatives with respect to u (coordinates in TM).

The vertical bundle is canonically defined; this is not the case for a horizontal bundle. We may describe horizontal bundles using three different points of view, each with its own advantages and disadvantages. The first two are easily introduced:

DEFINITION 1.1.3: A *horizontal bundle* is a subbundle $\mathcal{H}$ of $T(TM)$ such that

$$T(TM) = \mathcal{H} \oplus \mathcal{V}.$$

DEFINITION 1.1.4: A *horizontal map* is a bundle map $\Theta: \mathcal{V} \to T(TM)$ such that

$$\forall u \in TM \qquad (d\pi \circ \Theta)_u = \iota_u^{-1}. \tag{1.1.6}$$

We defined horizontal bundles (and horizontal maps) on TM, but it turns out that they are interesting only over $\tilde{M}$. In fact, let $o: M \to TM$ denote the zero section. It is easy to check that $do_p(\partial/\partial x^j) = \partial_j$; therefore we have the natural splitting

$$T_{o_p}(TM) = \mathcal{H}_{o_p} \oplus \mathcal{V}_{o_p},$$

where $\mathcal{H}_{o_p} = do_p(T_pM)$. *We shall then assume* that all our horizontal bundles coincide with $do_p(T_pM)$ over the zero section, and, analogously, that all our horizontal maps satisfy

$$\Theta_{o_p}(\dot{\partial}_h|_{o_p}) = \partial_h|_{o_p}.$$

Clearly, this may cause problems with the smoothness at the origin. We shall henceforth assume that our horizontal bundles and maps will be *smooth over* $\tilde{M}$, but they may be not smooth over the zero section. The reasons behind this approach will become clear in section 1.4, when we shall define the concept of Finsler metric.

As mentioned before, there is a third approach to horizontal bundles, via the notion of non-linear connection. But to describe it we need a digression on linear connections.

If $p: E \to M$ is any bundle over M, we shall denote by $\mathcal{X}(E)$ the space of sections of E.

Definition 1.1.5: A *linear connection* on a manifold M is a $\mathbb{R}$-linear map

$$\tilde{D}: \mathcal{X}(TM) \to \mathcal{X}(T^*M \otimes TM)$$

satisfying the derivation property

$$\forall \xi \in \mathcal{X}(TM)\, \forall f \in C^\infty(M)\ \tilde{D}(f\xi) = df \otimes \xi + f\tilde{D}\xi. \tag{1.1.7}$$

As a consequence, $\tilde{D}\xi$ at a point $p \in M$ depends only on the value of ξ and $d\xi$ at the point p. Indeed, let $\xi' \in \mathcal{X}(TM)$ be another vector field with $\xi(p) = \xi'(p)$ and $d\xi_p = d\xi'_p$. Then $\xi' = \xi + f\eta$ for suitable $\eta \in \mathcal{X}(TM)$ and $f \in C^\infty(M)$ with $f(p) = 0$ and $df_p = 0$. So

$$\tilde{D}\xi' = \tilde{D}\xi + \tilde{D}(f\eta) = \tilde{D}\xi + df \otimes \eta + f\tilde{D}\eta,$$

and $\tilde{D}\xi'_p = \tilde{D}\xi_p$.

There is another way of expressing this. Let ξ, $\xi' \in \mathcal{X}(TM)$ be such that $\xi(p) = \xi'(p) = u$; then $\xi' = \xi + f\eta$ with $f(p) = 0$. In particular,

$$\tilde{D}\xi'_p = \tilde{D}\xi_p + df_p \otimes \eta(p).$$

Now, for any $v \in T_pM$, writing $v = v^a(\partial/\partial x^a)$ and $\xi = \xi^b(\partial/\partial x^b)$, one has

$$d\xi_p(v) = v^a \partial_a|_u + v^a \frac{\partial \xi^b}{\partial x^a}(p)\dot{\partial}_b|_u. \tag{1.1.8}$$

So $d\xi'_p - d\xi_p$ maps T_pM into $\mathcal{V}_u$; furthermore, (1.1.8) also yields

$$\forall v \in T_pM\ \ d\xi'_p(v) - d\xi_p(v) = v(f)\iota_u(\eta(p)) = \iota_u(v(f)\eta(p)),$$

and thus

$$\iota_u^{-1} \circ (d\xi'_p - d\xi_p) = df_p \otimes \eta(p). \tag{1.1.9}$$

Summing up, if $\xi(p) = \xi'(p) = u$ we get

$$\tilde{D}\xi'_p - \tilde{D}\xi_p = \iota_u^{-1} \circ (d\xi'_p - d\xi_p), \tag{1.1.10}$$

which we may consider as an intrinsic way of saying that $\tilde{D}\xi_p$ depends only on $\xi(p)$ and $d\xi_p$.

There is another easy consequence of (1.1.7) worth remarking. If we apply (1.1.7) to the zero section o with $f \equiv 0$ we get

$$\tilde{D}o \equiv 0, \tag{1.1.11}$$

i.e., $\tilde{D}o_p(u) = o_p$ for all $p \in M$ and $u \in T_pM$.

We are now ready to introduce the third incarnation of horizontal bundles.

Definition 1.1.6: A *non-linear connection* is a map $\tilde{D}: \mathcal{X}(TM) \to \mathcal{X}(T^*M \otimes TM)$ satisfying (1.1.10) and (1.1.11). $\tilde{D}\xi$ is called the *covariant differential* of the vector field $\xi \in \mathcal{X}(TM)$, and $\tilde{D}\xi_p(u)$ (which we shall denote by $\tilde{\nabla}_u\xi$) is the *covariant derivative* of ξ in the direction of $u \in T_pM$.

Before stating the smoothness assumption we shall need, we remark that the covariant differential of a vector field is uniquely determined at its zeroes: indeed, if $\xi(p) = 0$ then (1.1.10) and (1.1.11) applied with $\xi' = o$ force

$$\tilde{D}\xi_p = \iota_{o_p}^{-1} \circ (d\xi_p - do_p).$$

So we shall say that a non-linear connection $\tilde{D}$ is *smooth* if for any $\xi \in \mathcal{X}(TM)$ and $p \in M$ with $\xi(p) \neq 0$ the section $\tilde{D}\xi$ is smooth in a neighborhood of p.

Equation (1.1.10) is enough to show that all non-linear connections enjoy two of the main properties of a differential operator: locality, and the dependence of $\tilde{\nabla}_u \xi$ only on the restriction of ξ to any curve with u as tangent vector. Indeed:

LEMMA 1.1.1: *Let $\tilde{D}$ be a non-linear connection on a manifold M. Then:*
(i) if $\xi, \xi' \in \mathcal{X}(TM)$ are such that $\xi \equiv \xi'$ in an open subset U of M, then $\tilde{D}\xi \equiv \tilde{D}\xi'$ in U;
(ii) take $p \in M$, $u \in T_pM$ and $\sigma: (-\varepsilon, \varepsilon) \to M$ a curve in M with $\sigma(0) = p$ and $\dot{\sigma}(0) = u$. Let ξ, ξ' be two vector fields with $\xi \circ \sigma = \xi' \circ \sigma$; then $\tilde{\nabla}_u \xi(p) = \tilde{\nabla}_u \xi'(p)$.

Proof: (i) Take $p \in U$; then $\xi(p) = \xi'(p)$, $d\xi_p = d\xi'_p$ and (1.1.10) yields $\tilde{D}\xi_p = \tilde{D}\xi'_p$.

(ii) Let $v = \xi(p) = \xi'(p)$; therefore (1.1.10) yields

$$\iota_v\big(\tilde{\nabla}_u \xi(p) - \tilde{\nabla}_u \xi'(p)\big) = d\xi_p(u) - d\xi'_p(u).$$

Since

$$d\xi_p(u)(g) = u(g \circ \xi) = \frac{d}{dt}(g \circ \xi \circ \sigma)\bigg|_{t=0}$$

for all $g \in C^\infty(TM)$, we get $d\xi_p(u) = d\xi'_p(u)$ and we are done. □

Now we may go back to horizontal bundles. We anticipated earlier that non-linear connections are just another way to look at horizontal bundles; let us now explain this assertion. Since, as already remarked, everything is determined over the zero section, we shall work in $\tilde{M}$.

Let $\mathcal{H}$ be a horizontal bundle; since $T\tilde{M} = \mathcal{H} \oplus \mathcal{V}$, we can associate to $\mathcal{H}$ a vertical projection $\kappa: T\tilde{M} \to \mathcal{V}$. Then we may define a non-linear connection $\tilde{D}_\mathcal{H}$ on M by setting

$$\tilde{D}_\mathcal{H}\xi_p = \iota_{\xi(p)}^{-1} \circ \kappa_{\xi(p)} \circ d\xi_p,$$

for any $p \in M$ and $\xi \in \mathcal{X}(TM)$. First of all, it is clear that $\tilde{D}_\mathcal{H}\xi$ is a linear map from $\mathcal{X}(TM)$ into itself, and thus an element of $\mathcal{X}(T^*M \otimes TM)$, smooth outside the zeroes of ξ, as required. Furthermore, if $\xi(p) = \xi'(p) = u$ then

$$\tilde{D}_\mathcal{H}\xi'_p - \tilde{D}_\mathcal{H}\xi_p = \iota_u^{-1} \circ \kappa_u \circ (d\xi'_p - d\xi_p) = \iota_u^{-1} \circ (d\xi'_p - d\xi_p),$$

because, as remarked before, $d\xi'_p - d\xi_p$ maps T_pM into $\mathcal{V}_u$. Finally, since by assumption we have $\mathcal{H}_{o_p} = do_p(T_pM)$, it follows that $\kappa_{o_p} \circ do_p \equiv 0$ and so $\tilde{D}_\mathcal{H} o \equiv 0$. Summing up, $\tilde{D}_\mathcal{H}$ is a non-linear connection as claimed.

Now let $\tilde{D}: \mathcal{X}(TM) \to \mathcal{X}(T^*M \otimes TM)$ be a non-linear connection. Take $u \in \tilde{M}_p$, and $\xi \in \mathcal{X}(TM)$ such that $\xi(p) = u$. Then define $\Theta_u^{\tilde{D}}: \mathcal{V}_u \to T_u\tilde{M}$ by

$$\Theta_u^{\tilde{D}} = d\xi_p \circ \iota_u^{-1} - \iota_u \circ \tilde{D}\xi_p \circ \iota_u^{-1}.$$

This definition does not depend on ξ, but only on u, thanks to (1.1.10). Furthermore, $\Theta^{\tilde{D}}$ is clearly smooth on $\tilde{M}$, and $(d\pi \circ \Theta)_u = \iota_u^{-1}$ because $d\pi_u \circ d\xi_p = \mathrm{id}$ and $d\pi_u \circ \iota_u = 0$. So $\Theta^{\tilde{D}}$ is a horizontal map.

Finally, to a horizontal map Θ we associate a horizontal bundle $\mathcal{H}^\Theta$ simply by setting

$$\mathcal{H}_u^\Theta = \Theta_u(\mathcal{V}_u);$$

indeed (1.1.6) implies that $\mathcal{H}^\Theta \cap \mathcal{V} = (0)$ and that $\mathcal{H}^\Theta$ has rank m.

In this way we have defined a correspondence among horizontal bundles, horizontal maps and non-linear connections. As announced before, this is one-to-one:

Proposition 1.1.2: *Let M be a manifold. Then the maps $\mathcal{H} \mapsto \tilde{D}_\mathcal{H}$, $\tilde{D} \mapsto \Theta^{\tilde{D}}$ and $\Theta \mapsto \mathcal{H}^\Theta$ define a one-to-one correspondence among horizontal bundles, non-linear connections and horizontal maps.*

Proof: It suffices to prove three facts.

(i) $\mathcal{H}^{\Theta^{\tilde{D}_\mathcal{H}}} = \mathcal{H}$. Indeed, if $\xi \in \mathcal{X}(TM)$ is such that $\xi(p) = u$ we have

$$\begin{aligned}\Theta_u^{\tilde{D}_\mathcal{H}}(\mathcal{V}_u) &= (d\xi_p \circ \iota_u^{-1} - \iota_u \circ \tilde{D}_\mathcal{H}\xi_p \circ \iota_u^{-1})(\mathcal{V}_u)\\ &= (d\xi_p \circ \iota_u^{-1} - \iota_u \circ (\iota_u^{-1} \circ \kappa_u \circ d\xi_p) \circ \iota_u^{-1})(\mathcal{V}_u)\\ &= (d\xi_p - \kappa_u \circ d\xi_p)(T_pM) = \mathcal{H}_u.\end{aligned}$$

(ii) $\tilde{D}_{\mathcal{H}^{\Theta^{\tilde{D}}}} = \tilde{D}$. Take $\xi \in \mathcal{X}(TM)$, $p \in M$ and let $u = \xi(p)$. By definition,

$$\mathcal{H}_u^{\Theta^{\tilde{D}}} = \Theta_u^{\tilde{D}}(\mathcal{V}_u) = (d\xi_p - \iota_u \circ \tilde{D}\xi_p)(T_pM).$$

Therefore the vertical projection associated to $\mathcal{H}^{\Theta^{\tilde{D}}}$ of $d\xi_p$ is $\iota_u \circ \tilde{D}\xi_p$, and so

$$\tilde{D}_{\mathcal{H}^{\Theta^{\tilde{D}}}}\xi_p = \iota_u^{-1} \circ (\iota_u \circ \tilde{D}\xi_p) = \tilde{D}\xi_p.$$

(iii) $\Theta^{\tilde{D}_{\mathcal{H}^\Theta}} = \Theta$. By definition, the vertical projection associated to $\mathcal{H}^\Theta$ is

$$\kappa_u = \mathrm{id} - \Theta_u \circ \iota_u \circ d\pi_u.$$

Then, taking any $\xi \in \mathcal{X}(TM)$ with $\xi(p) = u$, we get

$$\begin{aligned}\Theta_u^{\tilde{D}_{\mathcal{H}^\Theta}} &= d\xi_p \circ \iota_u^{-1} - \iota_u \circ \tilde{D}_{\mathcal{H}^\Theta}\xi_p \circ \iota_u^{-1} = d\xi_p \circ \iota_u^{-1} - \iota_u \circ (\iota_u^{-1} \circ \kappa_u \circ d\xi_p) \circ \iota_u^{-1}\\ &= d\xi_p \circ \iota_u^{-1} - (\mathrm{id} - \Theta_u \circ \iota_u \circ d\pi_u) \circ d\xi_p \circ \iota_u^{-1}\\ &= \Theta_u \circ \iota_u \circ (d\pi_u \circ d\xi_p) \circ \iota_u^{-1} = \Theta_u.\end{aligned}$$

□

So we have shown that non-linear connections, horizontal bundles and horizontal maps are just different aspects of the same object; in the rest of this work we shall need all of them.

Actually, we shall not always be interested in generic non-linear connections, but only in homogeneous ones.

Definition 1.1.7: A non-linear connection $\tilde{D}$ is *homogeneous* if

$$\tilde{D}(\lambda\xi) = \lambda\tilde{D}\xi$$

for all $\lambda \in \mathbb{R}$ and $\xi \in \mathcal{X}(TM)$.

It is easy to check that homogeneous connections enjoy the derivation property (1.1.7):

Lemma 1.1.3: *Let $\tilde{D}$ be a homogeneous connection. Then*

$$\tilde{D}(f\xi) = df \otimes \xi + f\tilde{D}\xi$$

for all $f \in C^\infty(M)$ and $\xi \in \mathcal{X}(TM)$.

Proof: Take $p \in M$ and apply (1.1.10) to $f\xi$ and $f(p)\xi$. We find

$$\begin{aligned}\tilde{D}(f\xi)_p - f(p)\tilde{D}\xi_p = \tilde{D}(f\xi)_p - \tilde{D}\big(f(p)\xi\big)_p &= \iota^{-1}_{f(p)\xi(p)} \circ \big(d(f\xi_p) - d(f(p)\xi)_p\big)\\ &= df_p \otimes \xi(p),\end{aligned}$$

where we used (1.1.9), and we are done. □

Accordingly to the philosophy expressed before, there should be a notion of homogeneity for horizontal bundles and maps.

Definition 1.1.8: For any $\lambda \in \mathbb{R}$, let $\mu_\lambda: TM \to TM$ denote the *multiplication map*

$$\mu_\lambda(p; u) = (p; \lambda u).$$

Since $\pi \circ \mu_\lambda = \pi$, the differential $d\mu_\lambda$ sends $\mathcal{V}_u$ onto $\mathcal{V}_{\mu_\lambda(u)}$ for all $u \in TM$. Clearly, a non-linear connection $\tilde{D}$ is homogeneous iff

$$\tilde{D}(\mu_\lambda \circ \xi) = \mu_\lambda \circ \tilde{D}\xi$$

for all $\lambda \in \mathbb{R}$ and $\xi \in \mathcal{X}(TM)$.

Definition 1.1.9: We say that a horizontal bundle $\mathcal{H}$ is *homogeneous* if

$$\mathcal{H}_{\mu_\lambda(u)} = d(\mu_\lambda)_u(\mathcal{H}_u)$$

for all $u \in TM$ and $\lambda \in \mathbb{R}$.

Definition 1.1.10: We say that a horizontal map Θ is *homogeneous* if

$$d(\mu_\lambda)_u \circ \Theta_u \circ \iota_u = \Theta_{\mu_\lambda(u)} \circ \iota_{\mu_\lambda(u)}$$

for all $u \in TM$ and $\lambda \in \mathbb{R}$.

The definition of homogeneous horizontal map seems slightly far-fetched, but it is the right one to get the following

Lemma 1.1.4: *Let M be a manifold. Then the correspondences defined in Proposition 1.1.2 preserve homogeneity.*

Proof: The proof depends on two remarks. First of all, it is not difficult to check that

$$\iota_{\mu_\lambda(u)} \circ \mu_\lambda = d(\mu_\lambda)_u \circ \iota_u \tag{1.1.12}$$

for all $u \in TM$ and $\lambda \in \mathbb{R}$.

Next, let κ be the vertical projection associated to a horizontal bundle $\mathcal{H}$. Then $\mathcal{H}$ is homogeneous iff

$$\kappa_{\mu_\lambda(u)} \circ d(\mu_\lambda)_u = d(\mu_\lambda)_u \circ \kappa_u \tag{1.1.13}$$

for all $u \in TM$ and $\lambda \in \mathbb{R}$.

Then $\mathcal{H}$ homogeneous implies, by (1.1.12) and (1.1.13), that $\tilde{D}_{\mathcal{H}}$ is homogeneous. If $\tilde{D}$ is homogeneous, then (1.1.12) yields

$$\begin{aligned} d(\mu_\lambda)_u \circ \Theta_u^{\tilde{D}} \circ \iota_u &= d(\mu_\lambda)_u \circ (d\xi_p \circ \iota_u^{-1} - \iota_u \circ \tilde{D}\xi_p \circ \iota_u^{-1}) \circ \iota_u \\ &= d(\mu_\lambda \circ \xi)_p - \iota_{\mu_\lambda(u)} \circ \mu_\lambda \circ \tilde{D}\xi_p \\ &= d(\mu_\lambda \circ \xi)_p - \iota_{\mu_\lambda(u)}\big(\tilde{D}(\mu_\lambda \circ \xi)_p\big) = \Theta^{\tilde{D}}_{\mu_\lambda(u)} \circ \iota_{\mu_\lambda(u)}, \end{aligned}$$

where as usual ξ is any vector field with $\xi(p) = u$, and $\Theta^{\tilde{D}}$ is homogeneous.

Finally, if Θ is homogeneous we have

$$d(\mu_\zeta)_u(\mathcal{H}_u^\Theta) = d(\mu_\lambda)_u \circ \Theta_u \circ \iota_u(T_pM) = \Theta_{\mu_\lambda(u)} \circ \iota_{\mu_\lambda(u)}(T_pM) = \mathcal{H}^\Theta_{\mu_\lambda(u)},$$

and $\mathcal{H}^\Theta$ is homogeneous. □

1.1.3. Local coordinates

Now let $\tilde{D}$ be a generic non-linear connection on a manifold M, with associated covariant differentiation $\tilde{\nabla}$; we would like to express $\tilde{\nabla}$ in local coordinates. Take $\xi \in \mathcal{X}(TM)$, $p \in M$ and set $u = \xi(p)$. Then (1.1.8) shows that

$$d\xi_p\left(\frac{\partial}{\partial x^h}\right) = \partial_h + \frac{\partial \xi^k}{\partial x^h}\dot{\partial}_k.$$

By definition of non-linear connection, the quantity $\iota_u(\tilde{\nabla}_{\partial/\partial x^h}\xi) - d\xi_p(\partial/\partial x^h)$ must depend only on u; being $\iota_u(\tilde{\nabla}_{\partial/\partial x^h}\xi)$ vertical, there exist $\Gamma_h^k = \Gamma_h^k(u) \in \mathbb{R}$ such that

$$\iota_u(\tilde{\nabla}_{\partial/\partial x^h}\xi) - d\xi_p\left(\frac{\partial}{\partial x^h}\right) = -\partial_h + \Gamma_h^k(u)\dot{\partial}_k.$$

Summing up we get

$$\tilde{\nabla}_v\xi(p) = v^h\left[\frac{\partial \xi^k}{\partial x^h}(p) + \Gamma_h^k(\xi(p))\right]\left.\frac{\partial}{\partial x^k}\right|_p. \tag{1.1.14}$$

The Γ_h^k are the *Christoffel symbols* of the non-linear connection. It is clear that the non-linear connection is smooth iff the Christoffel symbols are smooth on $\tilde{M}$ (where defined, of course).

Reading (1.1.14) for $\tilde{D}$, we get

$$\tilde{D}\xi = \left[\frac{\partial \xi^k}{\partial x^h} + \Gamma_h^k \circ \xi\right] dx^h \otimes \frac{\partial}{\partial x^k}.$$

Now it is easy to get the expression in local coordinates for the horizontal map $\Theta^{\tilde{D}}$. Take $u \in \tilde{M}_p$, and $\xi \in \mathcal{X}(TM)$ with $\xi(p) = u$. Now $\iota_u(\partial/\partial x^h) = \dot{\partial}_h$; therefore

$$\iota_u \circ \tilde{D}\xi \circ \iota_u^{-1}(\dot{\partial}_h) = \left[\frac{\partial \xi^k}{\partial x^h}(p) + \Gamma_h^k(u)\right]\dot{\partial}_k,$$

$$d\xi_p \circ \iota_u^{-1}(\dot{\partial}_h) = \partial_h + \frac{\partial \xi^k}{\partial x^h}(p)\dot{\partial}_k,$$

and so

$$\Theta_u^{\tilde{D}}(\dot{\partial}_h) = \partial_h - \Gamma_h^k(u)\dot{\partial}_k. \tag{1.1.15}$$

If Θ is any horizontal map, it is clear that Θ in local coordinates is expressed as in (1.1.15). We set

$$\delta_j|_u = \Theta_u(\dot{\partial}_j|_u) = \partial_j|_u - \Gamma_j^k(u)\dot{\partial}_k|_u \in \mathcal{H}_u^\Theta; \tag{1.1.16}$$

then $\{\delta_1, \ldots, \delta_m\}$ is a local frame for $\mathcal{H}^\Theta$, and $\{\delta_1, \ldots, \delta_m, \dot{\partial}_1, \ldots, \dot{\partial}_m\}$ is a local frame for $T\tilde{M}$. We shall denote by $\{dx^1, \ldots, dx^m, \psi^1, \ldots, \psi^m\}$ the dual coframe; clearly,

$$\psi^h = du^h + \Gamma_j^h\, dx^j.$$

DEFINITION 1.1.11: We set $\chi_u = \Theta_u \circ \iota_u : T_pM \to \mathcal{H}_u^\Theta$; then the *radial horizontal vector field* χ — the horizontal analogue of ι — is

$$\chi = \Theta \circ \iota \in \mathcal{X}(\mathcal{H}^\Theta).$$

Clearly,

$$\chi(\partial/\partial x^j) = \delta_j.$$

Now let $\mathcal{H}$ be a horizontal bundle, and $\{\delta_1, \ldots, \delta_m\}$ a local frame such that $d\pi(\delta_j) = \partial/\partial x^j$ for $j = 1, \ldots, m$. Then the δ_j's may be written as in (1.1.16) for suitable Γ_j^k's. If

$$X = X^i \partial_i + \dot{X}^j \dot{\partial}_j \in T_u \tilde{M},$$

then the local expression for the vertical projection associated to $\mathcal{H}$ is

$$\kappa_u(X) = (X^j \Gamma_j^k(u) + \dot{X}^k)\dot{\partial}_k,$$

and so

$$\tilde{D}_{\mathcal{H}}\xi_p\left(\frac{\partial}{\partial x^h}\right) = \left[\frac{\partial \xi^k}{\partial x^h}(p) + \Gamma_h^k(\xi(p))\right] \left.\frac{\partial}{\partial x^h}\right|_p,$$

completing a full circle.

Clearly, $\mathcal{H}$ is homogeneous iff $d(\mu_\lambda)_u(\delta_i) \in \mathcal{H}_{\mu_\lambda(u)}$ for $i = 1, \ldots, m$, $u \in \tilde{M}$ and $\lambda \in \mathbb{R}$. Since

$$\kappa_u(X^i\partial_i + \dot{X}^a\dot{\partial}_a) = [\dot{X}^a + X^i\Gamma_i^a(u)]\dot{\partial}_a|_u,$$
$$d(\mu_\lambda)_u(\dot{\partial}_a|_u) = \lambda\dot{\partial}_a|_{\mu_\lambda(u)},$$

and

$$d(\mu_\lambda)_u(\partial_i|_u) = \partial_i|_{\mu_\lambda(u)},$$

we find that $\mathcal{H}$ (and the associated non-linear connection and horizontal map) is homogeneous iff

$$\Gamma_i^a(\mu_\lambda(u)) = \lambda\Gamma_i^a(u), \tag{1.1.17}$$

that is iff

$$\forall i = 1, \ldots, m \qquad d(\mu_\lambda)_u(\delta_i|_u) = \delta_i|_{\mu_\lambda(u)}.$$

So a horizontal bundle (horizontal map, non-linear connection) is locally identified by the coefficients Γ_h^k. To see how they transform under a change of coordinates, we remark that

$$\begin{aligned}(\delta_i)_B &= (\partial_i)_B - (\Gamma_i^h)_B(\dot{\partial}_h)_B = \Theta((\partial_i)_B) = \Theta((\mathcal{J}_{BA}^{-1})_i^h(\partial_h)_A)\\ &= (\mathcal{J}_{BA}^{-1})_i^h(\partial_h)_A - (\mathcal{J}_{BA}^{-1})_i^h(\Gamma_h^k)_A(\dot{\partial}_k)_A = (\mathcal{J}_{BA}^{-1})_i^h(\delta_h)_A.\end{aligned} \tag{1.1.18}$$

Comparing (1.1.4) and (1.1.18) we find

$$(\Gamma_i^h)_B(\mathcal{J}_{BA}^{-1})_h^k + (H_{BA})_{jl}^h(\mathcal{J}_{BA}^{-1})_h^k(\mathcal{J}_{BA}^{-1})_i^j\, u_A^l = (\Gamma_h^k)_A(\mathcal{J}_{BA}^{-1})_i^h,$$

that is

$$\begin{aligned}(\Gamma_i^j)_B &= (\mathcal{J}_{BA})_k^j(\Gamma_h^k)_A(\mathcal{J}_{BA}^{-1})_i^h - (H_{BA})_{hl}^j(\mathcal{J}_{BA}^{-1})_i^h\, u_A^l\\ &= \frac{\partial x_B^j}{\partial x_A^k}(\Gamma_h^k)_A\frac{\partial x_A^h}{\partial x_B^i} - \frac{\partial^2 x_B^j}{\partial x_A^h \partial x_A^l}\frac{\partial x_A^h}{\partial x_B^i}\, u_A^l.\end{aligned} \tag{1.1.19}$$

Note that (1.1.18) together with (1.1.4) and (1.1.3) yield

$$(\psi^c)_B = (\mathcal{J}_{BA})_d^c(\psi^d)_A.$$

Finally, for the sake of future references we collect here the following easy computations:

$$
\begin{aligned}
&[\delta_i, \delta_j] = \{\delta_j(\Gamma_i^a) - \delta_i(\Gamma_j^a)\}\dot{\partial}_a,\\
&[\delta_i, \dot{\partial}_a] = \dot{\partial}_a(\Gamma_i^b)\dot{\partial}_b, \qquad (1.1.20)\\
&[\dot{\partial}_a, \dot{\partial}_b] = o.
\end{aligned}
$$

1.1.4. Geodesics

In what follows, horizontal bundles and maps will play a fundamental role, whereas the non-linear connection will somewhat fade in the background. But there is one instance where non-linear connections are essential: in defining parallel transport and geodesics.

If σ is a curve in M and ξ is a vector field defined only along (the trace of) σ, by Lemma 1.1.1 it makes sense to consider $\tilde{\nabla}_{\dot{\sigma}}\xi$.

DEFINITION 1.1.12: We say that ξ is *parallel* (or *horizontal*) if $\tilde{\nabla}_{\dot{\sigma}}\xi \equiv 0$; and σ itself is a *geodesic* if

$$\tilde{\nabla}_{\dot{\sigma}}\dot{\sigma} \equiv 0.$$

We may write the condition of parallelism and the equation of geodesics in local coordinates. Recalling (1.1.14), we see that a vector field ξ along a curve σ is parallel iff

$$\frac{d\xi^a}{dt} + \Gamma_b^a(\xi)\dot{\sigma}^b = 0,$$

for $a = 1, \ldots, m$; analogously, a curve σ is a geodesic iff

$$\ddot{\sigma}^a + \Gamma_b^a(\dot{\sigma})\dot{\sigma}^b = 0 \qquad (1.1.21)$$

for $a = 1, \ldots, m$. These equations are quasi-linear ODE; hence the local theory of geodesics and parallel transport for non-linear connections is more or less identical to the theory known for linear connections.

These equations may be interpreted in another way, as shown in the following

LEMMA 1.1.5: *Let $\tilde{D}: \mathcal{X}(TM) \to \mathcal{X}(T^*M \otimes TM)$ be a non-linear connection on a manifold M. Then a vector field ξ along a curve σ is parallel iff $\dot{\xi}$ is horizontal iff $\dot{\xi} = \chi_\xi(\dot{\sigma})$. Analogously, the curve σ itself is a geodesic iff $\ddot{\sigma}$ is horizontal iff $\ddot{\sigma} = \chi(\dot{\sigma})$.*

Proof: A vector field ξ along a curve σ is a curve in TM; analogously, $\dot{\xi}$ is a curve in $T(TM)$. A computation in local coordinates yields

$$\begin{aligned}\dot{\xi} &= \dot{\sigma}^j \partial_j + \frac{d\xi^\alpha}{dt}\dot{\partial}_\alpha \\ &= \dot{\sigma}^j \delta_j + \left[\frac{d\xi^\alpha}{dt} + \Gamma^\alpha_\beta(\xi)\dot{\sigma}^\beta\right]\dot{\partial}_\alpha, \\ &= \chi_\xi(\dot{\sigma}) + \iota_\xi(\tilde{\nabla}_{\dot{\sigma}}\xi),\end{aligned} \tag{1.1.22}$$

and the assertions follow. □

We shall return to geodesics and parallel transport in sections 1.5 and 1.6.

1.2. Vertical connections

1.2.1. Definitions

DEFINITION 1.2.1: A *vertical connection* on a manifold M is a linear connection on the vertical bundle, that is a $\mathbb{R}$-linear map $D: \mathcal{X}(\mathcal{V}) \to \mathcal{X}(T^*\tilde{M} \otimes \mathcal{V})$ such that

$$D(fV) = fDV + df \otimes V$$

for every $f \in C^\infty(\tilde{M})$ and $V \in \mathcal{X}(\mathcal{V})$. In other words, if V is a section of $\mathcal{V}$, then DV is a $\mathcal{V}$-valued 1-form on $\tilde{M}$. We set, as usual,

$$\nabla_X V = (DV)(X)$$

for all $X \in T_u\tilde{M}$ and $V \in \mathcal{X}(\mathcal{V})$. Note that D is not defined over the zero section of TM; we could have it defined there too but, in accord with the previous section and for reasons we shall see in section 1.4, we would not ask for any smoothness over the zero section. Since anyway we shall not need it, all our vertical connections will be undefined over the zero section.

In local coordinates, writing $V = V^a\dot{\partial}_a$, we get

$$DV = dV^a \otimes \dot{\partial}_a + V^a D\dot{\partial}_a,$$

and

$$D\dot{\partial}_a = \omega^b_a \otimes \dot{\partial}_b,$$

where (ω^b_a) is a locally defined matrix of 1-forms:

$$\omega^b_a = \tilde{\Gamma}^a_{b;i}\, dx^i + \tilde{\Gamma}^a_{bc}\, dv^c,$$

for suitable coefficients $\tilde{\Gamma}^a_{b;i}$ and $\tilde{\Gamma}^a_{bc}$.

So

$$\nabla_X V = [X(V^a) + \omega_b^a(X)V^b]\dot{\partial}_a \tag{1.2.1}$$

or, writing $X = X^j\partial_j + \dot{X}^b\dot{\partial}_b$,

$$\nabla_X V = \left\{ X^j \left[\frac{\partial V^a}{\partial x^j} + \tilde{\Gamma}^a_{b;j} V^b \right] + \dot{X}^c \left[\frac{\partial V^a}{\partial u^c} + \tilde{\Gamma}^a_{bc} V^b \right] \right\} \dot{\partial}_a.$$

The idea is that we may associate a non-linear connection (i.e., a horizontal bundle) to certain vertical connections. Let D be a vertical connection, and denote by $\Lambda: T\tilde{M} \to \mathcal{V}$ the bundle map defined by

$$\Lambda(X) = \nabla_X \iota.$$

In local coordinates, if $X = X^j\partial_j + \dot{X}^a\dot{\partial}_a$ we get

$$\Lambda(X) = [\dot{X}^a + \omega_b^a(X)u^b]\dot{\partial}_a.$$

DEFINITION 1.2.2: We say that D is a *good* vertical connection if $\Lambda|_{\mathcal{V}}: \mathcal{V} \to \mathcal{V}$ is a bundle isomorphism.

The importance of good vertical connections is explained by

LEMMA 1.2.1: *Let $D: \mathcal{X}(\mathcal{V}) \to \mathcal{X}(T^*\tilde{M} \otimes \mathcal{V})$ be a vertical connection. Then D is good iff*

$$T\tilde{M} = (\ker \Lambda) \oplus \mathcal{V},$$

that is iff $\mathcal{H} = \ker\Lambda$ is a horizontal bundle.

Proof: $\Lambda|_{\mathcal{V}}$ is an isomorphism iff $\mathcal{H}_u \cap \mathcal{V}_u = (0)$ for all $u \in \tilde{M}$; furthermore, if $\Lambda|_{\mathcal{V}}$ is an isomorphism then Λ is onto, $\mathcal{H}$ has rank $m = \dim M$ and the assertion follows. □

In local coordinates,

$$\Lambda(\dot{\partial}_b) = [\delta^a_b + \tilde{\Gamma}^a_{cb}u^c]\dot{\partial}_a \qquad \text{and} \qquad \Lambda(\partial_i) = \tilde{\Gamma}^a_{c;i}u^c\dot{\partial}_a,$$

where here δ^a_b is the Kronecker delta. Therefore D is good iff the matrix

$$L^a_b = \delta^a_b + \tilde{\Gamma}^a_{cb}u^c$$

is invertible. If $\left((L^{-1})^j_i\right)$ denotes the inverse matrix, a local frame for $\mathcal{H}$ is given by

$$\delta_i = \partial_i - (L^{-1})^b_a \tilde{\Gamma}^a_{c;i} u^c \dot{\partial}_b = \partial_i - \Gamma^b_i \dot{\partial}_b,$$

where

$$\Gamma^b_i = (L^{-1})^b_a \tilde{\Gamma}^a_{c;i} u^c$$

are the Christoffel symbols of the non-linear connection associated to $\mathcal{H}$.

1.2.2. Change of coordinates

So let again $D\colon \mathcal{X}(\mathcal{V}) \to \mathcal{X}(T^*\tilde{M} \otimes \mathcal{V})$ be a good vertical connection on M. We shall denote by $\{dz^i, \psi^a\}$ the local coframe dual to $\{\delta_i, \dot{\partial}_a\}$, where

$$\psi^a = du^a + \Gamma^a_j\, dx^j.$$

Correspondingly, we may decompose

$$\omega^a_b = \Gamma^a_{b;j}\, dx^j + \Gamma^a_{bc}\psi^c,$$

where

$$\Gamma^a_{bc} = \tilde{\Gamma}^a_{bc} \qquad \text{and} \qquad \Gamma^a_{b;j} = \tilde{\Gamma}^a_{b;j} - \tilde{\Gamma}^a_{bc}\Gamma^c_j.$$

In particular, if $X = X^j\delta_j + \dot{X}^b\dot{\partial}_b$ we have

$$\nabla_X V = \left\{X^j\left[\delta_j(V^a) + \Gamma^a_{b;j}V^b\right] + \dot{X}^c\left[\dot{\partial}_c(V^a) + \Gamma^a_{bc}V^b\right]\right\}\dot{\partial}_a. \tag{1.2.2}$$

Later on we shall need to know how ω^a_b, Γ^a_{bc} and $\Gamma^a_{b;j}$ behave under change of coordinates. We have

$$\begin{aligned}(\omega^b_a)_B \otimes (\mathcal{J}^{-1}_{BA})^k_b(\dot{\partial}_k)_A = (\omega^b_a)_B \otimes (\dot{\partial}_b)_B = D(\dot{\partial}_a)_B = D\big((\mathcal{J}^{-1}_{BA})^h_a(\dot{\partial}_h)_A\big)\\ = d(\mathcal{J}^{-1}_{BA})^k_a \otimes (\dot{\partial}_k)_A + (\mathcal{J}^{-1}_{BA})^h_a(\omega^k_h)_A \otimes (\dot{\partial}_k)_A.\end{aligned}$$

Now

$$d(\mathcal{J}^{-1}_{BA})^k_a = \frac{\partial}{\partial x^j_A}(\mathcal{J}^{-1}_{BA})^k_a\, dx^j_A = -(\mathcal{J}^{-1}_{BA})^h_a(\mathcal{J}^{-1}_{BA})^k_i(H_{BA})^i_{hj}\, dx^j_A,$$

and so

$$(\omega^b_a)_B = (\mathcal{J}^{-1}_{BA})^h_a(\omega^k_h)_A(\mathcal{J}_{BA})^b_k - (\mathcal{J}^{-1}_{BA})^h_a(H_{BA})^b_{hj}\, dx^j_A.$$

It follows that

$$\begin{aligned}(\Gamma^b_{a;i})_B &= (\mathcal{J}^{-1}_{BA})^h_a(\mathcal{J}^{-1}_{BA})^k_i(\Gamma^j_{h;k})_A(\mathcal{J}_{BA})^b_j - (\mathcal{J}^{-1}_{BA})^h_a(\mathcal{J}^{-1}_{BA})^k_i(H_{BA})^b_{hk},\\ (\Gamma^b_{ac})_B &= (\mathcal{J}^{-1}_{BA})^h_a(\mathcal{J}^{-1}_{BA})^k_c(\Gamma^d_{hk})_A(\mathcal{J}_{BA})^b_d.\end{aligned} \tag{1.2.3}$$

In the sequel it will be useful the following

LEMMA 1.2.2: *Let $D\colon \mathcal{X}(\mathcal{V}) \to \mathcal{X}(T^*\tilde{M} \otimes \mathcal{V})$ be a good vertical connection on M. Then*

$$\Gamma^a_j = \Gamma^a_{b;j}u^b.$$

Proof: Indeed

$$\begin{aligned}\Gamma^a_{b;j}u^b &= \tilde{\Gamma}^a_{b;j}u^b - \tilde{\Gamma}^a_{bc}u^b(L^{-1})^c_l\tilde{\Gamma}^l_{d;j}u^d = \left[L^a_c - \tilde{\Gamma}^a_{dc}u^d\right](L^{-1})^c_l\tilde{\Gamma}^l_{b;j}u^b\\ &= \Gamma^a_j.\end{aligned}$$

□

1.2.3. Linear connections on $\tilde{M}$

We have already seen that, if $\Theta: \mathcal{V} \to \mathcal{H}$ denotes the horizontal map associated to the horizontal bundle $\mathcal{H}$ associated to a good vertical connection D, in local coordinates we have

$$\Theta(\dot{\partial}_a) = \delta_a.$$

Using Θ, then, we may define a linear connection (still denoted by D) on $\mathcal{H}$ by setting

$$\forall H \in \mathcal{X}(\mathcal{H}) \qquad DH = \Theta\big(D\Theta^{-1}(H)\big),$$

that is

$$\forall H \in \mathcal{X}(\mathcal{H}) \quad \forall X \in T\tilde{M} \quad \nabla_X H = \Theta\big[\nabla_X(\Theta^{-1}H)\big].$$

In this way a good vertical connection gives rise to a linear connection D defined on $\tilde{M}$ by setting

$$\forall Y \in \mathcal{X}(T\tilde{M}) \qquad DY = D(\kappa Y) + D(Y - \kappa Y),$$

that is

$$\forall X, Y \in \mathcal{X}(T\tilde{M}) \quad \nabla_X Y = \nabla_X(\kappa Y) + \nabla_X(Y - \kappa Y),$$

where $\kappa: T\tilde{M} \to \mathcal{V}$ is the vertical projection associated to $\mathcal{H} = \ker \Lambda$. It will turn out that this linear connection (and in particular its horizontal part) is the right geometrical object to study. For the moment, we just remark that, by definition,

$$\forall H \in \mathcal{H} \quad \forall V \in \mathcal{V} \quad \nabla_H \mathcal{X} = 0 \quad \text{and} \quad \nabla_V \mathcal{X} = \Theta\Lambda(V).$$

Furthermore, by definition

$$D\delta_i = \omega_i^j \otimes \delta_j.$$

Clearly, the linear connection D on $\tilde{M}$ and the non-linear connection $\tilde{D}$ on M associated to $\mathcal{H}$ are closely related. To describe how, we need a definition.

Definition 1.2.3: Let $\xi \in \mathcal{X}(TM)$ be a tangent vector field; then the *vertical lift* $\xi^V \in \mathcal{X}(\mathcal{V})$ and the *horizontal lift* $\xi^H \in \mathcal{X}(\mathcal{H})$ are defined by

$$\xi^V(u) = \iota_u\big(\xi(\pi(u))\big) \in \mathcal{V}_u \qquad \text{and} \qquad \xi^H(u) = \chi_u\big(\xi(\pi(u))\big) \in \mathcal{H}_u$$

for all $u \in \tilde{M}$.

Then

Proposition 1.2.3: *Let $D: \mathcal{X}(\mathcal{V}) \to \mathcal{X}(T^*\tilde{M} \otimes \mathcal{V})$ be a good vertical connection. Let $\mathcal{H}$ be the horizontal bundle associated to D and let $\tilde{D}: \mathcal{X}(TM) \to \mathcal{X}(T^*M \otimes TM)$ be the non-linear connection associated to $\mathcal{H}$. Take $\xi \in \mathcal{X}(TM)$ and $p \in M$. Then*

$$\forall u \in T_pM \qquad \chi_{\xi(p)}(\tilde{\nabla}_u \xi) = \nabla_{\chi_{\xi(p)}(u)} \xi^H,$$

that is

$$\chi_{\xi(p)} \circ \tilde{D}\xi = D\xi^H \circ \chi_{\xi(p)}.$$

Proof: In local coordinates, (1.1.14) yields

$$\chi_{\xi(p)}(\tilde{\nabla}_u \xi) = u^h \left[\frac{\partial \xi^k}{\partial x^h}(p) + \Gamma^k_h(\xi(p)) \right] \delta_k|_{\xi(p)}.$$

On the other hand,

$$\begin{aligned}\nabla_{\chi_{\xi(p)}(u)} \xi^H &= u^h \left[\delta_h(\xi^k \circ \pi) + \Gamma^k_{b;h}(\xi(p))\xi^b(p) \right] \delta_k|_{\xi(p)} \\ &= u^h \left[\frac{\partial \xi^k}{\partial x^h}(p) + \Gamma^k_h(\xi(p)) \right] \delta_k|_{\xi(p)},\end{aligned}$$

by (1.2.2) and Lemma 1.2.2, and the assertion follows. □

1.3. Torsion and curvature

1.3.1. Connections and differentials on tensor bundles

DEFINITION 1.3.1: Let $D: \mathcal{X}(\mathcal{V}) \to \mathcal{X}(T^*\tilde{M} \otimes \mathcal{V})$ be a good vertical connection, and let D also denote the associated linear connection on $\tilde{M}$. We now define a linear connection on $T^*\tilde{M}$, which we shall denote by $\nabla: \mathcal{X}(T^*\tilde{M}) \to \mathcal{X}(T^*\tilde{M} \otimes T^*\tilde{M})$, simply requiring that

$$\nabla\varphi(X) + \varphi(DX) = d(\varphi(X)),$$

for every $\varphi \in \mathcal{X}(T^*\tilde{M})$ and $X \in \mathcal{X}(T\tilde{M})$. In particular, if $u \in \tilde{M}$ and $Y \in T_u\tilde{M}$, then $\nabla_Y\varphi \in T^*_u\tilde{M}$ is such that

$$(\nabla_Y\varphi)(X) + \varphi(\nabla_Y X) = Y\big(\varphi(X)\big).$$

This defines uniquely $\nabla\varphi$.

In local coordinates, if $\varphi = \varphi_i\, dx^i + \dot{\varphi}_a \psi^a$, then

$$\nabla_Y \varphi = [Y(\varphi_i) - \omega^j_i(Y)\varphi_j]\, dx^i + [Y(\dot{\varphi}_a) - \omega^b_a(Y)\dot{\varphi}_a]\psi^a,$$

that is

$$\nabla\varphi = [d\varphi_i - \varphi_j\,\omega^j_i] \otimes dx^i + [d\dot{\varphi}_a - \dot{\varphi}_b\,\omega^b_a] \otimes \psi^a.$$

Now let $T^{(r,s)}\tilde{M}$ denote the bundle of (r,s)-tensors over $\tilde{M}$, that is

$$T^{(r,s)}\tilde{M} = \underbrace{T^*\tilde{M} \otimes \cdots \otimes T^*\tilde{M}}_{r \text{ times}} \otimes \underbrace{T\tilde{M} \otimes \cdots \otimes T\tilde{M}}_{s \text{ times}}.$$

Then we can extend ∇ to a linear connection $\nabla: \mathcal{X}(T^{(r,s)}\tilde{M}) \to \mathcal{X}(T^*\tilde{M} \otimes T^{(r,s)}\tilde{M})$ just by setting

$$\nabla(R \otimes S) = \nabla R \otimes S + R \otimes \nabla S$$

for any pair of tensors R and S. For instance, if $R \in \mathcal{X}(T^{(1,1)}\tilde{M})$ is locally given by

$$R = R^i_{;j}\, \delta_i \otimes dx^j + R^i_b\, \delta_i \otimes \psi^b + \dot{R}^a_{;j}\, \dot{\partial}_a \otimes dx^j + \dot{R}^a_b\, \dot{\partial}_a \otimes \psi^b,$$

then

$$\begin{aligned}\nabla R = {} & \left[dR^i_{;j} - R^i_{;h}\omega^h_j + R^k_{;j}\omega^i_k\right] \delta_i \otimes dx^j + \left[dR^i_b - R^i_c\omega^c_b + R^k_b\omega^i_k\right] \delta_i \otimes \psi^b \\ & + \left[d\dot{R}^a_{;j} - \dot{R}^a_{;h}\omega^h_j + \dot{R}^d_{;j}\omega^a_d\right] \dot{\partial}_a \otimes dx^j + \left[d\dot{R}^a_b - \dot{R}^a_c\omega^c_b + \dot{R}^d_b\omega^a_d\right] \dot{\partial}_a \otimes \psi^b.\end{aligned}$$

We now introduce an exterior differential for $T^{(r,s)}\tilde{M}$-valued p-form coinciding with ∇ on $T^{(r,s)}\tilde{M}$-valued 0-form, i.e., on sections of $T^{(r,s)}\tilde{M}$.

DEFINITION 1.3.2: Let $\varphi \in \mathcal{X}(\bigwedge^p T^*\tilde{M} \otimes T^{(r,s)}\tilde{M})$ be a $T^{(r,s)}\tilde{M}$-valued p-form; locally, $\varphi = \varphi^a \otimes e_a$, where $\{e_a\}$ is a local frame for $T^{(r,s)}\tilde{M}$ composed by tensor products of $\{\delta_i, \dot{\partial}_a\}$ and $\{dx^j, \psi^b\}$. Then we set

$$D\varphi = d\varphi^a \otimes e_a + (-1)^p \varphi^a \wedge \nabla e_a. \tag{1.3.1}$$

For instance, if $\varphi \in \mathcal{X}(\bigwedge^p T^*\tilde{M} \otimes T\tilde{M})$, then

$$D\varphi = \left[d\varphi^i + (-1)^p\varphi^j \wedge \omega^i_j\right] \otimes \delta_i + \left[d\dot{\varphi}^a + (-1)^p\dot{\varphi}^b \wedge \omega^a_b\right] \otimes \dot{\partial}_a,$$

where we have written

$$\varphi = \varphi^i \otimes \delta_i + \dot{\varphi}^a \otimes \dot{\partial}_a. \tag{1.3.2}$$

We should check that $D\varphi$ is well-defined. For the sake of simplicity (and because it is standard), here we write the proof for $T\tilde{M}$-valued forms only; the general case is identical.

Let $\varphi \in \mathcal{X}(\bigwedge^p T^*\tilde{M} \otimes T\tilde{M})$ be expressed as in (1.3.2). It is then easy to see that

$$(\varphi^i)_B = (\mathcal{J}_{BA})^i_h(\varphi^h)_A \qquad \text{and} \qquad (\dot{\varphi}^a)_B = (\mathcal{J}_{BA})^a_b(\dot{\varphi}^b)_A,$$

and so we get

$$\begin{aligned}&\left[d(\varphi^i)_B + (-1)^p(\varphi^j)_B \wedge (\omega^i_j)_B\right] \otimes (\delta_i)_B \\ &= \Big[(H_{BA})^i_{hl}\, dx^l_A \wedge (\varphi^h)_A + (\mathcal{J}_{BA})^i_j d(\varphi^j)_A \\ &\quad + (-1)^p(\mathcal{J}_{BA})^j_k(\varphi^k)_A \wedge (\mathcal{J}^{-1}_{BA})^h_j\left[(\omega^s_h)_A(\mathcal{J}_{BA})^i_s - (H_{BA})^i_{hl}\, dx^l_A\right]\Big] \otimes (\mathcal{J}^{-1}_{BA})^j_i(\delta_j)_A \\ &= \left[d(\varphi^j)_A + (-1)^p(\varphi^k)_A \wedge (\omega^j_k)_A\right] \otimes (\delta_j)_A.\end{aligned}$$

An analogous computation works for the vertical components, and so $D\varphi$ is a globally well-defined $T\tilde{M}$-valued $(p+1)$-form. Finally, we explicitly remark that

$$D(f\varphi) = df \wedge \varphi + f D\varphi$$

for all $f \in C^\infty(\tilde{M})$.

1.3.2. The torsion

DEFINITION 1.3.3: The tangent bundle TM (and hence $\tilde{M}$) is naturally equipped with a 1-form, the *canonical form*

$$\eta = dx^i \otimes \partial_i + du^a \otimes \dot{\partial}_a \in \mathcal{X}(T^*\tilde{M} \otimes T\tilde{M}).$$

η is a well-defined $T\tilde{M}$-valued global 1-form on $\tilde{M}$, as easily checked using (1.1.3) and (1.1.4).

Now let $D: \mathcal{X}(\mathcal{V}) \to \mathcal{X}(T^*\tilde{M} \otimes \mathcal{V})$ be a good vertical connection, and let D also denote the associated linear connection on $\tilde{M}$. Then it is easy to see that

$$\eta = dx^i \otimes \delta_i + \psi^a \otimes \dot{\partial}_a;$$

in particular,

$$\eta \circ \iota = \iota \qquad \text{and} \qquad \eta \circ \chi = \chi.$$

DEFINITION 1.3.4: The *torsion* $\theta \in \mathcal{X}(\bigwedge^2 T^*\tilde{M} \otimes T\tilde{M})$ of a good vertical connection D is the $T\tilde{M}$-valued 2-form

$$\begin{aligned} \theta = D\eta &= (-dx^j \wedge \omega_j^i) \otimes \delta_i + (d\psi^a - \psi^b \wedge \omega_b^a) \otimes \dot{\partial}_a \\ &= \theta^i \otimes \delta_i + \dot{\theta}^a \otimes \dot{\partial}_a \in \mathcal{X}(\textstyle\bigwedge^2 T^*\tilde{M} \otimes T\tilde{M}), \end{aligned}$$

where D is the exterior differential defined in (1.3.1).

Locally, the $T\tilde{M}$-valued 2-form θ is given by

$$\begin{aligned} \theta^i &= \Gamma^i_{j;h}\, dx^h \wedge dx^j + \Gamma^i_{jc}\, \psi^c \wedge dx^j \\ &= \tfrac{1}{2}[\Gamma^i_{j;h} - \Gamma^i_{h;j}]\, dx^h \wedge dx^j + \Gamma^i_{jc}\, \psi^c \wedge dx^j; \end{aligned} \tag{1.3.3}$$

$$\dot{\theta}^a = \tfrac{1}{2}[\delta_j(\Gamma^a_i) - \delta_i(\Gamma^a_j)]\, dx^j \wedge dx^i + [\dot{\partial}_b(\Gamma^a_i) - \Gamma^a_{b;i}]\psi^b \wedge dx^i + \tfrac{1}{2}[\Gamma^a_{cb} - \Gamma^a_{bc}]\, \psi^b \wedge \psi^c. \tag{1.3.4}$$

The relationship between the torsion and the covariant derivative is the usual one:

PROPOSITION 1.3.1: *Let $D: \mathcal{X}(T\tilde{M}) \to \mathcal{X}(T^*\tilde{M} \otimes T\tilde{M})$ be the linear connection on $\tilde{M}$ induced by a good vertical connection. Then for any X, $Y \in \mathcal{X}(T\tilde{M})$ we have*

$$\nabla_X Y - \nabla_Y X = [X,Y] + \theta(X,Y).$$

Proof: Write $X = X^i\delta_i + \dot{X}^a\dot{\partial}_a$ and $Y = Y^j\delta_j + \dot{Y}^b\dot{\partial}_b$. Then

$$\begin{aligned} \nabla_X Y &= [X(Y^i) + \omega_j^i(X)Y^j]\delta_i + [X(\dot{Y}^a) + \omega_b^a(X)\dot{Y}^b]\dot{\partial}_a; \\ \nabla_Y X &= [Y(X^i) + \omega_j^i(Y)X^j]\delta_i + [Y(\dot{X}^a) + \omega_b^a(Y)\dot{X}^b]\dot{\partial}_a; \\ \theta(X,Y) &= [\omega_j^i(X)Y^j - \omega_j^i(Y)X^j]\delta_i + [\omega_b^a(X)\dot{Y}^b - \omega_b^a(Y)\dot{X}^b]\dot{\partial}_a + [d\psi^a(X,Y)]\dot{\partial}_a; \end{aligned}$$

$$d\psi^a(X,Y) = X(\dot{Y}^a) - Y(\dot{X}^a) - \psi^a([X,Y]);$$
$$[X,Y] = X(Y^i)\delta_i - Y(X^i)\delta_i + \psi^a([X,Y])\dot{\partial}_a,$$

where we used (1.1.20), and the assertion follows. □

1.3.3. The curvature

DEFINITION 1.3.5: The *curvature tensor* $R\colon \mathcal{X}(T\tilde{M}) \to \mathcal{X}(\bigwedge^2 T^*\tilde{M} \otimes T\tilde{M})$ associated to a good vertical connection D is

$$R = D \circ D,$$

that is

$$\forall X \in \mathcal{X}(T\tilde{M}) \qquad R_X = D(DX).$$

We called R a tensor for a good reason: it is $C^\infty(\tilde{M})$-linear. Indeed if we take $H = H^i\delta_i \in \mathcal{X}(\mathcal{H})$ and $f \in C^\infty(\tilde{M})$ we have

$$\begin{aligned} D(df \otimes H) &= D(H^i df \otimes \delta_i) = dH^i \wedge df \otimes \delta_i - H^i df \wedge \omega_j^i \otimes \delta_i \\ &= -df \wedge (dH^i + H^j\omega_j^i) \otimes \delta_i = -df \wedge DH. \end{aligned}$$

Analogously one shows that $D(df \otimes X) = -df \wedge DX$ for any $X \in \mathcal{X}(T\tilde{M})$, and hence

$$R(fX) = D\big(D(fX)\big) = D(df\otimes X + fDX) = -df\wedge DX + df\wedge DX + fR(X) = fR(X).$$

Locally, R is given by

$$\begin{aligned} R(\delta_i) &= D(\omega_i^j \otimes \delta_j) = [d\omega_i^j - \omega_i^h \wedge \omega_h^j] \otimes \delta_j, \\ R(\dot{\partial}_a) &= D(\omega_a^b \otimes \dot{\partial}_b) = [d\omega_a^b - \omega_a^c \wedge \omega_c^b] \otimes \dot{\partial}_b. \end{aligned}$$

DEFINITION 1.3.6: Set

$$\Omega_a^b = d\omega_a^b - \omega_a^c \wedge \omega_c^b,$$

and define the *curvature operator* $\Omega \in \mathcal{X}(\bigwedge^2 T^*\tilde{M} \otimes T^*\tilde{M} \otimes T\tilde{M})$ by

$$\Omega = \Omega_b^a \otimes [dx^b \otimes \delta_a + \psi^b \otimes \dot{\partial}_a].$$

The curvature operator Ω is a global $T^*\tilde{M}\otimes T\tilde{M}$-valued 2-form, that is $\Omega(X,Y)$ is a global $T\tilde{M}$-valued 1-form for any X, $Y \in \mathcal{X}(T\tilde{M})$. Indeed, we have

$$\Omega(X,Y)Z = R_Z(X,Y)$$

for any X, Y, $Z \in \mathcal{X}(T\tilde{M})$, and so Ω is well-defined.

The relationship between the curvature operator and the covariant derivative is as usual:

PROPOSITION 1.3.2: *Let $D: \mathcal{X}(T\tilde{M}) \to \mathcal{X}(T^*\tilde{M} \otimes T\tilde{M})$ be the linear connection on $\tilde{M}$ induced by a good vertical connection. Then for any $X, Y \in \mathcal{X}(T\tilde{M})$ we have*

$$\nabla_X \nabla_Y - \nabla_Y \nabla_X = \nabla_{[X,Y]} + \Omega(X,Y). \tag{1.3.5}$$

Proof: The assertion is proved as soon as we show that we get an identity both applying (1.3.5) to a vertical vector field $V \in \mathcal{X}(\mathcal{V})$ and applying it to a horizontal vector field $H \in \mathcal{X}(\mathcal{H})$. So take $V = V^a \dot{\partial}_a \in \mathcal{X}(\mathcal{V})$; then

$$\nabla_X V = [X(V^a) + \omega_l^a(X) V^l] \dot{\partial}_a;$$

$$\begin{aligned} \nabla_X(\nabla_Y V) &= \{X\big(Y(V^a) + \omega_b^a(Y)V^b\big) + \omega_l^a(X)[Y(V^l) + \omega_b^l(Y)V^b]\}\dot{\partial}_a \\ &= \{X\big(Y(V^a)\big) + X\big(\omega_b^a(Y)\big)V^b \\ &\quad + \omega_l^a(Y)X(V^l) + \omega_l^a(X)Y(V^l) + \omega_l^a(X)\omega_b^l(Y)V^b\}\dot{\partial}_a; \\ \nabla_Y(\nabla_X V) &= \{Y\big(X(V^a)\big) + Y\big(\omega_b^a(X)\big)V^b \\ &\quad + \omega_l^a(X)Y(V^l) + \omega_l^a(Y)X(V^l) + \omega_l^a(Y)\omega_b^l(X)V^b\}\dot{\partial}_a; \\ \nabla_{[X,Y]}V &= \{X\big(Y(V^a)\big) - Y\big(X(V^a)\big) + \omega_l^a([X,Y])V^l\}\dot{\partial}_a, \end{aligned}$$

and

$$d\omega_b^a(X,Y) = X\big(\omega_b^a(Y)\big) - Y\big(\omega_b^a(X)\big) - \omega_b^a([X,Y]).$$

So

$$\begin{aligned} \nabla_X \nabla_Y V - \nabla_Y \nabla_X V - \nabla_{[X,Y]} V &= [d\omega_b^a(X,Y) - (\omega_b^l \wedge \omega_l^a)(X,Y)]V^b \dot{\partial}_a \\ &= \Omega_b^a(X,Y) V^b \dot{\partial}_a = \Omega(X,Y)V. \end{aligned}$$

The same computation works for a horizontal vector field $H \in \mathcal{X}(\mathcal{H})$. □

We end this section recovering the Bianchi identities:

PROPOSITION 1.3.3: *Let $D: \mathcal{X}(T\tilde{M}) \to \mathcal{X}(T^*\tilde{M} \otimes T\tilde{M})$ be the linear connection on $\tilde{M}$ induced by a good vertical connection. Then*

$$\begin{cases} D\theta = \eta \wedge \Omega, \\ D\Omega = 0. \end{cases} \tag{1.3.6}$$

Proof: Recalling (1.3.3) and (1.3.4) we get

$$\begin{aligned} &d\theta^i + \theta^j \wedge \omega_j^i = dx^h \wedge d\omega_h^i - dx^h \wedge \omega_h^j \wedge \omega_j^i = dx^h \wedge \Omega_h^i, \\ &d\dot{\theta}^a + \dot{\theta}^b \wedge \omega_b^a = -d\psi^c \wedge \omega_c^a + \omega^c \wedge d\omega_c^a + d\psi^b \wedge \omega_b^a - \psi^c \wedge \omega_c^b \wedge \omega_b^a = \psi^c \wedge \Omega_c^a, \end{aligned}$$

and the first formula is proved. For the second one,

$$\begin{aligned}(D\Omega)^a_b &= d\Omega^a_b + \Omega^c_b \wedge \omega^a_c - \omega^c_b \wedge \Omega^a_c \\ &= -d\omega^c_b \wedge \omega^a_c + \omega^c_b \wedge d\omega^a_c + d\omega^c_b \wedge \omega^a_c - \omega^d_b \wedge \omega^c_d \wedge \omega^a_c \\ &\qquad - \omega^c_b \wedge d\omega^a_c + \omega^c_b \wedge \omega^d_c \wedge \omega^a_d \\ &= 0,\end{aligned}$$

and we are done. □

1.4. The Cartan connection

1.4.1. Finsler metrics

Finally we may start talking about Finsler metric.

DEFINITION 1.4.1: A *Finsler metric* on a manifold M is a function $F: TM \to \mathbb{R}^+$ satisfying the following properties:

(a) $G = F^2$ is smooth on $\tilde{M}$;
(b) $F(u) > 0$ for all $u \in \tilde{M}$;
(c) $F(\mu_\lambda(u)) = |\lambda| F(u)$ for all $u \in TM$ and $\lambda \in \mathbb{R}$;
(d) for any $p \in M$ the *indicatrix* $I_F(p) = \{u \in T_pM \mid F(u) < 1\}$ is strongly convex.

A manifold M endowed with a Finsler manifold will be called a *Finsler manifold.*

A couple of comments are in order. First of all, if (g_{ab}) is a Riemannian metric on M, it is clear that $F: TM \to \mathbb{R}^+$ given by

$$\forall u \in T_pM \qquad F(u) = \left(g_{ab}(p) u^a u^b\right)^{1/2}$$

is a Finsler metric; we shall say that F *comes from a Riemannian metric.* In this case, though, $G = F^2$ is smooth on the whole TM, and not only on $\tilde{M}$ as required in (a). As we shall see in a moment, this is not accidental: a Finsler metric F is smooth on TM iff it comes from a Riemannian metric. Condition (a) is the reason behind the somewhat anomalous smoothness assumptions in the previous sections.

The easiest example of Finsler metric not coming from a Riemannian metric is described in the following definition:

DEFINITION 1.4.2: A *real Minkowski space* is given by $\mathbb{R}^m$ endowed with the Finsler metric $F: \mathbb{R}^m \times \mathbb{R}^m \cong T\mathbb{R}^m \to \mathbb{R}^+$ defined by

$$\forall p \in \mathbb{R}^m \; \forall u \in T_p\mathbb{R}^m \cong \mathbb{R}^m \quad F(p; u) = \|u\|,$$

where $\|\cdot\|: \mathbb{R}^m \to \mathbb{R}^+$ is a norm with strongly convex unit ball on $\mathbb{R}^m$. If $\|\cdot\|$ is not the norm associated to a scalar product, then F does not come from a Riemannian metric.

A Riemannian structure on M is obtained defining an inner product on any tangent space T_pM varying smoothly with p. A Finsler structure, on the other hand, is obtained defining a norm on any T_pM varying smoothly with p; this is the content of conditions (a), (b) and (c). As we shall see in the next section, condition (c) will allow the measurement of length of curves on the manifold — and then to define geodesics. Finally, condition (d) is slightly stronger than convexity of the norm on each T_pM. To be precise, we are requiring that for every $u \in \tilde{M}$ the Hessian (with respect to the vector variables) of F^2 is positive definite. In symbols, we require that

$$\forall u \in \tilde{M} \qquad (G_{ab}(u)) > 0,$$

where (here and in the rest of the book) $G = F^2$ and subscripts denote derivatives. We shall use a semi-colon to distinguish between derivatives with respect to the point variables and derivatives with respect to the vector variables; for example,

$$G_{;i} = \frac{\partial G}{\partial x^i}, \qquad G_a = \frac{\partial G}{\partial u^a}, \qquad G_{a;i} = \frac{\partial^2 G}{\partial u^a \partial x^i},$$

and so on.

The main (and almost unique) property of the function G is its homogeneity, that is

$$G(p; \lambda u) = \lambda^2 G(p; u) \tag{1.4.1}$$

for all $p \in M$, $u \in T_pM$ and $\lambda \in \mathbb{R}$. Differentiating with respect to λ and setting $\lambda = 1$ we get

$$\forall (p; u) \in TM \qquad G_a(p; u)u^a = 2G(p; u). \tag{1.4.2}$$

On the other hand, differentiating (1.4.1) with respect to u^a we get

$$G_a(p; \lambda u) = \lambda G_a(p; u); \tag{1.4.3}$$

differentiating again with respect to λ and setting $\lambda = 1$ we get

$$G_{ab}(p; u)u^b = G_a(p; u) \qquad \text{and} \qquad G_{ab}(p; u)u^a u^b = 2G(p; u). \tag{1.4.4}$$

Differentiating another time (1.4.3) with respect to u^b we obtain

$$G_{ab}(p; \lambda u) = G_{ab}(p; u), \tag{1.4.5}$$

and hence

$$G_{abc}(p; u)u^c = 0. \tag{1.4.6}$$

This is enough to prove the claim about the smoothness of G:

Lemma 1.4.1: *Let $F: TM \to \mathbb{R}^+$ be a Finsler metric on a manifold M. Then $G = F^2$ is smooth on TM iff F comes from a Riemannian metric on M.*

Proof: One direction is clear. Conversely, assume G smooth (C^2 is enough) on TM. Take $p \in M$ and $u \in T_pM$, $u \neq o_p$; then (1.4.4) yields

$$2G(p; u) = G_{ab}(p; u)u^a u^b.$$

Take $t > 0$; by (1.4.5) $G_{ab}(p; tu) = G_{ab}(p; u)$. Hence

$$\forall t > 0 \qquad 2G(p; u) = G_{ab}(p; tu)u^a u^b.$$

Letting t go to zero, by smoothness we obtain

$$2G(p; u) = G_{ab}(p; o_p)u^a u^b.$$

It is then easy to check that setting $g_{ab}(p) = \frac{1}{2}G_{ab}(p; o_p)$ one gets a Riemannian metric on TM, whose associated norm is exactly F. □

A consequence of condition (d) is that the Hessian matrix (G_{ab}) is invertible. Later on we shall need derivatives of its inverse (G^{ab}); they are given by the following formula:

$$\partial(G^{ab}) = -G^{ar}G^{sb}\partial(G_{rs}), \tag{1.4.7}$$

where ∂ here denotes any first-order differential operator.

1.4.2. The Cartan connection

The aim of this section is to associate to any Finsler metric a good vertical connection which is, in some sense, the generalization of the Levi-Civita connection. The first observation is that condition (d) allows us to introduce a Riemannian structure on $\mathcal{V}$, by setting

$$\forall V, W \in \mathcal{V}_u \qquad \langle V \mid W \rangle_u = \tfrac{1}{2}G_{ab}(u)V^a W^b.$$

It is easy to check that $\langle \mid \rangle$ is well-defined, and it actually is a Riemannian metric on $\mathcal{V}$. Note that (1.4.4) implies that

$$G \equiv \langle \iota \mid \iota \rangle. \tag{1.4.8}$$

In other words, embedding $\tilde{M}$ in $\mathcal{V}$ by means of ι we recover the Finsler metric.

DEFINITION 1.4.3: The Riemannian structure $\langle \mid \rangle$ on $\mathcal{V}$ so defined is said to be *induced by* the Finsler metric.

The main result of this section is the construction of the good vertical connection associated to the Riemannian structure induced by F.

THEOREM 1.4.2: *Let $F: TM \to \mathbb{R}^+$ be a Finsler metric, and $\langle \mid \rangle$ the Riemannian structure on $\mathcal{V}$ induced by F. Then there is a unique vertical connection $D: \mathcal{X}(\mathcal{V}) \to \mathcal{X}(T^*\tilde{M} \otimes \mathcal{V})$ such that*

(i) *D is good;*
(ii) *for all $X \in T\tilde{M}$ and $V, W \in \mathcal{X}(\mathcal{V})$ one has*

$$X\langle V \mid W \rangle = \langle \nabla_X V \mid W \rangle + \langle V \mid \nabla_X W \rangle; \tag{1.4.9}$$

(iii) $\theta(V,W)=0$ *for all* V, $W\in\mathcal{V}$, *where* θ *is the torsion of the linear connection on* $\tilde{M}$ *induced by* D;
(iv) $\theta(H,K)\in\mathcal{V}$ *for all* H, $K\in\mathcal{H}$.

Proof: Assume such a connection exists; we shall recover the connection forms ω^a_b, showing its uniqueness.

First of all, (1.4.9) yields

$$\begin{aligned} G_{ars} &= \dot\partial_a(G_{rs}) = 2\dot\partial_a\langle\dot\partial_r \mid \dot\partial_s\rangle = 2\langle\nabla_{\dot\partial_a}\dot\partial_r \mid \dot\partial_s\rangle + 2\langle\dot\partial_r \mid \nabla_{\dot\partial_a}\dot\partial_s\rangle \\ &= 2\langle\omega^h_r(\dot\partial_a)\dot\partial_h \mid \dot\partial_s\rangle + 2\langle\dot\partial_r \mid \omega^k_s(\dot\partial_a)\dot\partial_k\rangle = G_{hs}\Gamma^h_{ra} + G_{rh}\Gamma^h_{sa}. \end{aligned}$$

Analogously we find

$$\begin{aligned} G_{rsa} &= G_{ha}\Gamma^h_{sr} + G_{sh}\Gamma^h_{ar}, \\ G_{sar} &= G_{hr}\Gamma^h_{as} + G_{ah}\Gamma^h_{rs}. \end{aligned}$$

Now, by (1.3.4) and (1.3.3) condition (iii) is equivalent to

$$\Gamma^h_{rs} = \Gamma^h_{sr};$$

so we obtain

$$\Gamma^h_{rs} = \tfrac{1}{2}G^{hk}G_{rsk}, \tag{1.4.10}$$

where (G^{hk}) denotes the inverse matrix of (G_{hk}). In particular, $\Gamma^h_{rs}u^r=0$; so $\Lambda|_{\mathcal{V}}=\mathrm{id}\,|_{\mathcal{V}}$ (where we recall that $\Lambda(X)=\nabla_X\iota$), and (i) turns out to be a consequence of (ii) and (iii).

Using again (ii) we get

$$\begin{aligned} \delta_i(G_{rs}) &= 2\delta_i\langle\dot\partial_r \mid \dot\partial_s\rangle = 2\langle\nabla_{\delta_i}\dot\partial_r \mid \dot\partial_s\rangle + 2\langle\dot\partial_r \mid \nabla_{\delta_i}\dot\partial_s\rangle \\ &= G_{hs}\Gamma^h_{r;i} + G_{rh}\Gamma^h_{s;i}. \end{aligned}$$

Analogously,

$$\begin{aligned} \delta_r(G_{si}) &= G_{hi}\Gamma^h_{s;r} + G_{sh}\Gamma^h_{i;r}; \\ \delta_s(G_{ir}) &= G_{hr}\Gamma^h_{i;s} + G_{ih}\Gamma^h_{r;s}. \end{aligned}$$

Now, by (1.3.3) and (1.3.4) condition (iv) is equivalent to

$$\Gamma^h_{i;j} = \Gamma^h_{j;i}; \tag{1.4.11}$$

hence we get

$$\begin{aligned} \Gamma^h_{i;j} &= \tfrac{1}{2}G^{hk}\left[\delta_j(G_{ki}) + \delta_i(G_{kj}) - \delta_k(G_{ij})\right] \\ &= \gamma^h_{ij} - \tfrac{1}{2}G^{hl}\left[G_{ilk}\Gamma^k_j + G_{jlk}\Gamma^k_i - G_{ijk}\Gamma^k_l\right], \end{aligned} \tag{1.4.12}$$

where

$$\gamma^h_{ij} = \tfrac{1}{2}G^{hk}\left[G_{ki;j} + G_{kj;i} - G_{ij;k}\right] = \gamma^h_{ji}.$$

This is not enough to determine uniquely the connection, because the coefficients Γ^k_i are still unknown. To recover them, we contract with ι, that is we compute the following:

$$\gamma_{ij}^h u^j = \tfrac{1}{2} G^{hk} \left[G_{ki;j} u^j + G_{k;i} - G_{i;k} \right],$$

and

$$\gamma_{ij}^h u^i u^j = G^{hk} \left[G_{k;j} u^j - G_{;k} \right],$$

where we used (1.4.1)–(1.4.5). Therefore Lemma 1.2.2 now yields

$$\Gamma_j^h = \Gamma_{i;j}^h u^i = \gamma_{ij}^h u^i - \Gamma_{jk}^h \Gamma_i^k u^i, \tag{1.4.13}$$

and so

$$\Gamma_j^h u^j = \gamma_{ij}^h u^i u^j = G^{hk} \left[G_{k;j} u^j - G_{;k} \right].$$

Thus we get

$$\Gamma_j^h = \tfrac{1}{2} G^{hk} \left[G_{kj;i} u^i + G_{k;j} - G_{j;k} \right] - \Gamma_{jk}^h G^{kl} \left[G_{l;i} u^i - G_{;l} \right]. \tag{1.4.14}$$

Hence we have determined the coefficients Γ_j^h and so, by (1.4.12), the connection forms ω_b^a.

For the existence, a long but straightforward computation shows that, under change of coordinates, the coefficients Γ_{bc}^a given by (1.4.10) behave as in (1.2.3), the coefficients Γ_i^a given by (1.4.14) behave as in (1.1.19), and the coefficients $\Gamma_{b;i}^a$ given by (1.4.12) behave as in (1.2.3). Therefore the forms

$$\omega_b^a = \Gamma_{b;i}^a \, dx^i + \Gamma_{bc}^a \psi^c = \tilde{\Gamma}_{b;i}^a \, dx^i + \Gamma_{bc}^a \, du^c,$$

are the connection forms of a good vertical connection satisfying (i)–(iv), and we are done. □

DEFINITION 1.4.4: The good vertical connection whose existence we have just proved is the *Cartan connection* associated to the Finsler metric F.

This is the Cartan connection introduced (in local coordinates) by Cartan and usually studied in classical Finsler geometry. Indeed, an easy computation shows that

$$\dot{\partial}_r(\gamma_{ij}^h) u^i u^j = -2 \Gamma_{kr}^h \gamma_{ij}^k u^i u^j;$$

so setting

$$G^h = \tfrac{1}{2} \gamma_{ij}^h u^i u^j = \tfrac{1}{2} G^{hk} \left[G_{k;i} u^i - G_{;k} \right],$$

from (1.4.13) we get

$$\Gamma_i^h = \dot{\partial}_i (G^h);$$

this is the way the Cartan connection is usually presented.

It should be remarked that the Cartan connection is not the only linear connection on $T\tilde{M}$ canonically associated to a Finsler metric; we just recall the Berwald connection (see [Rd1], [B]) and the connection recently defined by Bao and Chern (see [BC]). It would be interesting to know whether it is possible to give a global characterization for these connections similar to the one we just presented for the Cartan connection.

Coming back to our main concern, it is worthwhile to remark that the non-linear connection associated to the Cartan connection is homogeneous. Indeed, (1.4.5) and (1.4.10) show that

$$\Gamma^a_{bc}(\mu_\lambda(u)) = \lambda^{-1}\Gamma^a_{bc}(u),$$

and hence the assertion follows from (1.4.14), (1.4.1), (1.4.3) and (1.1.17).

In defining the Cartan connection, we asked only for the vanishing of part of the torsion. We cannot ask much more; for instance, we cannot ask for the vanishing of the whole horizontal part of the torsion without trivializing the theory. Indeed,

PROPOSITION 1.4.3: *Let $D: \mathcal{X}(\mathcal{V}) \to \mathcal{X}(T^*\tilde{M} \otimes \mathcal{V})$ be the Cartan connection associated to a Finsler metric $F: TM \to \mathbb{R}^+$ on a manifold M. Then $\theta(X,Y) \in \mathcal{V}$ for all X, $Y \in T\tilde{M}$ iff F comes from a Riemannian metric on M. In this case, the non-linear connection associated to the Cartan connection coincides with the Levi-Civita connection induced by the Riemannian metric.*

Proof: Indeed $\theta(X,Y) \in \mathcal{V}$ for all X, $Y \in T\tilde{M}$ iff $\Gamma^a_{bc} \equiv 0$ for all a, b, $c = 1, \ldots, m$, thanks to (1.3.3). This happens iff $G_{abc} \equiv 0$ iff $G_{ab}(p; u)$ depends only on the point p and not on the vector u. But this is the case iff

$$G(p; u) = \tfrac{1}{2}G_{ab}(p)u^a u^b$$

is a Riemannian metric.

Finally, if F comes from a Riemannian metric then (1.4.12) yields $\Gamma^h_{i;j} = \gamma^i_{ij}$, and the final assertion follows from Proposition 1.2.3. □

So let D be the Cartan connection associated to a Finsler metric F. Since D is good, we have a horizontal bundle $\mathcal{H}$ and a horizontal map $\Theta: \mathcal{V} \to \mathcal{H}$. Using Θ we may transfer the Riemannian structure from $\mathcal{V}$ to $\mathcal{H}$, by setting

$$\forall H, K \in \mathcal{H} \qquad \langle H \mid K \rangle = \langle \Theta^{-1}(H) \mid \Theta^{-1}(K) \rangle.$$

We can then define a Riemannian metric on the whole $T\tilde{M}$, just by stating that $\mathcal{H}$ is orthogonal to $\mathcal{V}$, that is

$$\forall H \in \mathcal{H} \quad \forall V \in \mathcal{V} \qquad \langle H \mid V \rangle = 0.$$

We have already seen how to extend a good connection to a linear connection on $T\tilde{M}$. It is then easy to check that the definitions imply

$$X\langle Y \mid Z \rangle = \langle \nabla_X Y \mid Z \rangle + \langle Y \mid \nabla_X Z \rangle$$

for all X, Y, $Z \in \mathcal{X}(T\tilde{M})$. Note that D is not the Levi-Civita connection associated to this Riemannian structure on $\tilde{M}$, because its torsion is not identically zero. But, on the other hand, the existence and the definition of this Riemannian structure depends on the Cartan connection; so it is sensible to say that the Cartan connection is a good generalization to the Finsler situation of the Levi-Civita connection.

1.4.3. The horizontal flag curvature

As we shall see more clearly in the next section, the aim of our approach to Finsler geometry is to provide a setting where, broadly speaking, the arguments used in Riemannian geometry can be used in Finsler geometry just rephrasing them in terms of the horizontal bundle $\mathcal{H}$ and of the radial horizontal field χ. As a consequence, we shall not be interested in the whole curvature tensor, but only in a particular contraction of its horizontal part, which we now define.

DEFINITION 1.4.5: Take $u \in \tilde{M}$; then the *horizontal flag curvature* at u is the bilinear form $R_u: \mathcal{H}_u \times \mathcal{H}_u \to \mathbb{R}$ given by

$$\forall H, K \in \mathcal{H}_u \qquad R_u(H,K) = \langle \Omega(\chi(u), H)K \mid \chi(u) \rangle_u.$$

To work with R_u we need at least to know that it is symmetric. This is the content of the following

PROPOSITION 1.4.4: *Let $F: TM \to \mathbb{R}^+$ be a Finsler metric on a manifold M, and D the Cartan connection associated to F. Then:*

(i) *for all $X, Y \in T\tilde{M}$ we have*

$$\langle \theta(X,Y) \mid \iota \rangle = 0; \tag{1.4.15}$$

(ii) *for any $u \in \tilde{M}$ the horizontal flag curvature R_u is symmetric, that is*

$$\forall H, K \in \mathcal{H} \qquad \langle \Omega(\chi, H)K \mid \chi \rangle = \langle \Omega(\chi, K)H \mid \chi \rangle.$$

Proof: (i) By (1.3.4) and (1.4.4) we see that (1.4.15) is equivalent to

$$G_a \delta_i(\Gamma^a_j) = G_a \delta_j(\Gamma^a_i). \qquad \text{and} \qquad G_a \dot{\partial}_b(\Gamma^a_i) = G_a \Gamma^a_{b;i}.$$

Now, (1.4.4) also implies $G_a G^{ab} = u^b$; hence (1.4.6), (1.4.10) and (1.4.14) show that

$$G_a \Gamma^a_{bc} = 0 \qquad \text{and} \qquad G_a \Gamma^a_i = G_{;i}. \tag{1.4.16}$$

Therefore

$$\begin{aligned} G_a \delta_i(\Gamma^a_j) &= \delta_i(G_a \Gamma^a_j) - \delta_i(G_a)\Gamma^a_j = \delta_i(G_{;j}) - \Gamma^a_j G_{a;i} + \Gamma^a_j \Gamma^b_i G_{ab} \\ &= G_{;ji} - \Gamma^a_i G_{a;j} - \Gamma^a_j G_{a;i} + \Gamma^a_j \Gamma^b_i G_{ab} = G_a \delta_j(\Gamma^a_i). \end{aligned}$$

Furthermore,

$$\begin{aligned} G_a \dot{\partial}_b(\Gamma^a_i) &= \dot{\partial}_b(G_a \Gamma^a_i) - G_{ab}\Gamma^a_i = G_{b;i} - G_{ab}\Gamma^a_i \\ &= \tfrac{1}{2}[G_{b;i} + G_{i;b} - G_{bi;l}u^l] + \tfrac{1}{2}G_{ibc}\Gamma^c_l u^l = G_a \Gamma^a_{b;i}, \end{aligned}$$

and (1.4.15) follows.

(ii) First of all,

$$\omega_b^a(H) = \Gamma^a_{b;s} H^s \qquad \text{and} \qquad \omega_b^a(\chi) = \Gamma^a_{b;s} u^s = \Gamma^a_{s;b} u^s = \Gamma^a_b,$$

where we used (1.4.11) and Lemma 1.2.2. This yields

$$\begin{aligned} d\omega_r^a(\chi, H) &= \left[\delta_d(\Gamma^a_{r;s})u^d - \delta_s(\Gamma^a_r) - \Gamma^c_s \Gamma^a_{r;c} - \Gamma^a_{rc}\big(\delta_d(\Gamma^c_s) - \delta_s(\Gamma^c_d)\big)u^d\right] H^s, \\ (\omega_r^c \wedge \omega_c^a)(\chi, H) &= \left[\Gamma^c_r \Gamma^a_{c;s} - \Gamma^a_c \Gamma^c_{s;r}\right] H^s, \end{aligned}$$

and so

$$\begin{aligned} \langle \Omega(\chi, H)K, \chi\rangle &= G_a\left[d\omega_r^a - \omega_r^c \wedge \omega_c^a\right](\chi, H) K^r \\ &= G_a\left[\delta_d(\Gamma^a_{r;s})u^d - \delta_s(\Gamma^a_r) - \Gamma^c_s\Gamma^a_{r;c} - \Gamma^c_r\Gamma^a_{c;s} + \Gamma^a_c\Gamma^c_{r;s}\right] K^r H^s \\ &= G_a\left[\delta_d(\Gamma^a_{s;r})u^d - \delta_r(\Gamma^a_s) - \Gamma^c_r\Gamma^a_{s;c} - \Gamma^c_s\Gamma^a_{c;r} + \Gamma^a_c\Gamma^c_{s;r}\right] K^r H^s \\ &= \langle \Omega(\chi, K)H, \chi\rangle, \end{aligned}$$

where we used (1.4.11) and part (i), and the assertion follows. □

DEFINITION 1.4.6: So we may safely say that a Finsler metric F has *positive* (or *negative*) *curvature* if the horizontal flag curvature R_u is positive (resp., negative) definite for any $u \in \tilde{M}$.

By the way, it is easy to check that

$$\forall H, K \in \mathcal{H}_u \;\; R_{\mu_\lambda(u)}\big(d(\mu_\lambda)_u(H), d(\mu_\lambda)_u(K)\big) = \lambda^2 R_u(H, K).$$

As we shall see in the next section, the horizontal flag curvature is the correct contraction of the curvature operator to consider, because it is the one appearing in the second variation formula. This is a tipical instance of the phenomenon we described in the preface: it is easier to define and manipulate formally objects on the tangent-tangent level, but they are geometrically meaningful only when brought by contractions at least half-way down toward the tangent level.

We end this section by proving that the curvature operator is antisymmetric:

PROPOSITION 1.4.5: *Let* $D: \mathcal{X}(T\tilde{M}) \to \mathcal{X}(T^*\tilde{M} \otimes T\tilde{M})$ *be the linear connection on* $\tilde{M}$ *induced by a good vertical connection. Then for any* X, Y, Z, $W \in T\tilde{M}$ *we have*

$$\langle \Omega(X, Y)Z \mid W\rangle = -\langle Z \mid \Omega(X, Y)W\rangle = -\langle \Omega(X, Y)W \mid Z\rangle.$$

In particular, $\langle \Omega(X, Y)Z \mid Z\rangle = 0$ *for all* X, Y, $Z \in T\tilde{M}$.

Proof: Indeed, Proposition 1.3.2 yields

$$\begin{aligned}\langle\Omega(X,Y)Z \mid W\rangle &= \langle\nabla_X\nabla_Y Z \mid W\rangle - \langle\nabla_Y\nabla_X Z \mid W\rangle - \langle\nabla_{[X,Y]}Z \mid W\rangle\\ &= X\langle\nabla_Y Z \mid W\rangle - \langle\nabla_Y Z \mid \nabla_X W\rangle\\ &\quad - Y\langle\nabla_X Z \mid W\rangle + \langle\nabla_X Z \mid \nabla_Y W\rangle\\ &\quad - [X,Y]\langle Z \mid W\rangle + \langle Z \mid \nabla_{[X,Y]}W\rangle\\ &= X\big(Y\langle Z \mid W\rangle\big) - X\langle Z \mid \nabla_Y W\rangle - \langle\nabla_Y Z \mid \nabla_X W\rangle\\ &\quad - Y\big(X\langle Z \mid W\rangle\big) + Y\langle Z \mid \nabla_X W\rangle + \langle\nabla_X Z \mid \nabla_Y W\rangle\\ &\quad - [X,Y]\langle Z \mid W\rangle + \langle Z \mid \nabla_{[X,Y]}W\rangle\\ &= -\langle Z \mid \nabla_X\nabla_Y W\rangle + \langle Z \mid \nabla_Y\nabla_X W\rangle + \langle Z \mid \nabla_{[X,Y]}W\rangle\\ &= -\langle Z \mid \Omega(X,Y)W\rangle.\end{aligned}$$

□

1.5. First and second variations

1.5.1. The setting

So far we have introduced a lot of formal objects more or less associated to a Finsler metric, claiming that we shall need them to study the geometry of a Finsler manifold; in this section we start doing exactly this.

As mentioned before, a Finsler metric can be used to measure the length of curves.

DEFINITION 1.5.1: A *regular* curve $\sigma\colon [a,b] \to M$ is a C^1-curve such that

$$\forall t \in [a,b] \qquad \dot\sigma(t) = d\sigma_t\left(\frac{d}{dt}\right) \neq 0.$$

The *length*, with respect to the Finsler metric $F\colon TM \to \mathbb{R}^+$, of the regular curve σ is then given by

$$L(\sigma) = \int_a^b F\big(\dot\sigma(t)\big)\,dt.$$

It is clear that, thanks to condition (c), the length of a curve does not depend on the parametrization; moreover, if σ is a regular curve we can always assume that, up to a reparametrization, $F(\dot\sigma)$ is constant. Needless to say, we can also measure the length of piecewise regular curves, just by adding the lengths of the regular pieces.

A geodesic for the Finsler metric F is a curve which is a critical point of the length functional. To be more precise:

DEFINITION 1.5.2: Let $\sigma_0: [a, b] \to M$ be a regular curve with $F(\dot{\sigma}_0) \equiv c_0$. A *regular variation* of σ_0 is a C^1-map $\Sigma: (-\varepsilon, \varepsilon) \times [a, b] \to M$ such that
(i) $\sigma_0(t) = \Sigma(0, t)$ for all $t \in [a, b]$;
(ii) for every $s \in (-\varepsilon, \varepsilon)$ the curve $\sigma_s(t) = \Sigma(s, t)$ is a regular curve in M;
(iii) $F(\dot{\sigma}_s) \equiv c_s > 0$ for every $s \in (-\varepsilon, \varepsilon)$.
A regular variation Σ is *fixed* if it moreover satisfies
(iv) $\sigma_s(a) = \sigma_0(a)$ and $\sigma_s(b) = \sigma_0(b)$ for all $s \in (-\varepsilon, \varepsilon)$.

If Σ is a regular variation of σ_0, we define the function $\ell_\Sigma: (-\varepsilon, \varepsilon) \to \mathbb{R}^+$ by

$$\ell_\Sigma(s) = L(\sigma_s).$$

DEFINITION 1.5.3: We shall say that a regular curve σ_0 is a *geodesic* for F iff

$$\frac{d\ell_\Sigma}{ds}(0) = 0$$

for all fixed regular variations Σ of σ_0.

Our first aim is to write the first variation of the length functional; we shall then find the differential equation satisfied by the geodesics, and we shall show that every geodesic for F is a geodesic for the non-linear connection associated to the Cartan connection, and conversely.

To write the first variation formula, we shall need to pull-back the Cartan connection along a curve. Unfortunately, the Cartan connection does not live on the tangent bundle, but on $\tilde{TM}$; for this reason the pull-back procedure we shall presently describe is a bit more involved than the usual one in Riemannian geometry.

Let $\Sigma: (-\varepsilon, \varepsilon) \times [a, b] \to M$ be a regular variation of a curve $\sigma_0: [a, b] \to M$. Let

$$p: \Sigma^*(TM) \to (-\varepsilon, \varepsilon) \times [a, b]$$

be the pull-back bundle, and let $\gamma: \Sigma^*(TM) \to TM$ be the fiber map such that the diagram

$$\begin{array}{ccc} \Sigma^*(TM) & \xrightarrow{\gamma} & TM \\ {\scriptstyle p}\downarrow & & \downarrow{\scriptstyle \pi} \\ (-\varepsilon, \varepsilon) \times [a, b] & \xrightarrow{\Sigma} & M \end{array}$$

commutes; γ is the map identifying each $\Sigma^*(TM)_{(s,t)}$ with $T_{\Sigma(s,t)}M$. In particular, a local frame for $\Sigma^*(TM)$ is given by the local fields

$$\left.\frac{\partial}{\partial x^i}\right|_{(s,t)} = \gamma^{-1}\left(\left.\frac{\partial}{\partial x^i}\right|_{\Sigma(s,t)}\right),$$

for $i = 1, \ldots, m$. So an element $\xi \in \mathcal{X}(\Sigma^*(TM))$ can be written locally as

$$\xi(s,t) = u^i(s,t) \left. \frac{\partial}{\partial x^i} \right|_{(s,t)} .$$

Accordingly, a local frame on $T(\Sigma^*(TM))$ is then given by $\{\partial_s, \partial_t, \dot{\partial}_i\}$, where $\partial_s = \partial/\partial s$, $\partial_t = \partial/\partial t$ and $\dot{\partial}_i = \partial/\partial u^i$.

Two particularly important sections of $\Sigma^*(TM)$ are

$$T = \gamma^{-1}\left(d\Sigma\left(\frac{\partial}{\partial t}\right)\right) = \frac{\partial \Sigma^i}{\partial t} \frac{\partial}{\partial x^i},$$

and

$$U = \gamma^{-1}\left(d\Sigma\left(\frac{\partial}{\partial s}\right)\right) = \frac{\partial \Sigma^i}{\partial s} \frac{\partial}{\partial x^i}.$$

DEFINITION 1.5.4: The section U is the *transversal vector* of Σ.

We remark explicitly that, setting $\Sigma^*\tilde{M} = \gamma^{-1}(\tilde{M})$, by assumption we have $T \in \mathcal{X}(\Sigma^*\tilde{M})$; furthermore,

$$T(s,t) = \gamma^{-1}(\dot{\sigma}_s(t)). \tag{1.5.1}$$

Now we may pull-back $T\tilde{M}$ over $\Sigma^*\tilde{M}$ by using γ, obtaining the commutative diagram

$$\begin{array}{ccc} \gamma^*(T\tilde{M}) & \xrightarrow{\tilde{\gamma}} & T\tilde{M} \\ \tilde{p}\big\downarrow & & \big\downarrow\tilde{\pi} \\ \Sigma^*\tilde{M} & \xrightarrow{\gamma} & \tilde{M} \\ p\big\downarrow & & \big\downarrow\pi \\ (-\varepsilon,\varepsilon)\times[a,b] & \xrightarrow{\Sigma} & M \end{array} .$$

In other words, for any $u \in \Sigma^*\tilde{M}_{(s,t)} = \gamma^{-1}(\tilde{M}_{\Sigma(s,t)})$ the map $\tilde{\gamma}$ identifies $\gamma^*(T\tilde{M})_u$ with $T_{\gamma(u)}\tilde{M}$.

In the previous section we showed how to define a Riemannian structure on $T\tilde{M}$. This induces a Riemannian structure on $\gamma^*(T\tilde{M})$ by

$$\forall X, Y \in \gamma^*(T\tilde{M})_u \quad \langle X \mid Y\rangle_u = \langle \tilde{\gamma}(X) \mid \tilde{\gamma}(Y)\rangle_{\gamma(u)}.$$

Analogously, on $T\tilde{M}$ we have the Cartan connection D; this induces a linear connection

$$D^*\colon \mathcal{X}(\gamma^*(T\tilde{M})) \to \mathcal{X}(T^*(\Sigma^*\tilde{M}) \otimes \gamma^*(T\tilde{M}))$$

by setting

$$\nabla^*_X Y = \tilde{\gamma}^{-1}(\nabla_{d\gamma(X)}\tilde{\gamma}(Y))$$

for all $X \in T(\Sigma^*\tilde{M})$ and $Y \in \mathcal{X}(\gamma^*(T\tilde{M}))$. In particular,

$$X\langle Y \mid Z\rangle = \langle \nabla^*_X Y \mid Z\rangle + \langle Y \mid \nabla^*_X Z\rangle$$

for all $X \in T(\Sigma^*\tilde{M})$ and $Y, Z \in \mathcal{X}(\gamma^*(T\tilde{M}))$.

Now take $u \in (\Sigma^*\tilde{M})_{(s,t)}$; then

$$d\gamma_u(T(\Sigma^*\tilde{M})) \subset T_{\gamma(u)}\tilde{M} \qquad \text{and} \qquad \tilde{\gamma}(\gamma^*(T\tilde{M})_u) = T_{\gamma(u)}\tilde{M}.$$

Therefore we get a fiber map $\Xi\colon T(\Sigma^*\tilde{M}) \to \gamma^*(T\tilde{M})$ such that the diagram

$$\begin{array}{ccc} T(\Sigma^*\tilde{M}) & \xrightarrow{\Xi} & \gamma^*(T\tilde{M}) \\ & {\scriptstyle d\gamma}\searrow & \downarrow{\scriptstyle \tilde{\gamma}} \\ & & T\tilde{M} \end{array}$$

commutes, that is $\tilde{\gamma} \circ \Xi = d\gamma$. Since

$$d\gamma(\partial_t) = \frac{\partial \Sigma^i}{\partial t}\partial_i, \qquad d\gamma(\partial_s) = \frac{\partial \Sigma^i}{\partial s}\partial_i \qquad \text{and} \qquad d\gamma(\dot{\partial}_a) = \dot{\partial}_a,$$

it follows that $d\gamma(T(\Sigma^*\tilde{M})) \supset \mathcal{V}$. Moreover, setting $\mathcal{V}^* = \ker dp$, where p is the canonical projection of $\Sigma^*\tilde{M}$ onto $(-\varepsilon, \varepsilon)\times[a, b]$, it is clear that $d\gamma$ is an isomorphism between $\mathcal{V}^*$ and $\mathcal{V}$. Therefore setting

$$\mathcal{H}^* = (d\gamma)^{-1}\left(\mathcal{H} \cap d\gamma(T(\Sigma^*\tilde{M}))\right)$$

we have decomposed $T(\Sigma^*\tilde{M})$ as

$$T(\Sigma^*\tilde{M}) = \mathcal{H}^* \oplus \mathcal{V}^*.$$

A local frame for $\mathcal{H}^*$ is given by

$$\delta_t = \partial_t - (\Gamma_i^a \circ \gamma)\frac{\partial \Sigma^i}{\partial t}\dot{\partial}_a, \qquad \delta_s = \partial_s - (\Gamma_i^a \circ \gamma)\frac{\partial \Sigma^i}{\partial s}\dot{\partial}_a,$$

so that, setting $T^H = d\gamma(\delta_t)$ and $U^H = d\gamma(\delta_s)$, we find

$$T^H(u) = \frac{\partial \Sigma^i}{\partial t}(s,t)\delta_i|_{\gamma(u)} = \chi_{\gamma(u)}(\gamma(T(s,t))) = \chi_{\gamma(u)}(\dot{\sigma}_s(t)) \in \mathcal{H}_{\gamma(u)},$$

and

$$U^H(u) = \frac{\partial \Sigma^i}{\partial s}(s,t)\delta_i|_{\gamma(u)} = \chi_{\gamma(u)}(\gamma(U(s,t))) \in \mathcal{H}_{\gamma(u)},$$

for all $u \in \Sigma^*\tilde{M}_{(s,t)}$. So T^H and U^H are the horizontal lifts of $\gamma(T)$ and $\gamma(U)$ respectively. Moreover, they are tangent horizontal vector fields along γ, and

$$T^H(T(s,t)) = \chi(\dot{\sigma}_s(t)). \tag{1.5.2}$$

We end this preparatory subsection with two final formulas:

$$\begin{aligned} \tilde{\gamma}(\nabla^*_X\Xi(Y) - \nabla^*_Y\Xi(X)) &= \nabla_{d\gamma(X)}d\gamma(Y) - \nabla_{d\gamma(Y)}d\gamma(X) \\ &= [d\gamma(X), d\gamma(Y)] + \theta(d\gamma(X), d\gamma(Y)) \\ &= d\gamma[X,Y] + \theta(d\gamma(X), d\gamma(Y)), \end{aligned} \tag{1.5.3}$$

and

$$\tilde{\gamma}\circ(\nabla^*_X\nabla^*_Y - \nabla^*_Y\nabla^*_X - \nabla^*_{[X,Y]}) = (\nabla_{d\gamma(X)}\nabla_{d\gamma(Y)} - \nabla_{d\gamma(Y)}\nabla_{d\gamma(X)} - \nabla_{d\gamma[X,Y]})\circ\tilde{\gamma}$$
$$= \Omega\big(d\gamma(X), d\gamma(Y)\big)\circ\tilde{\gamma}, \tag{1.5.4}$$

for all $X, Y \in \mathcal{X}\big(T(\Sigma^*\tilde{M})\big)$.

1.5.2. The first variation of the length integral

We are now able to prove the first variation formula for Finsler metrics:

THEOREM 1.5.1: *Let $F\colon TM \to \mathbb{R}^+$ be a Finsler metric on a manifold M. Take a regular curve $\sigma_0\colon [a,b] \to M$, with $F(\dot{\sigma}_0) \equiv c_0 > 0$, and let $\Sigma\colon (-\varepsilon,\varepsilon)\times[a,b] \to M$ be a regular variation of σ_0. Then*

$$\frac{d\ell_\Sigma}{ds}(0) = \frac{1}{c_0}\left\{ \langle U^H \mid T^H\rangle_{\dot{\sigma}_0}\Big|_a^b - \int_a^b \langle U^H \mid \nabla_{T^H}T^H\rangle_{\dot{\sigma}_0}\,dt \right\}.$$

In particular, if the variation Σ is fixed we have

$$\frac{d\ell_\Sigma}{ds}(0) = -\frac{1}{c_0}\int_a^b \langle U^H \mid \nabla_{T^H}T^H\rangle_{\dot{\sigma}_0}\,dt.$$

Proof: By definition,

$$\ell_\Sigma(s) = \int_a^b \big(G(\dot{\sigma}_s)\big)^{1/2}\,dt;$$

therefore

$$\frac{d\ell_\Sigma}{ds} = \frac{d}{ds}\int_a^b \big(G(\dot{\sigma}_s)\big)^{1/2}\,dt = \int_a^b \frac{\partial}{\partial s}\Big[\big(G(\dot{\sigma}_s)\big)^{1/2}\Big]\,dt$$
$$= \frac{1}{2c_s}\int_a^b \frac{\partial}{\partial s}[G(\dot{\sigma}_s)]\,dt = \frac{1}{2c_s}\int_a^b \frac{\partial}{\partial s}\langle \Xi(\delta_t) \mid \Xi(\delta_t)\rangle_T\,dt,$$

where $c_s \equiv F(\dot{\sigma}_s)$, which is constant by definition of regular variation, and we used

$$G(\dot{\sigma}_s) = \langle \chi(\dot{\sigma}_s) \mid \chi(\dot{\sigma}_s)\rangle_{\dot{\sigma}_s} = \big\langle \tilde{\gamma}^{-1}\big(\chi(\dot{\sigma}_s)\big) \mid \tilde{\gamma}^{-1}\big(\chi(\dot{\sigma}_s)\big)\big\rangle_T$$
$$= \langle \Xi(\delta_t) \mid \Xi(\delta_t)\rangle_T,$$

which holds by (1.5.1) and (1.5.2). Now, lifting $\langle \Xi(\delta_t)|\Xi(\delta_t)\rangle_T$ to $\Sigma^*\tilde{M}$ in the obvious way, we get a function which does not depend on the vector variables; therefore

$$\frac{1}{2}\frac{\partial}{\partial s}\langle \Xi(\delta_t) \mid \Xi(\delta_t)\rangle_T = \frac{1}{2}\delta_s\langle \Xi(\delta_t) \mid \Xi(\delta_t)\rangle_T = \langle \nabla^*_{\delta_s}\Xi(\delta_t) \mid \Xi(\delta_t)\rangle_T$$
$$= \langle \nabla^*_{\delta_t}\Xi(\delta_s) \mid \Xi(\delta_t)\rangle_T - \langle \Xi([\delta_t,\delta_s]) \mid \Xi(\delta_t)\rangle_T - \langle \theta(T^H,U^H) \mid T^H\rangle_{\dot{\sigma}_s},$$

by (1.5.3). Now, $\theta(T^H, U^H) \in \mathcal{V}_{\dot\sigma_s}$ because we are working with the Cartan connection, and hence it is orthogonal to T^H. Analogously, $[\delta_t, \delta_s] \in \mathcal{V}^*$ and so $\Xi([\delta_t, \delta_s])$ is orthogonal to $\Xi(\delta_t)$. Thus

$$\begin{aligned}\frac{1}{2}\frac{\partial}{\partial s}\langle \Xi(\delta_t) \mid \Xi(\delta_t)\rangle_T &= \langle \nabla^*_{\delta_t}\Xi(\delta_s) \mid \Xi(\delta_t)\rangle_T \\ &= \delta_t\langle \Xi(\delta_s) \mid \Xi(\delta_t)\rangle_T - \langle \Xi(\delta_s) \mid \nabla^*_{\delta_t}\Xi(\delta_t)\rangle_T \qquad (1.5.5) \\ &= \frac{\partial}{\partial t}\langle U^H \mid T^H\rangle_{\dot\sigma_s} - \langle U^H \mid \nabla_{T^H}T^H\rangle_{\dot\sigma_s}.\end{aligned}$$

Summing up we have found

$$\frac{d\ell_\Sigma}{ds}(s) = \frac{1}{c_s}\left\{\langle U^H \mid T^H\rangle_{\dot\sigma_s}\Big|_a^b - \int_a^b \langle U^H \mid \nabla_{T^H}T^H\rangle_{\dot\sigma_s}\,dt\right\},$$

and the assertion follows. □

As a corollary we get the equation of geodesics:

COROLLARY 1.5.2: *Let $F\colon TM \to \mathbb{R}^+$ be a Finsler metric on a manifold M, and $\sigma\colon [a,b] \to M$ a regular curve. Then σ is a geodesic for F iff*

$$\nabla_{T^H}T^H \equiv 0,$$

where $T^H(u) = \chi_u(\dot\sigma(t)) \in \mathcal{H}_u$ for all $u \in \tilde{M}_{\sigma(t)}$.

Proof: The curve σ is a geodesic iff $(d\ell_\Sigma/ds)(0) = 0$ for any fixed regular variation Σ of σ. Since Σ is fixed, we have $U^H(0,a) = U^H(0,b) = 0$. By Theorem 1.5.1, it follows that σ is a geodesic iff $\nabla_{T^H}T^H = 0$. □

In local coordinates,

$$\nabla_{T^H}T^H = \dot\sigma^j\left[\delta_j(\dot\sigma^i \circ \pi) + (\Gamma^i_{k;j} \circ \dot\sigma)\dot\sigma^k\right]\delta_i = \left[\ddot\sigma^i + (\Gamma^i_j \circ \dot\sigma)\dot\sigma^j\right]\delta_i, \qquad (1.5.6)$$

and so we have proved

COROLLARY 1.5.3: *Let $F\colon TM \to \mathbb{R}^+$ be a Finsler metric on a manifold M. Then a regular curve $\sigma\colon [a,b] \to M$ is a geodesic for F iff it is a geodesic for the non-linear connection associated to the Cartan connection iff*

$$\ddot\sigma = \chi(\dot\sigma).$$

Proof: It follows from (1.5.6), (1.1.21) and (1.1.22). □

In particular, we have shown that a curve σ is a geodesic for F iff $\dot\sigma$ is an integral curve of the radial horizontal vector field χ.

1.5.3. The second variation of the length integral

Our final aim for this section is the second variation formula. It is contained in

THEOREM 1.5.4: *Let $F\colon TM \to \mathbb{R}^+$ be a Finsler metric on a manifold M. Take a geodesic $\sigma_0\colon [a,b] \to M$, with $F(\dot\sigma_0) \equiv 1$, and let $\Sigma\colon (-\varepsilon,\varepsilon)\times[a,b] \to M$ be a regular variation of σ_0. Then*

$$\frac{d^2\ell_\Sigma}{ds^2}(0) = \langle \nabla_{U^H} U^H \mid T^H\rangle_{\dot\sigma_0}\Big|_a^b + \int_a^b \left[\|\nabla_{T^H} U^H\|^2_{\dot\sigma_0} - \langle \Omega(T^H,U^H)U^H \mid T^H\rangle_{\dot\sigma_0} - \left|\frac{\partial}{\partial t}\langle U^H \mid T^H\rangle_{\dot\sigma_0}\right|^2\right] dt,$$

where $\|H\|^2_u = \langle H \mid H\rangle_u$ for all $u \in \tilde M$ and $H \in \mathcal{H}_u$. In particular, if the variation Σ is fixed we have

$$\frac{d^2\ell_\Sigma}{ds^2}(0) = \int_a^b \left[\|\nabla_{T^H} U^H\|^2_{\dot\sigma_0} - \langle \Omega(T^H,U^H)U^H \mid T^H\rangle_{\dot\sigma_0} - \left|\frac{\partial}{\partial t}\langle U^H \mid T^H\rangle_{\dot\sigma_0}\right|^2\right] dt.$$

Proof: During the proof of the first variation formula we saw that

$$\frac{d\ell_\Sigma}{ds}(s) = \int_a^b \frac{\langle \nabla^*_{\delta_t}\Xi(\delta_s) \mid \Xi(\delta_t)\rangle_T}{(\langle \Xi(\delta_t) \mid \Xi(\delta_t)\rangle_T)^{1/2}}\, dt.$$

The integrand is a function defined over $(-\varepsilon,\varepsilon)\times[a,b]$; we may lift it over $\Sigma^*\tilde M$ and then compute

$$\begin{aligned}\frac{\partial}{\partial s}\left[\frac{\langle \nabla^*_{\delta_t}\Xi(\delta_s) \mid \Xi(\delta_t)\rangle_T}{(\langle \Xi(\delta_t) \mid \Xi(\delta_t)\rangle_T)^{1/2}}\right] &= \delta_s\left[\frac{\langle \nabla^*_{\delta_t}\Xi(\delta_s) \mid \Xi(\delta_t)\rangle_T}{(\langle \Xi(\delta_t) \mid \Xi(\delta_t)\rangle_T)^{1/2}}\right]\\ &= \frac{\delta_s\langle \nabla^*_{\delta_t}\Xi(\delta_s) \mid \Xi(\delta_t)\rangle_T}{(\langle \Xi(\delta_t) \mid \Xi(\delta_t)\rangle_T)^{1/2}} - \frac{1}{2}\frac{\langle \nabla^*_{\delta_t}\Xi(\delta_s) \mid \Xi(\delta_t)\rangle_T}{(\langle \Xi(\delta_t) \mid \Xi(\delta_t)\rangle_T)^{3/2}}\,\delta_s\langle \Xi(\delta_t) \mid \Xi(\delta_t)\rangle_T.\end{aligned}$$

Now we already saw in (1.5.5) that $\frac{1}{2}\delta_s\langle \Xi(\delta_t) \mid \Xi(\delta_t)\rangle_T = \langle \nabla^*_{\delta_t}\Xi(\delta_s) \mid \Xi(\delta_t)\rangle_T$; hence

$$\begin{aligned}\frac{1}{2}\frac{\langle \nabla^*_{\delta_t}\Xi(\delta_s) \mid \Xi(\delta_t)\rangle_T}{(\langle \Xi(\delta_t) \mid \Xi(\delta_t)\rangle_T)^{3/2}}\,\delta_s\langle \Xi(\delta_t) \mid \Xi(\delta_t)\rangle_T &= \frac{\left|\langle \nabla^*_{\delta_t}\Xi(\delta_s) \mid \Xi(\delta_t)\rangle_T\right|^2}{(\langle \Xi(\delta_t) \mid \Xi(\delta_t)\rangle_T)^{3/2}}\\ &= \frac{\left|\delta_t\langle \Xi(\delta_s) \mid \Xi(\delta_t)\rangle_T - \langle \Xi(\delta_s) \mid \nabla^*_{\delta_t}\Xi(\delta_t)\rangle_T\right|^2}{(\langle \Xi(\delta_t) \mid \Xi(\delta_t)\rangle_T)^{3/2}}\\ &= \frac{\left|\frac{\partial}{\partial t}\langle U^H \mid T^H\rangle_{\dot\sigma_s} - \langle U^H \mid \nabla_{T^H} T^H\rangle_{\dot\sigma_s}\right|^2}{(\langle T^H \mid T^H\rangle_{\dot\sigma_s})^{3/2}}.\end{aligned}$$

Furthermore

$$\begin{aligned}\delta_s\langle \nabla^*_{\delta_t}\Xi(\delta_s) \mid \Xi(\delta_t)\rangle_T &= \langle \nabla^*_{\delta_s}\nabla^*_{\delta_t}\Xi(\delta_s) \mid \Xi(\delta_t)\rangle_T + \langle \nabla^*_{\delta_t}\Xi(\delta_s) \mid \nabla^*_{\delta_s}\Xi(\delta_t)\rangle_T\\ &= \langle \nabla^*_{\delta_t}\nabla^*_{\delta_s}\Xi(\delta_s) \mid \Xi(\delta_t)\rangle_T - \langle \nabla^*_{[\delta_t,\delta_s]}\Xi(\delta_s) \mid \Xi(\delta_t)\rangle_T\\ &\quad - \langle \Omega(T^H,U^H)U^H \mid T^H\rangle_{\dot\sigma_s} + \langle \nabla^*_{\delta_t}\Xi(\delta_s) \mid \nabla^*_{\delta_t}\Xi(\delta_s)\rangle_T\\ &\quad - \langle \nabla^*_{\delta_t}\Xi(\delta_s) \mid \Xi([\delta_t,\delta_s])\rangle_T - \langle \nabla_{T^H}U^H \mid \theta(T^H,U^H)\rangle_{\dot\sigma_s},\end{aligned}$$

by (1.5.3) and (1.5.4). Now: $\nabla^*_{\delta_t}\Xi(\delta_s)$ and $\nabla_{T^H}U^H$ are horizontal; $\theta(T^H,U^H)$ and $\Xi([\delta_t,\delta_s])=[T^H,U^H]$ are vertical; furthermore for any $V\in\mathring{\mathcal{V}}^*$ one has

$$\begin{aligned}\langle\nabla^*_V\Xi(\delta_s)\mid\Xi(\delta_t)\rangle_T &= \langle\nabla_{d\gamma(V)}U^H\mid T^H\rangle_{\dot\sigma_s}\\ &= G_{ab}(\dot\sigma_s)V^c\left[\dot\partial_c\left(\frac{\partial\Sigma^a}{\partial s}\circ p\right)+\Gamma^a_{dc}(\dot\sigma_s)\left(\frac{\partial\Sigma^d}{\partial s}\circ p\right)\right]\dot\sigma^b_s\\ &= G_a(\dot\sigma_s)\Gamma^a_{dc}(\dot\sigma_s)V^c\left(\frac{\partial\Sigma^d}{\partial s}\circ p\right)=0,\end{aligned}$$

by (1.4.16). Therefore

$$\begin{aligned}&\delta_s\langle\nabla^*_{\delta_t}\Xi(\delta_s)\mid\Xi(\delta_t)\rangle_T\\ &\quad=\langle\nabla^*_{\delta_t}\nabla^*_{\delta_s}\Xi(\delta_s)\mid\Xi(\delta_t)\rangle_T-\langle\Omega(T^H,U^H)U^H\mid T^H\rangle_{\dot\sigma_s}+\langle\nabla^*_{\delta_t}\Xi(\delta_s)\mid\nabla^*_{\delta_t}\Xi(\delta_s)\rangle_T\\ &\quad=\delta_t\langle\nabla^*_{\delta_s}\Xi(\delta_s)\mid\Xi(\delta_t)\rangle_T-\langle\nabla^*_{\delta_s}\Xi(\delta_s)\mid\nabla^*_{\delta_t}\Xi(\delta_t)\rangle_T\\ &\qquad-\langle\Omega(T^H,U^H)U^H\mid T^H\rangle_{\dot\sigma_s}+\|\nabla^*_{\delta_t}\Xi(\delta_s)\|^2_T\\ &\quad=\frac{\partial}{\partial t}\langle\nabla_{U^H}U^H\mid T^H\rangle_{\dot\sigma_s}-\langle\nabla_{U^H}U^H\mid\nabla_{T^H}T^H\rangle_{\dot\sigma_s}\\ &\qquad-\langle\Omega(T^H,U^H)U^H\mid T^H\rangle_{\dot\sigma_s}+\|\nabla_{T^H}U^H\|^2_{\dot\sigma_s}.\end{aligned}$$

Recalling the hypotheses on σ, that is $\nabla_{T^H}T^H=0$ and $\|T^H\|=1$ for $s=0$, we finally get the assertion. □

We remark that, by (1.5.2),

$$\begin{aligned}\langle\Omega(T^H,U^H)U^H\mid T^H\rangle_{\dot\sigma_0}&=\langle\Omega(\chi(\dot\sigma_0),U^H)U^H\mid\chi(\dot\sigma_0)\rangle_{\dot\sigma_0}\\ &=R_{\dot\sigma_0}(U^H,U^H).\end{aligned}$$

So, as announced at the end of the previous section, from the point of view of variation formulas the only geometrically meaningful curvature is the horizontal flag curvature.

1.6. The exponential map

1.6.1. Parallel transport

Let $\sigma\colon[0,a]\to M$ be a regular curve, and ξ a vector field along σ. In subsection 1.1.4 we said that ξ is parallel if $\tilde\nabla_{\dot\sigma}\xi\equiv0$, where $\tilde\nabla$ is the non-linear connection associated to the Cartan connection. Recalling Proposition 1.2.3 we find that ξ is parallel iff

$$\nabla_{T^H}\xi^H\equiv0,\tag{1.6.1}$$

that is, by Lemma 1.1.5, iff

$$\dot\xi=\chi_\xi(\dot\sigma).$$

In local coordinates, (1.6.1) is expressed by

$$\dot{\xi}^a = -\Gamma^a_b(\xi)\dot{\sigma}^b, \qquad \text{for } a = 1, \ldots, m;$$

therefore the Cauchy problem

$$\begin{cases} \nabla_{T^H} \xi^H = 0, \\ \xi(0) = \xi_0 \in \tilde{M}_{\sigma(0)}, \end{cases} \tag{1.6.2}$$

always admits a unique solution $\xi: [0, \varepsilon) \to \tilde{M}$. A priori, ε could depend on ξ_0; however, momentarily we shall see that this is not the case. Indeed, we know that $G(\xi) = \langle \xi^H, \xi^H \rangle_\xi$; so if ξ is parallel one has

$$\frac{d}{dt}(G(\xi)) = T^H(\langle \xi^H, \xi^H \rangle_\xi) = 2\langle \nabla_{T^H} \xi^H, \xi^H \rangle_\xi = 0. \tag{1.6.3}$$

So the length of a parallel vector field is constant. Being the Christoffel symbols of the Cartan connection homogeneous, it follows that ε could be chosen independent of ξ_0 — and depending continuously on $\sigma(0)$. Since $[0, a]$ is a compact interval, the uniqueness of the solution of the Cauchy problem (1.6.2) implies that we can actually choose $\varepsilon = a$. Summing up, we have proved most of the following

PROPOSITION 1.6.1: *Let $\sigma: [0, a] \to M$ be a regular curve on a Finsler manifold M, with $\sigma(0) = p$ and $\sigma(a) = q$. Define a map $A_\sigma: \tilde{M}_p \to \tilde{M}_q$ by setting $A_\sigma(u) = \xi(a)$, where ξ is the unique parallel vector field along σ such that $\xi(0) = u$. Then A_σ is a well-defined bijective homogeneous isometry between $\tilde{M}_p$ and $\tilde{M}_q$.*

Proof: We are only left to show that A_σ is bijective. Let $\sigma_-: [0, a] \to M$ be the curve defined by $\sigma_-(t) = \sigma(a - t)$, so that $\sigma_-(0) = q$, $\sigma_-(a) = p$ and $\dot{\sigma}_-(t) = -\dot{\sigma}(a - t)$ for all $t \in [0, a]$. Take $u \in \tilde{M}_p$, and let ξ be the parallel vector field along σ such that $\xi(0) = u$. If we set $\xi_-(t) = \xi(a - t)$, then ξ_- is clearly parallel along σ_-; hence $A_{\sigma_-}(A_\sigma(u)) = u$. Analogously one proves that $A_\sigma \circ A_{\sigma_-}$ is the identity. □

DEFINITION 1.6.1: The map A_σ so defined is the *parallel transport* along the curve σ.

1.6.2. Geodesics

Having a way to measure the length of curves, we can define a distance on any Finsler manifold.

DEFINITION 1.6.2: Let $F: TM \to \mathbb{R}$ be a Finsler metric on a manifold M. Then the *distance* $d_F: M \times M \to \mathbb{R}^+$ induced by F on M is given by

$$\forall p, q \in M \qquad d_F(p, q) = \inf\{L(\sigma)\},$$

where the infimum is taken with respect to the piecewise regular curves $\sigma: [a, b] \to M$ with $\sigma(a) = p$ and $\sigma(b) = q$.

The proof that d_F is a true distance inducing the manifold topology on M is identical to the corresponding one in Riemannian geometry.

Now let us look to geodesics. In local coordinates, the geodesic equation is

$$\ddot{\sigma}^a = -\Gamma^a_b(\dot{\sigma})\dot{\sigma}^b, \qquad \text{for } a = 1, \ldots, m;$$

hence given $p \in M$ and $u \in \tilde{M}_p$ there exists a unique geodesic $\sigma: [0, \varepsilon] \to M$ with $\sigma(0) = p$ and $\dot{\sigma}(0) = u$.

DEFINITION 1.6.3: Fix $p \in M$; then the unique geodesic $\sigma_u: [0, a] \to M$ with $\sigma_u(0) = p$ and $\dot{\sigma}_u(0) = u$ will be called the *radial geodesic* tangent to $u \in \tilde{M}_p$. If $q = \sigma_u(a)$, we shall say that σ_u *connects* p to q.

A couple of properties of geodesics are trivial. First of all, we have already remarked in (1.6.3) that the length of the tangent vector to a geodesic is constant, that is

$$F(\dot{\sigma}_u) \equiv F(\dot{\sigma}_u(0)) = F(u). \tag{1.6.4}$$

This readily implies that if σ_u is defined on $[0, a]$ then

$$L(\sigma_u) = a\,F(u). \tag{1.6.5}$$

Furthermore, the non-linear connection associated to the Cartan connection is homogeneous; then, exactly as in the Riemannian case, one proves that

$$\sigma_{tu}(s) = \sigma_u(ts) \tag{1.6.6}$$

for all s, $t \in \mathbb{R}$ such that one side (and hence the other too) is defined.

Fix $p \in M$, and denote by $\tilde{S}_p$ the set of vector in $\tilde{M}_p$ of unitary length, i.e.,

$$\tilde{S}_p = \{u \in \tilde{M}_p \mid F(u) = 1\}.$$

We know that there is $\varepsilon > 0$ so that $\sigma_u(t)$ is defined for all $t \in [0, \varepsilon]$ and $u \in \tilde{S}_p$; by (1.6.6), this implies that $\sigma_{\varepsilon u}(1)$ is defined. Setting $\sigma_{o_p}(1) = p$, we see that $\sigma_u(1)$ is defined in a neighborhood of the origin in T_pM; furthermore, it depends continuously on u, because

$$d_F\big(\sigma_u(1), p\big) \le L(\sigma_u|_{[0,1]}) = F(u), \tag{1.6.7}$$

by (1.6.5).

DEFINITION 1.6.4: The *exponential map* $\exp_p$ from T_pM into M is defined in a neighborhood of the origin by setting

$$\exp_p(u) = \sigma_u(1).$$

It is continuous where defined, and smooth outside the origin. As usual, we can patch together all these exponential maps getting the *exponential map* exp defined in a neighborhood of the zero section in TM.

The properties of $\exp_p$ are much the same as the ones enjoyed by its Riemannian analogue. To prove them, however, we need the notion of double of a differential manifold.

Let M be a differentiable manifold of dimension m, and take $p \in M$ any point. If B is a small ball centered at p and $M' = M \setminus B$, the double of M at p is the orientable manifold obtained gluing two copies of M' along the boundaries; roughly speaking, we are separating the directions at p. More formally:

DEFINITION 1.6.5: The *double* of M at p is a pair $(\hat{M}, \pi)$ such that $\hat{M}$ is a differentiable manifold of dimension m, and $\pi: \hat{M} \to M$ is a proper smooth surjective map such that

(a) $\pi^{-1}(p) = S$ is a compact submanifold of $\hat{M}$ diffeomorphic to the sphere S^{m-1};
(b) at any point $\hat{p} \in S$, and with respect to any system of coordinates near $\hat{p}$ on $\hat{M}$ and near p on M, the Jacobian determinant $\det(\partial\pi^h/\partial x^k)$ vanishes of order $m-1$ at $\hat{p}$;
(c) $\hat{M} \setminus S$ has exactly two connected components $\hat{M}_1$, $\hat{M}_2$, and moreover the restrictions $\pi: \hat{M}_j \to M \setminus \{p\}$ are diffeomorphisms for $j = 1, 2$.

These properties determine the double at a point up to diffeomorphisms commuting with the projection π. Furthermore, the double may be described near p as follows. Let U be a coordinate neighborhood on M centered at p, $B(\varepsilon) \subset U$ the ball of radius ε centered at p, and let $V = \pi^{-1}(B(\varepsilon))$. Let $\pi^0: (-\varepsilon, \varepsilon) \times S^{m-1} \to B(\varepsilon)$ be the map given by

$$\pi^0(t, b) = tb.$$

Moreover, define $\pi^\pm: B(\varepsilon) \setminus \{0\} \to (-\varepsilon, \varepsilon) \times S^{m-1}$ by

$$\pi^\pm(x) = \left(\pm\|x\|, \frac{x}{\|x\|}\right);$$

clearly,

$$\pi^0 \circ \pi^\pm = \pm\,\mathrm{id}.$$

Then for $\varepsilon > 0$ small enough there exists a diffeomorphism $f: V \to (-\varepsilon, \varepsilon) \times S^{m-1}$ such that

$$\pi(x) = \pi^0(f(x)).$$

The map f is called a *trivialization* of the double $(\hat{M}, \pi)$. For proofs and further details, see [St].

We need another remark. Take $p \in M$; if $j_p: T_pM \to TM$ is the inclusion, then the differential $d(j_p)_u: T_u(T_pM) \to T_u(TM)$ is a canonical isomorphism between $T_u(T_pM)$ and $\mathcal{V}_u$ for any $u \in T_pM$. So we can (and we shall) identify $T(T_pM)$ with $\mathcal{V}|_{T_pM}$. In particular, in this way the maps $\iota_u: T_pM \to \mathcal{V}_u$ play the same role as the usual identification maps $k_u: T_pM \to T_u(T_pM)$ — compare (1.1.5).

Using the double we are now able to describe the properties of $\exp_p$:

THEOREM 1.6.2: *Let M be a Finsler manifold, and take $p \in M$. Then the exponential map $\exp_p$ at p has the following properties:*

(i) the restriction of $\exp_p$ to any line through the origin in T_pM is a (necessarily smooth) geodesic;

(ii) $\exp_p$ is continuous where defined, and smooth outside the origin;

(iii) $\exp_p$ is Lipschitz in a neighborhood of the origin;

(iv) $\exp_p$ is differentiable at the origin, and $d(\exp_p)_{o_p} = \iota_{o_p}^{-1}$;

(v) there is a neighborhood U_0 of the origin in T_pM such that $\exp_p: U_0 \to \exp_p(U_0)$ is a homeomorphism, and a diffeomorphism restricted to $U_0 \setminus \{o_p\}$;

(vi) for any $u_0, u \in \tilde{M}_p$ the map $t \mapsto d(\exp_p)_{tu}(\iota_{tu}(u_0))$ is smooth wherever defined.

Proof: (i) is true by definition and (1.6.6), and we have already remarked (ii). From now on, we shall work in a coordinate neighborhood U centered at p; in particular, we shall directly identify $\mathcal{V}_u \cong T_u(T_pM)$ and T_pM, disregarding all references to the identification maps ι_u.

By (1.6.7), we can find $r_0 > 0$ so that $\exp_p(tu) \in U$ as soon as $t \in [0, r_0]$ and $u \in \tilde{S}_p$. Now the map $t \mapsto \exp_p(tu) = \sigma_u(t)$ is the solution of a ODE system with Lipschitz coefficients; therefore there is $c_0 > 0$ such that

$$\|\sigma_{u_1}(t) - \sigma_{u_2}(t)\|^2 + \|\dot\sigma_{u_1}(t) - \dot\sigma_{u_2}(t)\|^2 \le c_0^2 \|u_1 - u_2\|^2,$$

for $u_1, u_2 \in \tilde{S}_p$ and $t \in [0, r_0]$, where $\|\cdot\|$ is any fixed norm. In particular,

$$\|\sigma_{u_1}(t) - \sigma_{u_2}(t)\| \le |t| \int_0^1 \|\dot\sigma_{u_1}(st) - \dot\sigma_{u_2}(st)\|\, ds \le c_0 |t| \|u_1 - u_2\|.$$

Now take $v_1 = t_1 u_1$, $v_2 = t_2 u_2 \in T_pM$, with $0 \le t_1 \le t_2 \le r_0$ and $u_1, u_2 \in \tilde{S}_p$. Then

$$\begin{aligned} \|\exp_p(v_1) - \exp_p(v_2)\| &\le \|\sigma_{u_1}(t_1) - \sigma_{u_2}(t_1)\| + \|\sigma_{u_2}(t_1) - \sigma_{u_2}(t_2)\| \\ &\le c_0 |t_1| \|u_1 - u_2\| + c_1 |t_1 - t_2| \\ &\le \sqrt{2}(c_0 + c_1)\|v_1 - v_2\|, \end{aligned}$$

where $c_1 > 0$ is independent of u_j and t_j for $j = 1, 2$, and (iii) is proved.

Next, by definition,

$$\exp_p(tu) = tu + o(|t|)$$

for all $u \in \tilde{S}_p$; since the remainder term is uniform in u, (iv) follows.

To prove (v) and (vi), let (L, π_1) be the double of T_pM at the origin, and $(\hat{M}, \pi_2)$ the double of M at p. Let $f: V \to (-\varepsilon, \varepsilon) \times S^{m-1}$ be a trivialization of $(\hat{M}, \pi_2)$ with respect to our chart U. Again, there is $\rho > 0$ such that $\exp_p(tu) \in \pi_2^0(f(V))$ for all $|t| < \rho$ and $u \in \tilde{S}_p$.

Define the map $F: (-\rho, \rho) \times S^{m-1} \to (-\varepsilon, \varepsilon) \times S^{m-1}$ by setting

$$F(t, b) = \begin{cases} \pi_2^+\left[\exp_p\left(\pi_1^0(t, b)\right)\right] & \text{if } t > 0; \\ (0, b) & \text{if } t = 0; \\ \pi_2^-\left[\exp_p\left(\pi_1^0(t, b)\right)\right] & \text{if } t < 0. \end{cases}$$

Now, F is smooth everywhere, by (i) and (ii). Furthermore, $dF_{(0,b)} = \mathrm{id}$ for every $b \in S^{m-1}$, by (iv); therefore F is a diffeomorphism in a neighborhood of $\{0\} \times S^{m-1}$. Moreover,

$$\exp_p = \pi_2 \circ f^{-1} \circ F \circ g \circ \pi_1^{-1},$$

where $g\colon V' \to (-\rho, \rho) \times S^{n-1}$ is a trivialization of (L, π_1), and so (v) follows.

Finally, $\exp_p \circ \pi_1 \circ g^{-1} = \pi_2 \circ f^{-1} \circ F$ is smooth, and $\pi_1 \circ g^{-1} = \pi_1^0$; so

$$\begin{aligned} d(\exp_p)_{tu}(u_0) &= d(\exp_p \circ \pi_1^0)_{(t\|u\|, u/\|u\|)}(\|u_0\|, u_0) \\ &= d(\pi_2 \circ f^{-1} \circ F)_{(t\|u\|, u/\|u\|)}(\|u_0\|, u_0), \end{aligned}$$

and we are done. □

DEFINITION 1.6.6: A continuous map $f\colon M \to M'$ between two manifolds which is smooth outside a point $p \in M$, a homeomorhism in a neighborhood U of p and a diffeomorphism in $U \setminus \{p\}$ will be called a *local E-diffeomorphism* at p; if moreover $U = M$ we shall briefly say that f is an *E-diffeomorphism* at p.

So $\exp_p$ is a local E-diffeomorphism at the origin. For $r > 0$ set

$$\tilde{B}_p(r) = \{u \in T_pM \mid F(u) < r\}, \qquad \tilde{S}_p(r) = \{u \in \tilde{M}_p \mid F(u) = r\};$$

in particular, $\tilde{B}_p(1)$ is the F-indicatrix at p, and $\tilde{S}_p(1) = \tilde{S}_p$.

We shall moreover set

$$B_p(r) = \exp_p(\tilde{B}_p(r)), \qquad S_p(r) = \exp_p(\tilde{S}_p(r)).$$

We shall see later on that for $r > 0$ small enough we have

$$B_p(r) = \{q \in M \mid d_F(p,q) < r\}, \qquad S_p(r) = \{q \in M \mid d_F(p,q) = r\}.$$

Using this terminology, Theorem 1.6.2.(v) says that there is a $r > 0$ so that $\exp_p$ is a local E-diffeomorphism at o_p between $\tilde{B}_p(r)$ and $B_p(r)$. Then the following definition comes as no surprise:

DEFINITION 1.6.7: The *injectivity radius* $\mathrm{ir}(p)$ of M at p is defined by

$$\mathrm{ir}(p) = \sup\{r > 0 \mid \exp_p \text{ is a local } E\text{-diffeomorphism at } o_p \text{ between } \tilde{B}_p(r) \text{ and } B_p(r)\}.$$

1.6.3. The Gauss lemma

Let M be a Finsler manifold, and fix $p \in M$. For $r < \mathrm{ir}(p)$ we can define a vector field T over $B_p(r)$ by

$$T(\exp_p u) = \frac{1}{F(u)}\, d(\exp_p)_u\big(\iota(u)\big) \in T_{\exp_p(u)}M; \tag{1.6.8}$$

since we have identified $T_u(T_pM)$ with $\mathcal{V}_u$, the vector field T is well-defined. Moreover, as we shall momentarily see,

$$T(\exp_p u) = \frac{1}{F(u)}\, \dot\sigma_u(1); \tag{1.6.9}$$

so T yields the unit tangent vector to the geodesics issuing from p, and thus it is sensible to denote it using the same letter used in section 1.5 for the tangent vector to a curve.

Let us prove (1.6.9).

LEMMA 1.6.3: *Let $F: TM \to \mathbb{R}^+$ be a Finsler metric, and fix $p \in M$. Then*

$$T(\exp_p u) = \frac{\dot\sigma_u(1)}{F(u)}$$

for all $u \in \tilde M_p$ with $F(u) < \mathrm{ir}(p)$. In particular,

$$F(T) \equiv 1. \tag{1.6.10}$$

Proof: A curve σ in $\tilde M_p$ with $\sigma(0) = u$ and $\dot\sigma(0) = \iota(u)$ is given by

$$\sigma(s) = u + su.$$

Hence

$$\begin{aligned} T(\exp_p u) &= \frac{1}{F(u)}\, d(\exp_p)_u\big(\iota(u)\big) = \frac{1}{F(u)}\, \frac{d}{ds}\big\{\exp_p\big((1+s)u\big)\big\}\Big|_{s=0} \\ &= \frac{1}{F(u)}\, \frac{d}{dt}\{\exp_p(tu)\}\Big|_{t=1} = \frac{1}{F(u)}\, \frac{d}{dt}\big(\sigma_u(t)\big)\Big|_{t=1} = \frac{\dot\sigma_u(1)}{F(u)}. \end{aligned}$$

Since, by (1.6.4),

$$F\big(\dot\sigma_u(1)\big) = F(u),$$

(1.6.10) follows. □

So $t \mapsto T\big(\sigma_u(t)\big)$ is the unit tangent vector to the geodesic σ_u; in particular,

$$\nabla_{T^H} T^H = 0. \tag{1.6.11}$$

Now take $r < \mathrm{ir}(p)$, $u_0 \in \tilde S_p(r)$, set $q = \exp_p(u_0) \in S_p(r)$ and take

$$u \in T_q\big(S_p(\delta)\big) \subset T_{q_0}M.$$

Choose a curve $\gamma = \exp_p \tilde{\gamma}: (-\varepsilon, \varepsilon) \to M$ in $S_p(r)$ with $\gamma(0) = q$ and $\dot{\gamma}(0) = u$, where $\tilde{\gamma}$ is a curve in $\tilde{S}_p(r)$ with $\tilde{\gamma}(0) = u_0$. Now we define a variation of σ_{u_0} by setting

$$\Sigma(s,t) = \sigma_{\tilde{\gamma}(s)}(t) = \exp_p\bigl(t\tilde{\gamma}(s)\bigr),$$

for $t \in [0,1]$. Since $\ell_\Sigma(s) = L(\sigma_{\tilde{\gamma}(s)}) = F(\tilde{\gamma}(s)) \equiv r$, the first variation formula Theorem 1.5.1 yields

$$\langle U^H \mid T^H \rangle_{T(q)} = 0,$$

where U is the transversal vector, because σ_{u_0} is a geodesic and $U^H(0) = o$. But

$$U^H(T(q)) = \chi_{T(q)}(U(1)) = \chi_{T(q)}(u) = u^H(T(q)),$$

and we have proved a Finsler version of the classical Gauss lemma:

Proposition 1.6.4: *Let $F: TM \to \mathbb{R}^+$ be a Finsler metric, fix $p \in M$ and take $r < \mathrm{ir}(p)$ and $q \in S_p(r)$. Then $u \in T_qM$ belongs to $T_q(S_p(r))$ iff*

$$\langle u^H \mid T^H \rangle_{T(q)} = 0.$$

Proof: We have seen that if $v \in T_q(S_p(r))$ then v^H is orthogonal to T^H in $\mathcal{H}_{T(q)}$; being both $T_q(S_p(r))$ and $(T^H)^\perp$ vector spaces of dimension $m-1$, we are done.□

Note that, by the homogeneity properties of G_{ab}, in the statement of Proposition 1.6.4 we may replace $T(q)$ by any non-zero multiple of $\dot{\sigma}_u(1)$, where $u \in \tilde{S}_p(r)$ is such that $q = \exp_p(u)$.

1.6.4. Geodesics are locally minimizing

Our next aim is to show that geodesics are locally minimizing. This can be proved by calculus of variations' arguments (see for instance [Rd1]); here instead we shall give a differential geometric proof, due to Bao and Chern (see [BC]). We begin with the following

Lemma 1.6.5: *Let $F: TM \to \mathbb{R}^+$ be a Finsler metric. Take $p \in M$ and $u_0 \in \tilde{M}_p$. Then:*

(i) *$F_{ab}(u_0)U^aU^b \geq 0$ for all $U \in \mathcal{V}_{u_0}$, with equality iff U and $\iota(u_0)$ are linearly dependent;*

(ii) *$F_a(u_0)u^a \leq F(u)$ for all $u \in \tilde{M}_p$, with equality iff u and u_0 are linearly dependent;*

(iii) *$\langle \iota_{u_0}(u) \mid \iota(u_0)\rangle_{u_0} \leq F(u)F(u_0)$ for all $u \in \tilde{M}_p$, with equality iff u and u_0 are linearly dependent.*

Proof: By assumption,

$$\tfrac{1}{2}G_{ab}(u_0)U^aU^b \geq 0$$

with equality iff $U = O$. Now,

$$\tfrac{1}{2}G_{ab} = FF_{ab} + F_aF_b;$$

therefore

$$\begin{aligned} F_a(u_0)F_b(u_0)U^aU^b &= \left(F_a(u_0)U^a\right)^2 \\ &= \left[\frac{1}{2F(u_0)}G_a(u_0)U^a\right]^2 = \frac{1}{G(u_0)}|\langle U \mid \iota(u_0)\rangle_{u_0}|^2 \\ &\le \frac{1}{G(u_0)}\|U\|^2_{u_0}G(u_0) = \|U\|^2_{u_0} \\ &= [F(u_0)F_{ab}(u_0) + F_a(u_0)F_b(u_0)]U^aU^b, \end{aligned}$$

with equality iff U and $\iota(u_0)$ are linearly dependent, and (i) follows.

Now, expanding $F(u)$ at the second order near u_0 (p is fixed) we get

$$F(u) = F(u_0) + F_a(u_0)(u - u_0)^a + F_{ab}(u_0 + \varepsilon(u - u_0))(u - u_0)^a(u - u_0)^b,$$

for a suitable $\varepsilon \in [0, 1]$. Therefore (i) yields

$$F(u_0) + F_a(u_0)(u - u_0)^a \le F(u),$$

with equality iff u and u_0 are linearly dependent. Since

$$F_a(u_0)(u - u_0)^a = F_a(u_0)u^a - F(u_0),$$

this yields (ii).

Finally, (iii) follows from

$$F_a(u_0)u^a = \frac{1}{2F(u_0)}G_a(u_0)u^a = \frac{1}{F(u_0)}\langle \iota_{u_0}(u) \mid \iota(u_0)\rangle_{u_0}.$$

□

We can now prove the usual statement about the local minimizing property of geodesics:

THEOREM 1.6.6: *Let M be a Finsler manifold, and fix $p \in M$ and $r < \mathrm{ir}(p)$. Take a point $q \in S_p(r)$; let $\gamma: [0, 1] \to M$ be any piecewise smooth regular curve in M joining p to q, and let σ_0 be the radial geodesic in $B_p(r)$ connecting p to q. Then*

$$L(\sigma_0) \le L(\gamma),$$

and the equality holds iff γ is a monotone reparametrization of σ_0.

Proof: Take $u_0 \in \tilde{M}_p$ such that $\exp_p(u_0) = q$; then $\sigma_0(t) = \exp_p(tu_0)$. Assume first that the image of γ is contained in $\overline{B_p(r)}$. Then we may write

$$\gamma(s) = \exp_p\bigl(t(s)u(s)\bigr),$$

where $s \in [0,1]$, $F(u(s)) \equiv F(u_0) = r$, $t(0) = 0$, $t(1) = 1$, $t(s) \in [0,1]$ and $u(1) = u_0$.

We now define a variation $\Sigma: [0,1] \times [0,1] \to M$ of σ_0 by setting

$$\Sigma(s,t) = \exp_p(tu(s)),$$

so that $\gamma(s) = \Sigma(s, t(s))$; in particular

$$\dot\gamma(s) = U(s, t(s)) + T(s, t(s))\frac{dt}{ds}, \tag{1.6.12}$$

where U is the transversal vector of Σ, and here $T = \partial\Sigma/\partial t$. Now, Lemma 1.6.5.(iii) applied with $u = \dot\gamma(s)$ and $u_0 = T(s, t(s))$ yields

$$L(\gamma) = \int_0^1 F(\gamma(s); \dot\gamma(s))\,ds \geq \int_0^1 \frac{1}{F(T(s,t(s)))}\langle \dot\gamma(s)^H \mid T^H(s,t(s))\rangle_{T(s,t(s))}\,ds. \tag{1.6.13}$$

The curves $t \mapsto \exp_p(tu(s))$ are geodesics of constant length; so first of all

$$F(T(s,t(s))) = F(T(s,0)) = F(u(s)) \equiv r.$$

Furthermore, for each $s \in [0,1]$ the vector $U(s_0, t(s_0))$ is tangent to the curve

$$s \mapsto \exp_p(t(s_0)u(s)) \in S_p(t(s_0)r);$$

by Proposition 1.6.4 (and the remark immediately after)

$$\langle U^H(s,t(s)) \mid T^H(s,t(s))\rangle_{T(s,t(s))} = 0.$$

So (1.6.12) and (1.6.13) yield

$$L(\gamma) \geq \frac{1}{r}\int_0^1 \langle T^H(s,t(s))\frac{dt}{ds} \mid T^H(s,t(s))\rangle_{T(s,t(s))}\,ds = \frac{1}{r}\int_0^1 r^2\frac{dt}{ds}\,ds = r = L(\sigma_0),$$

with equality iff $U(s,t(s)) \equiv 0$, that is iff $u(s) \equiv u_0$, and so the equality holds iff γ is a monotone reparametrization of σ_0.

Finally, if the image of γ is not contained in $\overline{B_p(r)}$, let $s_0 \in [0,1]$ be the first s such that $\gamma(s) \in S_p(r)$. Then the previous argument yields $L(\gamma|_{[0,s_0]}) \geq r = L(\sigma_0)$; since $L(\gamma) > L(\gamma|_{[0,s_0]})$, we are done. □

As a corollary we have

COROLLARY 1.6.7: *Let M be a Finsler manifold, and fix $p \in M$. Let $r_0 = \mathrm{ir}(p)$; then*

$$\forall u \in \tilde B_p(r_0) \qquad d_F(p, \exp_p(u)) = F(u).$$

In particular,

$$B_p(r) = \{q \in M \mid d_F(p,q) < r\} \quad \text{and} \quad S_p(r) = \{q \in M \mid d_F(p,q) = r\} = \partial B_p(r)$$

for all $r < \mathrm{ir}(p)$.

Proof: Indeed, for every $u \in \tilde{B}_p(r_0)$ we have

$$d_F(p, \exp_p(u)) = L(\sigma_u|_{[0,1]}) = F(u).$$

□

Another consequence of Theorem 1.6.6 is:

LEMMA 1.6.8: *Let M be a Finsler manifold, fix $p \in M$ and take $r < \mathrm{ir}(p)$. If $q \notin B_p(r)$, then there exists $q' \in S_p(r)$ such that $d_F(p,q) = r + d_F(q', q)$.*

Proof: Let $\gamma\colon [0,1] \to M$ be a curve from p to q. Since $q \notin B_p(r)$, there is a first $s_0 \in (0,1]$ such that $\gamma(s_0) \in S_p(r)$. Since, by Corollary 1.6.7, $d_F(p, \gamma(s_0)) = r$, we have

$$L(\gamma) \geq r + d_F(\gamma(s_0), q) \geq r + d_F(S_p(r), q),$$

where as usual

$$d_F(S_p(r), q) = \inf_{q' \in S_p(r)} \{d_F(q', q)\}.$$

Since γ is generic, it follows that

$$d_F(p,q) \geq r + d_F(S_p(r), q).$$

But by the triangle inequality the reverse inequality is true too, and so

$$d_F(p,q) = r + d_F(S_p(r), q).$$

Since $S_p(r)$ is compact, we are done. □

We are now able to state and prove the classical Hopf-Rinow theorem for Finsler metrics:

THEOREM 1.6.9: *Let $F\colon TM \to \mathbb{R}^+$ be a Finsler metric on a manifold M. Then the following assertions are equivalent:*

(i) *the distance d_F is complete;*
(ii) *closed bounded subsets of M are compact;*
(iii) *there is $p \in M$ such that $\exp_p$ is defined on all of T_pM;*
(iv) *$\exp$ is defined on all of TM.*

In addition, any of the previous statements implies that

(v) *for any p, $q \in M$ there is a geodesic σ connecting p to q such that*

$$L(\sigma) = d_F(p,q).$$

Proof: Thanks to Theorem 1.6.6 and Corollary 1.6.8 we can repeat word by word the proof of the Hopf-Rinow theorem for Riemannian manifolds described in [CE].□

DEFINITION 1.6.8: A Finsler metric F and the underlying Finsler manifold M are *complete* iff the induced distance d_F is.

There is a final consequence of Theorem 1.6.6 and of our version of Gauss' lemma that shall be useful in the third chapter.

Let $F: TM \to \mathbb{R}^+$ a Finsler metric, and fix $p \in M$. Define $\rho = \rho_p: M \to \mathbb{R}^+$ by

$$\rho_p(q) = d_F(p, q).$$

By Corollary 1.6.7,

$$\rho = F \circ (\exp_p)^{-1}$$

in $B_p(r)$ for any $r < \mathrm{ir}(p)$; in particular, it is smooth there (but possibly at p, of course).

Now take $r < \mathrm{ir}(p)$ and $q \in S_p(r)$. Then a vector $u \in T_qM$ belongs to $T_q(S_p(r))$ iff

$$0 = u(\rho) = d\rho(u);$$

so Proposition 1.6.4 says that

$$d\rho(u) = 0 \qquad \Longleftrightarrow \qquad \langle u^H \mid T^H \rangle_T = 0.$$

We can do better than this:

PROPOSITION 1.6.10: *Let $F: TM \to \mathbb{R}^+$ be a Finsler metric, fix $p \in M$ and take $r < \mathrm{ir}(p)$ and $q \in S_p(r)$. Then for all $u \in T_qM$ we have*

$$\langle u^H \mid T^H \rangle_{T(q)} = d\rho(u) = u^H(\rho \circ \pi). \tag{1.6.14}$$

Proof: If we write $u = u^a(\partial/\partial x^a)$, then $u^H = u^a \delta_a$; but $\delta_a(\rho \circ \pi) = \partial\rho/\partial x^a$, and so

$$u^H(\rho \circ \pi) = u(\rho) = d\rho(u).$$

We have already proved the first equality if u^H is orthogonal to T^H. On the other hand, if we take $u_0 \in T_pM$ so that $q = \exp_p u_0$, we have

$$\begin{aligned} d\rho(T(q)) &= \frac{1}{F(u_0)}\, d\rho \circ d(\exp_p)_{u_0}(\iota(u_0)) = \frac{1}{F(u_0)}\, dF_{u_0}(\iota(u_0)) \\ &= \frac{1}{F(u_0)}\, u_0^a F_a(u_0) = \frac{1}{F(u_0)}\, u_0^a \left(\frac{1}{2F(u_0)} G_a(u_0)\right) = \frac{G(u_0)}{G(u_0)} \\ &= 1 = G(T(q)) = \langle T^H \mid T^H \rangle_{T(q)}, \end{aligned}$$

and (1.6.14) is verified for $u = T(q)$.

Finally, let $u \in T_qM$ be generic, and write $u = u_0 + \lambda T(q)$, with u_0^H orthogonal to T^H and $\lambda \in \mathbb{R}$. Then

$$d\rho(u) = d\rho(u_0) + \lambda d\rho(T(q)) = \lambda = \langle u_0^H + \lambda T^H \mid T^H \rangle_{T(q)} = \langle u^H \mid T^H \rangle_{T(q)}.$$

□

COROLLARY 1.6.11: *Let* $F\colon TM \to \mathbb{R}^+$ *be a Finsler metric on a manifold* M, *fix* $p \in M$ *and take* $r < \mathrm{ir}(p)$, $u \in \tilde{S}_p(r)$, *and let* $q = \exp_p(u)$. *Then for all* $U \in \mathcal{V}_u$ *we have*

$$\langle U \mid \iota \rangle_u = \langle d(\exp_p)_u(U)^H \mid d(\exp_p)_u(\iota)^H \rangle_{T(q)}.$$

Proof: Indeed by (1.6.14) and Corollary 1.6.7, together with the definition of $T(q)$, we have

$$\begin{aligned}\langle d(\exp_p)_u(U)^H \mid d(\exp_p)_u(\iota)^H \rangle_{T(q)} &= F(u)\langle d(\exp_p)_u(U)^H \mid T^H \rangle_{T(q)} \\ &= F(u) d\rho\big(d(\exp_p)_u(U)\big) \\ &= F(u)\, d(\rho \circ \exp_p)_u(U) \\ &= F(u)\, dF_u(U) = \frac{1}{2} dG_u(U) \\ &= \frac{1}{2} G_a(u) U^a = \langle U \mid \iota \rangle_u.\end{aligned}$$

□

1.7. Jacobi fields and Auslander's theorems

1.7.1. Jacobi fields

In this section we go on with our recasting of classical Riemannian geometry in a Finsler setting. After discussing variation formulas and geodesics, we start dealing with Jacobi fields.

DEFINITION 1.7.1: A *geodesic variation* $\Sigma\colon (-\varepsilon, \varepsilon) \times [0, a] \to M$ of a geodesic $\sigma_0\colon [0, a] \to M$ is a regular variation of σ_0 such that $\sigma_s = \Sigma(s, \cdot)$ is a geodesic for any $s \in (-\varepsilon, \varepsilon)$.

This means that if we put

$$\forall u \in \tilde{M}_{\sigma_s(t)} \qquad T^H(u) = \chi_u\big(\dot\sigma_s(t)\big)$$

then we have

$$\forall s \in (-\varepsilon, \varepsilon) \qquad \nabla_{T^H} T^H|_{\dot\sigma_s} \equiv 0.$$

Setting, as in section 1.5,

$$U^H(u) = \chi_u\left(\frac{\partial \Sigma^a}{\partial s} \frac{\partial}{\partial x^a}\bigg|_{\sigma_s(t)} \right),$$

we may then compute

$$0 = \nabla_{U^H}\nabla_{T^H}T^H = \nabla_{T^H}\nabla_{U^H}T^H + \nabla_{[U^H,T^H]}T^H + \Omega(U^H,T^H)T^H$$
$$= \nabla_{T^H}\nabla_{T^H}U^H + \nabla_{T^H}\big([U^H,T^H] + \theta(U^H,T^H)\big) + \nabla_{[U^H,T^H]}T^H - \Omega(T^H,U^H)T^H. \tag{1.7.1}$$

Now,

$$[U^H,T^H] = \left[\frac{\partial\Sigma^a}{\partial s}\delta_a, \frac{\partial\Sigma^b}{\partial t}\delta_b\right]$$
$$= \frac{\partial\Sigma^a}{\partial s}\frac{\partial\Sigma^b}{\partial t}[\delta_a,\delta_b] + \left\{\frac{\partial\Sigma^a}{\partial s}\delta_a\left(\frac{\partial\Sigma^b}{\partial t}\right) - \frac{\partial\Sigma^a}{\partial t}\delta_a\left(\frac{\partial\Sigma^b}{\partial s}\right)\right\}\delta_b.$$

Since $\partial\Sigma^a/\partial s$ and $\partial\Sigma^a/\partial t$ do not depend on the vector variables, we get

$$\frac{\partial\Sigma^a}{\partial s}\delta_a\left(\frac{\partial\Sigma^b}{\partial t}\right) = \frac{\partial\Sigma^a}{\partial s}\frac{\partial}{\partial x^a}\left(\frac{\partial\Sigma^b}{\partial t}\right) = \frac{\partial^2\Sigma^b}{\partial s\partial t} = \frac{\partial^2\Sigma^b}{\partial t\partial s} = \frac{\partial\Sigma^a}{\partial t}\delta_a\left(\frac{\partial\Sigma^b}{\partial s}\right),$$

and so

$$[U^H,T^H] = \frac{\partial\Sigma^a}{\partial s}\frac{\partial\Sigma^b}{\partial t}[\delta_a,\delta_b] = \frac{\partial\Sigma^a}{\partial s}\frac{\partial\Sigma^b}{\partial t}\{\delta_b(\Gamma^c_{;a}) - \delta_a(\Gamma^c_{;b})\}\dot\partial_c = -\theta(U^H,T^H),$$

because we are working with the Cartan connection.

In particular, then, $[U^H,T^H]$ is vertical; but for any $V \in \mathcal{V}_{\dot\sigma_s}$ we have

$$(\nabla_V T^H)(\dot\sigma_s) = \left[V\left(\frac{\partial\Sigma^a}{\partial t}\right) + \Gamma^a_{b;j}(\dot\sigma_s)V^j(\dot\sigma_s)^b\right]\delta_a = 0,$$

because $V(\partial\Sigma^a/\partial t) = 0$ and $\Gamma^a_{b;j}(\dot\sigma_s)(\dot\sigma_s)^b = 0$. Summing up, we have proved

Proposition 1.7.1: *Let $\Sigma: (-\varepsilon,\varepsilon)\times[0,a] \to M$ be a geodesic variation of a geodesic $\sigma_0: [0,a] \to M$ in a Finsler manifold M. Set*

$$J(t) = \frac{\partial\Sigma^a}{\partial s}(0,t)\left.\frac{\partial}{\partial x^a}\right|_{\sigma_0(t)} \in T_{\sigma_0(t)}M$$

and

$$J^H(t) = \chi_{\dot\sigma_0(t)}\big(J(t)\big) \in \mathcal{H}_{\dot\sigma_0(t)}$$

for all $t \in [0,a]$. Then

$$\nabla_{T^H}\nabla_{T^H}J^H - \Omega(T^H,J^H)T^H \equiv 0. \tag{1.7.2}$$

Note that, since $T^H(\dot\sigma_0(t)) = \chi(\dot\sigma_0(t))$, the equation (1.7.2) can be expressed as

$$\nabla_\chi\nabla_\chi J^H - \Omega(\chi,J^H)\chi \equiv 0$$

along $\dot\sigma_0$.

DEFINITION 1.7.2: Let $\sigma\colon[0,a]\to M$ be a geodesic. A vector field J along σ is called a *Jacobi field* if it satisfies the Jacobi equation (1.7.2) for all $t\in[0,a]$, where $J^H(t)=\chi_{\dot\sigma(t)}\bigl(J(t)\bigr)$. Note that $\dot\sigma$ and $t\dot\sigma$ are Jacobi fields; the first one never vanishing, the second one vanishing only at $t=0$. The set of all Jacobi fields along σ will be denoted by $\mathcal{J}(\sigma)$.

In local coordinates, the Jacobi equation (1.7.2) is a second order linear differential system; hence, given the initial conditions $J(0)$ and $(\nabla_{T^H}J^H)(0)$, there exists a unique solution of the system, defined on $[0,a]$. Our first aim is to express it in a compact form, at least for Jacobi fields vanishing at zero.

PROPOSITION 1.7.2: *Let $\sigma\colon[0,a]\to M$ be a geodesic in a Finsler manifold M, with $p=\sigma(0)$ and $u=\dot\sigma(0)$. Take $H\in\mathcal{H}_u$, and set*

$$W_t=\iota_{tu}\circ\iota_u^{-1}\bigl(\Theta^{-1}(H)\bigr)\in\mathcal{V}_{tu},\tag{1.7.3}$$

where $\Theta\colon\mathcal{V}\to\mathcal{H}$ is the horizontal map induced by the Cartan connection. Then the unique Jacobi field J along σ such that $J(0)=0$ and $\nabla_{T^H}J^H(0)=H$ is given by

$$\forall t\in[0,a]\qquad J(t)=d(\exp_p)_{tu}(tW_t),$$

where we are again identifying $T_{tu}(T_pM)$ with $\mathcal{V}_{tu}$.

Proof: Put $u(s)=u+s\iota_u^{-1}(W_1)$ for $s\in(-\varepsilon,\varepsilon)$, where $\varepsilon>0$ is so small that $u(s)$ is never vanishing. Define $\Sigma\colon(-\varepsilon,\varepsilon)\times[0,a]\to M$ by

$$\Sigma(s,t)=\exp_p\bigl(tu(s)\bigr).$$

Σ is a geodesic variation, and so

$$J=\frac{\partial\Sigma^a}{\partial s}(0,t)\frac{\partial}{\partial x^a}$$

is a Jacobi field. Clearly, $J(0)=0$; moreover,

$$J(t)=\frac{\partial\Sigma}{\partial s}(0,t)=d(\exp_p)_{tu}(tW_t)=t\,d(\exp_p)_{tu}(W_t).$$

Therefore

$$\nabla_{T^H}J^H=\nabla_{T^H}\left[t\bigl(d(\exp_p)_{tu}(W_t)\bigr)^H\right]=\bigl(d(\exp_p)_{tu}(W_t)\bigr)^H+t\nabla_{T^H}\bigl(d(\exp_p)_{tu}(W_t)\bigr)^H,$$

and, when $t=0$,

$$\nabla_{T^H}J^H(0)=\bigl(\iota_u^{-1}(W_1)\bigr)^H=\Theta(W_1)=H,$$

because, by Theorem 1.6.2, $t\mapsto d(\exp_p)_{tu}(W_t)$ is smooth and $d(\exp_p)_{o_p}=\iota_{o_p}^{-1}$. □

In particular, then, zeroes of Jacobi fields and critical points of $\exp_p$ are one and the same thing, exactly as in Riemannian geometry. To express it better, let us recall a customary definition:

DEFINITION 1.7.3: Let $\sigma\colon [0,a] \to M$ be a geodesic. The point $\sigma(t_0)$ is said to be *conjugate* to $\sigma(0)$ along σ, where $t_0 \in (0,a]$, if there exists a non-identically zero Jacobi field J along σ such that $J(0) = 0 = J(t_0)$.

Then:

PROPOSITION 1.7.3: *Let $\sigma\colon [0,a] \to M$ be a geodesic in a Finsler manifold M, and set $p = \sigma(0)$. The point $q = \sigma(t_0)$, with $t_0 \in (0,a]$, is conjugate to p along σ iff $u_0 = t_0\dot\sigma(0)$ is a critical point of* $\exp_p$.

Proof: The point q is a conjugate point of p along σ iff there is a non-identically zero Jacobi field J along σ with $J(0) = J(t_0) = 0$. Let $u = \dot\sigma(0)$ and $H = \nabla_{T^H} J^H(0)$; by Proposition 1.7.2,

$$J(t) = d(\exp_p)_{tu}(tW_t),$$

where $W_t \in \mathcal{V}_{tu}$ is given by (1.7.3); note that J is non-identically zero iff $H \neq 0$. Therefore $q = \sigma(t_0)$ is conjugate to p along σ iff

$$0 = J(t_0) = d(\exp_p)_{u_0}(t_0 W_{t_0}),$$

that is iff u_0 is a critical point of $\exp_p$. □

In particular, it is important to notice that the zeroes of a non-trivial Jacobi field J are discrete: indeed, if they were not, we would have $J(t_0) = 0$ and $\nabla_{T^H} J^H(t_0) = 0$ for some $t_0 \in [0,a]$, and the uniqueness theorem for solutions of a Cauchy problem would imply $J \equiv 0$.

We end this section with two results concerning the behavior of $\langle J^H \mid T^H\rangle_{\dot\sigma}$ along a geodesic σ.

PROPOSITION 1.7.4: *Let $J \in \mathcal{J}(\sigma)$ be a Jacobi field along a geodesic $\sigma\colon [0,a] \to M$ in a Finsler manifold M. Then*

$$\langle J^H \mid T^H\rangle_{\dot\sigma}(t) = t\langle \nabla_{T^H} J^H \mid T^H\rangle_{\dot\sigma}(0) + \langle J^H \mid T^H\rangle_{\dot\sigma}(0).$$

Proof: We have

$$\begin{aligned}\frac{d}{dt}\langle \nabla_{T^H} J^H \mid T^H\rangle_{\dot\sigma} &= T^H\langle \nabla_{T^H} J^H \mid T^H\rangle_{\dot\sigma}\\ &= \langle \nabla_{T^H}\nabla_{T^H} J^H \mid T^H\rangle_{\dot\sigma} = \langle \Omega(T^H, J^H)T^H \mid T^H\rangle_{\dot\sigma} = 0,\end{aligned}$$

by Proposition 1.4.5. In particular,

$$\langle \nabla_{T^H} J^H \mid T^H\rangle_{\dot\sigma} \equiv \langle \nabla_{T^H} J^H \mid T^H\rangle_{\dot\sigma}(0). \tag{1.7.4}$$

Moreover,

$$\frac{d}{dt}\langle J^H \mid T^H\rangle_{\dot\sigma} = T^H\langle J^H \mid T^H\rangle_{\dot\sigma} = \langle \nabla_{T^H} J^H \mid T^H\rangle_{\dot\sigma} \equiv \langle \nabla_{T^H} J^H \mid T^H\rangle_{\dot\sigma}(0),$$

and we are done. □

COROLLARY 1.7.5: *Let $J \in \mathcal{J}(\sigma)$ be a Jacobi field along a geodesic $\sigma\colon [0,a] \to M$ in a Finsler manifold M. Assume that $\langle J^H \mid T^H\rangle_{\dot\sigma}(0) = \langle J^H \mid T^H\rangle_{\dot\sigma}(a)$. Then $\langle J^H \mid T^H\rangle_{\dot\sigma} \equiv \langle J^H \mid T^H\rangle_{\dot\sigma}(0)$, and $\langle \nabla_{T^H} J^H \mid T^H\rangle_{\dot\sigma} \equiv 0$.*

Proof: It follows from Proposition 1.7.4 and (1.7.4). □

DEFINITION 1.7.4: Let $\sigma\colon [0,a] \to M$ be a geodesic in a Finsler manifold M. A *proper Jacobi field* along σ is a Jacobi field $J \in \mathcal{J}(\sigma)$ such that

$$\langle J^H \mid T^H\rangle_{\dot\sigma} \equiv 0.$$

For instance, by Corollary 1.7.5 any Jacobi field J with $J(0) = J(a) = o$ is proper. The space of all proper Jacobi fields along σ will be denoted by $\mathcal{J}_0(\sigma)$.

1.7.2. The Morse index form

DEFINITION 1.7.5: Let $\sigma\colon [a,b] \to M$ be a geodesic in a Finsler manifold M; we say that σ is a *normal geodesic* if it is parametrized by arc-length, i.e., if $F(\dot\sigma) \equiv 1$. In particular, $T(\sigma) = \dot\sigma$.

Let $\sigma\colon [a,b] \to M$ be a normal geodesic in a Finsler manifold M; we shall denote by $\mathcal{X}[a,b]$ the space of all piecewise smooth vector fields ξ along σ such that

$$\langle \xi^H \mid T^H\rangle_T \equiv 0.$$

Moreover, we shall denote by $\mathcal{X}_0[a,b]$ the subspace of all $\xi \in \mathcal{X}[a,b]$ such that $\xi(a) = \xi(b) = o$.

DEFINITION 1.7.6: The *Morse index form* $I = I_a^b\colon \mathcal{X}[a,b] \times \mathcal{X}[a,b] \to \mathbb{R}$ of the normal geodesic $\sigma\colon [a,b] \to M$ is the symmetric (by Proposition 1.4.4) bilinear form

$$I(\xi,\eta) = \int_a^b \Big[\langle \nabla_{T^H}\xi^H \mid \nabla_{T^H}\eta^H\rangle_T - \langle \Omega(T^H,\xi^H)\eta^H \mid T^H\rangle_T\Big]\,dt$$

for all ξ, $\eta \in \mathcal{X}[a,b]$. Note that, by Theorem 1.5.4, if $U \in \mathcal{X}_0[a,b]$ is the transversal vector of a fixed variation Σ of σ then

$$I(U,U) = \frac{d^2\ell_\Sigma}{ds^2}(0). \tag{1.7.5}$$

There is another formula expressing the Morse index form:

Lemma 1.7.6: *Let $\sigma\colon[a,b]\to M$ be a normal geodesic in a Finsler manifold M, and take $\xi\in\mathcal{X}[a,b]$ smooth. Then*

$$I(\xi,\eta)=\langle\nabla_{T^H}\xi^H\mid\eta^H\rangle_T\Big|_a^b-\int_a^b\langle\nabla_{T^H}\nabla_{T^H}\xi^H-\Omega(T^H,\xi^H)T^H\mid\eta^H\rangle_T\,dt,\quad(1.7.6)$$

for all $\eta\in\mathcal{X}[a,b]$.

Proof: Clearly we can assume η smooth (up to breaking σ up in a finite number of pieces). Then

$$\frac{d}{dt}\langle\nabla_{T^H}\xi^H\mid\eta^H\rangle_T=T^H\langle\nabla_{T^H}\xi^H\mid\eta^H\rangle_T=\langle\nabla_{T^H}\nabla_{T^H}\xi^H\mid\eta^H\rangle_T+\langle\nabla_{T^H}\xi^H\mid\nabla_{T^H}\eta^H\rangle_T,$$
$$\langle\Omega(T^H,\xi^H)\eta^H\mid T^H\rangle_T=-\langle\Omega(T^H,\xi^H)T^H\mid\eta^H\rangle_T,$$

and we are done. □

In particular, then, the kernel of the Morse index form is composed by proper Jacobi fields:

Corollary 1.7.7: *Let $\sigma\colon[a,b]\to M$ be a normal geodesic in a Finsler manifold M, and take $\xi\in\mathcal{X}[a,b]$. Then $I(\xi,\mathcal{X}_0[a,b])=\{0\}$ iff ξ is a proper Jacobi field. In particular,*

$$\ker I|_{\mathcal{X}_0[a,b]}=\mathcal{X}_0[a,b]\cap\mathcal{J}_0(\sigma).$$

Proof: Assume that $I(\xi,\mathcal{X}_0[a,b])=\{0\}$. Then by (1.7.6) for every $\eta\in\mathcal{X}_0[a,b]$ smooth we have

$$0=I(\xi,\eta)=-\int_a^b\langle\nabla_{T^H}\nabla_{T^H}\xi^H-\Omega(T^H,\xi^H)T^H\mid\eta^H\rangle_T\,dt,$$

and so $\xi\in\mathcal{J}_0(\sigma)$. The converse statement is obvious. □

As we have already remarked in the previous subsection, there is a strong connection between Jacobi fields and conjugate points. Our next aim is to exploit this connection to describe a relation between the Morse index form and the minimizing properties of geodesics.

Definition 1.7.7: Let $\sigma\colon[a,b]\to M$ be a normal geodesic in a Finsler manifold M. We shall say that σ *contains no conjugate points* if $\sigma(t)$ is not conjugate to $\sigma(a)$ along σ for all $t\in(a,b]$. We shall also say that $\sigma(b)$ is the *first conjugate point* to $\sigma(a)$ along σ if $\sigma(b)$ is conjugate to $\sigma(a)$ and all previous points of the image of σ are not.

PROPOSITION 1.7.8: *Let $\sigma: [a,b] \to M$ be a normal geodesic in a Finsler manifold M containing no conjugate points. Then the Morse index form I_a^b is positive definite on $\mathcal{X}_0[a,b]$.*

Proof: The assumption is that $\exp_{\sigma(a)}$ is a local E-diffeomorphism in a neighborhood of the image of σ; so we can argue as in the proof of Theorem 1.6.6 to show that σ is length-minimizing among nearby curves. By (1.7.5), then, I is positive semidefinite on $\mathcal{X}_0[a,b]$.

Now assume that $\xi \in \mathcal{X}_0[a,b]$ is such that $I(\xi,\xi) = 0$; we claim that $\xi \in \ker I$. Indeed, take $\eta \in \mathcal{X}_0[a,b]$; then for any $\varepsilon \in \mathbb{R}^*$

$$0 \leq I(\xi + \varepsilon\eta, \xi + \varepsilon\eta) = \varepsilon\big[2I(\xi,\eta) + \varepsilon I(\eta,\eta)\big].$$

Dividing by ε and letting $\varepsilon \to 0^+$ (respectively, $\varepsilon \to 0^-$) we get $I(\xi,\eta) \geq 0$ (respectively, $I(\xi,\eta) \leq 0$), and so $I(\xi,\eta) = 0$, as claimed.

It follows, by Corollary 1.7.7, that ξ is a Jacobi field with $J(a) = J(b) = o$; being $\sigma(b)$ not conjugated to $\sigma(a)$ along σ this implies $\xi \equiv 0$, and we are done. □

As a consequence, the proper Jacobi fields realize the minimum of the Morse index form among vector fields with given boundary values:

COROLLARY 1.7.9: *Let $\sigma: [a,b] \to M$ be a normal geodesic in a Finsler manifold M containing no conjugate points. Let $\xi \in \mathcal{X}[a,b]$ and $J \in \mathcal{J}_0(\sigma)$ be such that $\xi(a) = J(a)$ and $\xi(b) = J(b)$. Then*

$$I(J,J) \leq I(\xi,\xi),$$

with equality iff $\xi \equiv J$.

Proof: By (1.7.6)

$$I(J,\xi) = \langle \nabla_{T^H} J^H \mid \xi^H \rangle_T \Big|_a^b = \langle \nabla_{T^H} J^H \mid J^H \rangle_T \Big|_a^b = I(J,J).$$

Then if $\xi \not\equiv J$ we have

$$0 < I(\xi - J, \xi - J) = I(\xi,\xi) - 2I(\xi,J) + I(J,J) = I(\xi,\xi) - I(J,J).$$

□

Of course, the Morse index form becomes positive semidefinite at the first conjugate point:

PROPOSITION 1.7.10: *Let $\sigma: [a,b] \to M$ be a normal geodesic in a Finsler manifold M such that $\sigma(b)$ is the first conjugate point to $\sigma(a)$ along σ. Then I_a^b is positive semidefinite on $\mathcal{X}_0[a,b]$ and*

$$\ker I_a^b|_{\mathcal{X}_0[a,b]} = \mathcal{X}_0[a,b] \cap \mathcal{J}_0(\sigma) \neq \{0\}.$$

Proof: By Corollary 1.7.7, it is enough to show that $I_a^b \geq 0$. Take $b' \in (a,b)$ and define a map $T_{b'}: \mathcal{X}_0[a,b] \to \mathcal{X}_0[a,b']$ by setting

$$T_{b'}(\xi)(t) = \xi(bt/b').$$

The map $T_{b'}$ is clearly an isomorphism; so we can define a symmetric bilinear form $I_{b'}: \mathcal{X}_0[a,b] \times \mathcal{X}_0[a,b] \to \mathbb{R}$ by

$$I_{b'}(\xi,\eta) = I_a^{b'}\left(T_{b'}(\xi), T_{b'}(\eta)\right).$$

Then

$$I_a^b(\xi,\xi) = \lim_{b' \to b} I_{b'}(\xi,\xi) \geq 0$$

for all $\xi \in \mathcal{X}_0[a,b]$, by Proposition 1.7.8. □

To complete the picture, we prove the following:

PROPOSITION 1.7.11: *Let $\sigma: [a,b] \to M$ be a normal geodesic in a Finsler manifold M. Then there is $t_0 \in (a,b)$ such that $\sigma(t_0)$ is conjugate to $\sigma(a)$ along σ iff there exists $\xi \in \mathcal{X}_0[a,b]$ such that $I_a^b(\xi,\xi) < 0$.*

Proof: If there exists such a ξ, by Propositions 1.7.8 and 1.7.10 there must be a $t_0 \in (a,b)$ such that $\sigma(t_0)$ is conjugate to $\sigma(a)$ along σ.

Conversely, let $t_0 \in (a,b)$ such that $\sigma(t_0)$ is conjugated to $\sigma(a)$ along σ. Then there is a non-identically zero Jacobi field $J \in \mathcal{X}_0[a,t_0]$. Take $t' \in (a,t_0)$ and $t'' \in (t_0,b)$ such that $J(t') \neq 0$ and $d_F(\sigma(t'),\sigma(t'')) < \mathrm{ir}(\sigma(t''))$. In particular, $\sigma|_{[t',t'']}$ contains no conjugate points to $\sigma(t'')$.

Let $\gamma: (-\varepsilon,\varepsilon) \to M$ be a curve in M such that $\gamma(0) = \sigma(t')$ and $\dot\gamma(0) = J(t')$. If we write $\gamma = \exp_{\sigma(t'')}(\tilde\gamma)$, then we may consider the geodesic variation Σ given by

$$\Sigma(s,t) = \exp_{\sigma(t'')}\left(t\tilde\gamma(s)\right),$$

as usual. The transversal vector U of Σ then is a proper Jacobi field (by Corollary 1.7.5) belonging to $\mathcal{X}[t',t'']$ such that $U(t') = J(t')$ and $U(t'') = 0$.

We now define $\xi \in \mathcal{X}_0[a,b]$ by

$$\xi(t) = \begin{cases} J(t) & \text{if } t \in [a,t']; \\ U(t) & \text{if } t \in [t',t'']; \\ 0 & \text{if } t \in [t'',b]. \end{cases}$$

We also denote by $J' \in \mathcal{X}[a,t'']$ the extension of J obtained by setting $J'(t) = 0$ if $t \in [t_0,t'']$. Clearly, J' is not smooth at t_0, and so it is not a Jacobi field on $[t',t'']$. Therefore

$$\begin{aligned} I_a^b(\xi,\xi) &= I_a^{t'}(\xi,\xi) + I_{t'}^{t''}(\xi,\xi) = I_a^{t'}(J,J) + I_{t'}^{t''}(U,U) \\ &< I_a^{t'}(J,J) + I_{t'}^{t''}(J',J') = I_a^{t'}(J,J) + I_{t'}^{t_0}(J,J) \\ &= I_a^{t_0}(J,J) = 0, \end{aligned}$$

by Corollary 1.7.9 and Corollary 1.7.7. □

In particular, a geodesic containing a conjugate point cannot be length minimizing:

COROLLARY 1.7.12: *Let $\sigma: [a, b] \to M$ be a normal geodesic in a Finsler manifold M. Assume that there exists $t_0 \in (a, b)$ such that $\sigma(t_0)$ is conjugated to $\sigma(a)$ along σ. Then σ is not length-minimizing, that is $d_F(\sigma(a), \sigma(b)) < L(\sigma)$.*

Proof: If σ were length-minimizing, then by (1.7.5) the Morse index form I_a^b along σ would have been positive semidefinite, against Proposition 1.7.11. □

A final corollary, which is a major step in the proof of the Finsler version of the Cartan-Hadamard theorem:

COROLLARY 1.7.13: *Let $\sigma: [a, b] \to M$ be a normal geodesic in a Finsler manifold M. Assume that the Morse index form I_a^b is positive definite on $\mathcal{X}_0[a, b]$. Then σ contains no conjugate points.*

Proof: By Propositions 1.7.10 and 1.7.11. □

1.7.3. The Cartan-Hadamard and Bonnet theorems

In this subsection we complete our presentation of real Finsler metrics providing a proof for the Finsler versions of the classical Cartan-Hadamard and Bonnet theorems. The theorems were originally proved by Auslander in [Au1, 2]; our proofs, however, are devised so to stress the similarities with the Riemannian case.

We start with a consequence of Corollary 1.7.13:

LEMMA 1.7.14: *Let M be a complete Finsler manifold, and fix $p \in M$. Assume that the Morse index form I_0^a is positive definite on $\mathcal{X}_0[0, a]$ for all $a > 0$ and along every radial normal geodesic issuing from p. Then the exponential map $\exp_p: T_pM \to M$ is a local diffeomorphism on $\tilde{M}_p$ — and an E-diffeomorphism at the origin.*

Proof: Indeed, by Corollary 1.7.13 no radial normal geodesic contains conjugate points, and thus, by Proposition 1.7.3, $\exp_p$ has no critical points. □

The assumptions in this Lemma are slightly weaker than the usual hypotheses of the Cartan-Hadamard theorem, as shown in

LEMMA 1.7.15: *Let M be a complete Finsler manifold of nonpositive (horizontal flag) curvature. Then the Morse index form I_0^a is positive definite on $\mathcal{X}_0[0, a]$ for all $a > 0$ and along every radial normal geodesic issuing from p for all $p \in M$.*

Proof: Let $\sigma: [0,a] \to M$ be a radial normal geodesic issuing from p, and take $\xi \in \mathcal{X}_0[0,a]$. Then

$$I(\xi,\xi) = \int_0^a \left[\|\nabla_{T^H}\xi^H\|^2 - R_T(\xi^H,\xi^H)\right] dt \geq 0,$$

because the horizontal flag curvature is nonpositive. Furthermore, $I(\xi,\xi) = 0$ implies $\nabla_{T^H}\xi^H \equiv 0$, that is that ξ is parallel along σ. But $\xi(0) = o_p$; hence, by (1.6.3), ξ is identically zero, and we are done. □

We do not directly assume here that the horizontal flag curvature is nonpositive because in chapter 3 we shall encounter a situation where we shall be able to control the sign of the Morse index form without controlling directly the sign of the horizontal flag curvature.

Anyway, from now on we can proceed with the standard proof of the Cartan-Hadamard theorem. First of all we need the following result:

LEMMA 1.7.16: *Let $F_j: TM_j \to \mathbb{R}^+$ be a Finsler metric on a manifold M_j, for $j = 1, 2$, and assume that F_1 is complete. Fix a point $p_0 \in M_1$, and let $f: M \to N$ be a local diffeomorphism of M_1 onto M_2 (and a local E-diffeomorphism at p_0, but still differentiable there) such that*

$$F_2\big(df_p(u)\big) \geq F_1(u)$$

for all $p \in M_1$ and $u \in T_pM_1$. Then f is a covering map.

Proof: We can repeat word by word the proof of Lemma 3.3 in [DoC]. □

And finally:

THEOREM 1.7.17: *Let M be a complete Finsler manifold, and fix $p \in M$. Assume that the Morse index form I_0^a is positive definite on $\mathcal{X}_0[0,a]$ for all $a > 0$ and along every radial normal geodesic issuing from p (e.g., assume that the horizontal flag curvature is nonpositive). Then $\exp_p: T_pM \to M$ is a covering map, smooth outside the origin. In particular, if M is simply connected then $\exp_p$ is a E-diffeomorphism at the origin.*

Proof: By Lemma 1.7.14, $f = \exp_p: T_pM \to M$ is a local diffeomorphism on $\tilde{M}_p$, an E-diffeomorphism at the origin, where it is still differentiable, and it is onto, because M is complete. We can then use it to pull-back a Finsler metric onto T_pM, which coincides with the original F at the origin. This metric is complete, because the geodesics through the origin are straight lines. Hence we can apply the Lemma 1.7.16, and we are done. □

We end this chapter by proving the Bonnet theorem for Finsler manifolds by using our second variation formula:

Theorem 1.7.18: *Let M be a complete Finsler manifold of dimension at least 2 such that*

$$R_u(H,H) \ge c\langle H \mid H\rangle_u \tag{1.7.7}$$

for some $c > 0$ and all $u \in \tilde{M}$ and $H \in \mathcal{H}_u$. Then the diameter $\operatorname{diam}(M)$ *is bounded above, namely*

$$\operatorname{diam}(M) \le \frac{\pi}{\sqrt{c}}.$$

In particular, M is compact with finite fundamental group.

Proof: Let $S \subset \mathbb{R}^3$ be the euclidean 2-sphere of radius $1/\sqrt{c}$; then it has constant sectional curvature c and (intrinsic) diameter $\pi/\sqrt{c}$. Then for any $a > \pi/\sqrt{c}$ there is a non-minimizing geodesic $\tau_a\colon [0,a] \to S$, and so the usual proof of the Bonnet theorem for the 2-sphere yields a vector field $U_a = f_a E_a$ — where E_a is a parallel unit vector field along τ_a orthogonal to $\dot{\tau}_a$ — such that $U_a(0) = U_a(a) = 0$ and

$$0 > \int_0^a \left[\|\nabla_{\dot{\tau}_a} U_a\|^2 - c\|U_a\|^2\right] dt = \int_0^a \left(|\dot{f}_a|^2 - cf_a^2\right) dt.$$

Now, let $\sigma\colon [0,a] \to M$ be a length-minimizing normal geodesic; we must prove that $a \le \pi/\sqrt{c}$. Assume, by contradiction, that $a > \pi/\sqrt{c}$. Let ξ be a parallel vector field along σ such that ξ^H is orthogonal to T^H and $\|\xi^H\|_{\dot{\sigma}} \equiv 1$ along $\dot{\sigma}$; since

$$\frac{d}{dt}\langle \xi^H \mid T^H\rangle_{\dot{\sigma}} = T^H\langle \xi^H \mid T^H\rangle_{\dot{\sigma}} = \langle \nabla_{T^H}\xi^H \mid T^H\rangle_{\dot{\sigma}} + \langle \xi^H \mid \nabla_{T^H}T^H\rangle_{\dot{\sigma}} = 0$$

and

$$\frac{d}{dt}\langle \xi^H \mid \xi^H\rangle_{\dot{\sigma}} = T^H\langle \xi^H \mid \xi^H\rangle_{\dot{\sigma}} = 2\langle \nabla_{T^H}\xi^H \mid \xi^H\rangle_{\dot{\sigma}} = 0,$$

such a ξ exists. Put $U = f_a\xi$, where f_a is the function defined above. Being σ a globally minimizing geodesic, we should have

$$\frac{d^2\ell_\Sigma}{ds^2}(0) \ge 0$$

for any fixed variation Σ of σ. But on the other hand the second variation formula applied to a variation Σ with U as transversal vector yields

$$\begin{aligned}\frac{d^2\ell_\Sigma}{ds^2}(0) &= \int_0^a \left[\|\nabla_{T^H}U^H\|_{\dot{\sigma}}^2 - R_{\dot{\sigma}}(U^H,U^H)\right] dt\\ &\le \int_0^a \left[|\dot{f}_a|^2 - c\langle U^H \mid U^H\rangle_{\dot{\sigma}}\right] dt\\ &= \int_0^a \left[|\dot{f}_a|^2 - cf_a^2\right] dt < 0,\end{aligned}$$

(because $\langle U^H \mid T^H\rangle_{\dot{\sigma}} \equiv 0$, $\nabla_{T^H}\xi^H \equiv 0$ and $\|\xi^H\|_{\dot{\sigma}} \equiv 1$ along $\dot{\sigma}$), contradiction.

Finally, let $\pi\colon \hat{M} \to M$ be the universal covering space of M. Since π is a local diffeomorphism, we can pull-back the Finsler metric on M to a Finsler metric on $\hat{M}$. Clearly the horizontal flag curvature on $\hat{M}$ satisfies again (1.7.7), and so $\hat{M}$ is still compact. Hence for any $p \in M$ the set $\pi^{-1}(p)$ has finite cardinality, and so $\pi_1(M)$ must be finite. □

In this way it is possible to recover even the Rauch comparison theorem (see, e.g., [BC]); but for our aims this is enough, and in the next chapter we shall start dealing with the complex case.

CHAPTER 2

Complex Finsler geometry

2.0. Introduction

In this chapter we start dealing with complex Finsler metrics. The main idea is exactly the same as in the real case: replacing the Finsler metric on the tangent bundle by a Hermitian metric on a suitable bundle it becomes possible to get geometrical results on the Finsler manifold using Hermitian techniques. Contrarily to the real case, we did not find in the literature much references on complex Finsler geometry. We anyway would like to recall here the pioneering work of Rund [Rd2], Fukui's [Fu] examination of the Cartan connection in the complex case, Faran's [F] work on the equivalence problem for complex Finsler metrics, and above all Kobayashi's paper [K], which inspired our approach.

So let M be a complex manifold, and $F: T^{1,0}M \to \mathbb{R}^+$ a complex strongly pseudoconvex Finsler metric on M (see section 2.3 for the precise definitions). Then, as in the real case, using the Levi form of $G = F^2$ it is possible to define a Hermitian metric on the vertical bundle $\mathcal{V}$ in such a way that the canonical section $\iota: \tilde{M} \to \mathcal{V}$ (where this time $\tilde{M}$ stands for the complement of the zero section in $T^{1,0}M$) turns out again to be an isometric embedding of $\tilde{M}$ into $\mathcal{V}$.

Now to any Hermitian metric on a vector bundle it is associated the Chern connection, which is the unique complex linear connection (or connection of type $(1,0)$, as it is also called) for which the given Hermitian metric is parallel. It turns out that the Chern connection is automatically good, and so we get a complex horizontal bundle $\mathcal{H}$ such that $T^{1,0}\tilde{M} = \mathcal{H} \oplus \mathcal{V}$, and a bundle isomorphism $\Theta: \mathcal{V} \to \mathcal{H}$. Using Θ we can define an Hermitian metric on $\mathcal{H}$, the associated Chern connection and a canonical section $\chi = \Theta \circ \iota$; and this is the setting we need to work.

The torsion of the Chern connection is, in general, not zero. Generalizing the usual definition in Hermitian geometry, we shall say that a complex Finsler metric is Kähler if (a suitable contraction of) the horizontal part of the torsion of the Chern connection vanishes. As we shall see in sections 2.3 and 2.4, Kähler Finsler metrics enjoy properties similar to the ones of Hermitian Kähler metrics. In particular, we can compute the first and second variation of the length integral in complex terms, and prove the local existence and uniqueness of geodesics for Kähler Finsler metrics even when the associated real Finsler metric is not convex — that is in cases where the existence of geodesics is not a consequence of the real theory.

We shall describe geometrical (and function theoretical) applications of this theory in the next chapter; the finals sections of this one are instead devoted to the study of the holomorphic curvature of a complex Finsler metric, and to the comparison of the Chern-Finsler connection with the Cartan connection. In particular, we shall give a variational characterization of the holomorphic curvature.

More precisely, the content of this chapter is the following. In the first two sections we review the theory of horizontal bundles and maps, and of non-linear and vertical connections, adapting the presentation to the complex case. In the third section we introduce the vertical Chern connection associated to a complex Finsler metric, and we discuss the notion of Kähler Finsler metric. In the fourth section we compute the first and second variation of the length integral in this context, proving the existence of geodesics for strongly pseudoconvex weakly Kähler Finsler metrics. In the fifth section we examine the notion of holomorphic curvature of a strongly pseudoconvex Finsler metric. Finally, in the sixth section we compare the Chern-Finsler connection with the Cartan connection under various Kähler hypotheses.

2.1. Complex non-linear connections

2.1.1 Preliminaries

Let M be a complex manifold of complex dimension n. Let $\{z^1,\dots,z^n\}$ be a set of local complex coordinates, with $z^\alpha = x^\alpha + i x^{n+\alpha}$, so that $\{x^1,\dots,x^n,x^{n+1},\dots,x^{2n}\}$ are local real coordinates. Lowercase greek indices will run from 1 to n, whereas lowercase roman indices will run from 1 to $2n$. Let $T_{\mathbb{R}}M$ denote the *real tangent bundle* of M; it is a real bundle of rank $2n$ equipped with a complex structure J. With respect to the local frame $\{\partial/\partial x^1,\dots,\partial/\partial x^{2n}\}$, the complex structure J acts in the following way:

$$J(\partial/\partial x^a) = \begin{cases} \partial/\partial x^{a+n} & \text{if } 1 \le a \le n, \\ -\partial/\partial x^{a-n} & \text{if } n+1 \le a \le 2n. \end{cases}$$

In other words, writing $J(\partial/\partial x^a) = J_a^b\,\partial/\partial x^b$ we have

$$J_a^b = \begin{cases} \delta_{a+n}^b & \text{if } 1 \le a \le n, \\ -\delta_{a-n}^b & \text{if } n+1 \le a \le 2n, \end{cases}$$

(where δ_b^a is the Kronecker delta), that is

$$J = \begin{pmatrix} 0 & -I_n \\ I_n & 0 \end{pmatrix}.$$

Let $T_{\mathbb{C}}M = T_{\mathbb{R}}M \otimes_{\mathbb{R}} \mathbb{C}$ be the *complexified tangent bundle;* it is a complex bundle of rank $2n$. Set

$$\frac{\partial}{\partial z^\alpha} = \frac{1}{2}\left(\frac{\partial}{\partial x^\alpha} - i\frac{\partial}{\partial x^{\alpha+n}}\right) \qquad \text{and} \qquad \frac{\partial}{\partial \bar z^\alpha} = \frac{1}{2}\left(\frac{\partial}{\partial x^\alpha} + i\frac{\partial}{\partial x^{\alpha+n}}\right);$$

then $\{\partial/\partial z^1,\dots,\partial/\partial z^n,\partial/\partial\bar z^1,\dots,\partial/\partial\bar z^n\}$ is a local frame for $T_{\mathbb C}M$. The complex structure J acts complex linearly on $T_{\mathbb C}M$, and $J^2=-I$; so $T_{\mathbb C}M$ splits as the sum of the two eigenbundles

$$T_{\mathbb C}M = T^{1,0}M\oplus T^{0,1}M,$$

where $T^{1,0}M=\{v\in T_{\mathbb C}M\mid Jv=iv\}$ and $T^{0,1}M=\{v\in T_{\mathbb C}M\mid Jv=-iv\}$. The bundle $T^{1,0}M$ (which we shall sometimes simply denote by TM) is the *holomorphic* (or *complex*) *tangent bundle* of M. It is easy to check that $\{\partial/\partial z^1,\dots,\partial/\partial z^n\}$ is a local frame for $T^{1,0}M$, and $\{\partial/\partial\bar z^1,\dots,\partial/\partial\bar z^n\}$ one for $T^{0,1}M$.

The real splitting

$$T_{\mathbb C}M = T_{\mathbb R}M\oplus i(T_{\mathbb R}M)$$

induces a *conjugation* on $T_{\mathbb C}M$ commuting with J; since $\overline{\partial/\partial z^j}=\partial/\partial\bar z^j$, we have $\overline{T^{1,0}M}=T^{0,1}M$.

As well known, the bundles $T^{1,0}M$ and $T_{\mathbb R}M$ are isomorphic, and both are sensible representations of the tangent bundle of M. In this book we choose as explicit isomorphism the bundle map ${}^o\colon T^{1,0}M\to T_{\mathbb R}M$ given by

$$\forall v\in T^{1,0}M \qquad v^o=v+\bar v.$$

Note that $(\partial/\partial z^\alpha)^o=\partial/\partial x^\alpha$, $(i\partial/\partial z^\alpha)^o=\partial/\partial x^{\alpha+n}$ and that

$$Jv^o=J(v+\bar v)=Jv+\overline{Jv}=iv+\overline{iv}=(iv)^o=i(v-\bar v),$$

that is ^{o} is a real isomorphism preserving J. The inverse ${}_o\colon T_{\mathbb R}M\to T^{1,0}M$ is given by

$$\forall u\in T_{\mathbb R}M \qquad u_o=\frac{1}{2}(u-iJu). \tag{2.1.1}$$

Furthermore, locally an element of $T^{1,0}M$ could be written as

$$v=v^\alpha\frac{\partial}{\partial z^\alpha}. \tag{2.1.2}$$

Then, setting $v^\alpha=u^\alpha+iu^{\alpha+n}$, it is easy to check that

$$v^o=u^a\frac{\partial}{\partial x^a}. \tag{2.1.3}$$

Conversely, if $u=u^a\,\partial/\partial x^a\in T_{\mathbb R}M$, then

$$u_o=(u^\alpha+iu^{\alpha+n})\frac{\partial}{\partial z^\alpha}. \tag{2.1.4}$$

Let $T^*_{\mathbb R}M$ denote the *real cotangent bundle* of M; a local frame for it is $\{dx^a\}$. Also $T^*_{\mathbb R}M$ is endowed with a complex structure J^* given by

$$\forall\gamma\in T^*_{\mathbb R}M\quad \forall u\in T_{\mathbb R}M\qquad J^*\gamma(u)=\gamma(Ju).$$

It follows that, writing $J^*(dx^a)=(J^*)^a_b\,dx^b$, we have

$$(J^*)^a_b=\begin{cases}-\delta_b^{a+n} & \text{if } 1\le a\le n,\\ \delta_b^{a-n} & \text{if } n+1\le a\le 2n;\end{cases}$$

in other words,

$$J^*(dx^a) = \begin{cases} -dx^{a+n} & \text{if } 1 \le a \le n, \\ dx^{a-n} & \text{if } n+1 \le a \le 2n. \end{cases}$$

As before, let $T^*_{\mathbb{C}}M = T^*_{\mathbb{R}}M \otimes_{\mathbb{R}} \mathbb{C}$, and set

$$dz^\alpha = dx^\alpha + i\,dx^{\alpha+n} \qquad \text{and} \qquad d\bar{z}^\alpha = dx^\alpha - i\,dx^{\alpha+n};$$

the set $\{dz^\alpha, d\bar{z}^\alpha\}$ is a local coframe for $T^*_{\mathbb{C}}M$. The map J^* induces the complex splitting

$$T^*_{\mathbb{C}}M = \bigwedge^{1,0}M \oplus \bigwedge^{0,1}M, \tag{2.1.5}$$

where $\bigwedge^{1,0}M = \{\gamma \in T^*_{\mathbb{C}}M \mid J^*\gamma = i\gamma\}$ and $\bigwedge^{0,1}M = \{\gamma \in T^*_{\mathbb{C}}M \mid J^*\gamma = -i\gamma\}$. Note that $\bigwedge^{1,0}M = (T^{0,1}M)^\perp$ and $\bigwedge^{0,1}M = (T^{1,0}M)^\perp$; so $\bigwedge^{1,0}M$ is naturally isomorphic to $(T^{1,0}M)^*$ and $\{dz^\alpha\}$ is the dual frame of $\{\partial/\partial z^\alpha\}$.

Again, the real splitting $T^*_{\mathbb{C}}M = T^*_{\mathbb{R}}M \oplus iT^*_{\mathbb{R}}M$ induces a conjugation commuting with J^*; since $\overline{dz^\alpha} = d\bar{z}^\alpha$, clearly $\overline{\bigwedge^{1,0}M} = \bigwedge^{0,1}M$. The real isomorphism ${}^o\colon \bigwedge^{1,0}M \to T^*_{\mathbb{R}}M$ is given by

$$\forall \omega \in \bigwedge^{1,0}M \qquad\qquad \omega^o = \frac{1}{2}(\omega + \bar{\omega}),$$

so that $(dz^\alpha)^o = dx^\alpha$, $(i\,dz^\alpha)^o = dx^{\alpha+n}$ and

$$J^*\omega^o = (i\omega)^o = \frac{i}{2}(\omega - \bar{\omega}).$$

The inverse isomorphism ${}_o\colon T^*_{\mathbb{R}}M \to \bigwedge^{1,0}M$ is given by

$$\forall \gamma \in T^*_{\mathbb{R}}M \qquad\qquad \gamma_o = \gamma - iJ\gamma.$$

Finally, the splitting (2.1.5) induces a splitting

$$\bigwedge^r(T_{\mathbb{C}}M) = \bigoplus_{p+q=r} \bigwedge^{p,q}M$$

where $\bigwedge^{p,q}M = \bigwedge^p(\bigwedge^{1,0}M) \wedge \bigwedge^q(\bigwedge^{0,1}M)$. As before, $\overline{\bigwedge^{p,q}M} = \bigwedge^{q,p}M$, and $\bigwedge^r(T^*_{\mathbb{R}}M)$ is both a subbundle of $\bigwedge^r(T^*_{\mathbb{C}}M)$ and isomorphic (via ^{o}) to $\bigwedge^{r,0}M$. We shall denote by $A^{p,q}(M)$ the space of smooth sections of $\bigwedge^{p,q}M$, that is the space of smooth (p,q)-forms on M, and we shall set

$$A(M) = \bigoplus_{p,q \in \mathbb{N}} A^{p,q}(M).$$

Let us now look at the tangent-tangent level. Since $T_{\mathbb{R}}M \cong T^{1,0}M$ is a complex manifold, we have a complex structure J on it. Thus we have at least six spaces to consider: $T_{\mathbb{R}}(T_{\mathbb{R}}M)$, $T_{\mathbb{C}}(T_{\mathbb{R}}M)$, $T^{1,0}(T_{\mathbb{R}}M)$, $T_{\mathbb{R}}(T^{1,0}M)$, $T_{\mathbb{C}}(T^{1,0}M)$, $T^{1,0}(T^{1,0}M)$ and so on. Luckily, the situation is not so complicated as it seems at first glance. Indeed, (2.1.3) and (2.1.4) show that the isomorphisms ^{o} and ${}_o$ preserve the real coordinates of $T^{1,0}M$ and $T_{\mathbb{R}}M$. This means that, as far as only the tangent-tangent spaces are concerned, we can (and we shall) safely identify $T^{1,0}M$ and $T_{\mathbb{R}}M$. In particular, $\tilde{M}$ will denote either $T^{1,0}M$ or $T_{\mathbb{R}}M$ minus the zero section, depending on the actual situation, without any further warnings.

A local frame over $\mathbb{R}$ for $T_{\mathbb{R}}\tilde{M}$ is given by $\{\partial_1^o, \ldots, \partial_{2n}^o, \dot{\partial}_1^o, \ldots, \dot{\partial}_{2n}^o\}$, where

$$\partial_a^o = \frac{\partial}{\partial x^a} \qquad \text{and} \qquad \dot{\partial}_a^o = \frac{\partial}{\partial u^a};$$

analogously, a local frame over $\mathbb{C}$ for $T^{1,0}\tilde{M}$ is given by $\{\partial_1, \ldots, \partial_n, \dot{\partial}_1, \ldots, \dot{\partial}_n\}$, where

$$\partial_\alpha = \frac{\partial}{\partial z^\alpha} \qquad \text{and} \qquad \dot{\partial}_\alpha = \frac{\partial}{\partial v^\alpha},$$

and, as always, $z^\alpha = x^\alpha + ix^{\alpha+n}$ and $v^\alpha = u^\alpha + iu^{\alpha+n}$.

There is a natural isomorphism, still denoted by ${}^o\colon T^{1,0}\tilde{M} \to T_{\mathbb{R}}\tilde{M}$, given by

$$\forall X \in T^{1,0}\tilde{M} \qquad\qquad X^o = X + \overline{X},$$

with inverse ${}_o\colon T_{\mathbb{R}}\tilde{M} \to T^{1,0}\tilde{M}$ given again by (2.1.1). It is easy to check that

$$(\partial_\alpha)^o = \partial_\alpha^o, \qquad (i\partial_\alpha)^o = \partial_{\alpha+n}^o, \qquad (\dot{\partial}_\alpha)^o = \dot{\partial}_\alpha^o, \qquad \text{and} \qquad (i\dot{\partial}_\alpha)^o = \dot{\partial}_{\alpha+n}^o.$$

There also are $T_{\mathbb{R}}^*\tilde{M}$ and $\bigwedge^{1,0}\tilde{M}$, with dual coframes

$$\{dx^1, \ldots, dx^{2n}, du^1, \ldots, du^{2n}\} \qquad \text{and} \qquad \{dz^1, \ldots, dz^n, dv^1, \ldots, dv^n\}$$

respectively. Note that, as in the real case, $dz^\alpha|_v$ is different from $dz^\alpha|_p$.

2.1.2. Change of coordinates

Let us now summarize what happens under holomorphic change of coordinates. Notations, proofs and arguments needed are exactly as in the real case (cf. subsection 1.1.1). This time the relevant matrices are

$$(\mathcal{J}_{BA})^\alpha_\beta = \frac{\partial z_B^\alpha}{\partial z_A^\beta} \qquad \text{and} \qquad (H_{BA})^\alpha_{\beta\gamma} = \frac{\partial^2 z_B^\alpha}{\partial z_A^\beta \partial z_A^\gamma}.$$

Therefore the change of coordinates on $T^{1,0}M$ are given by

$$dz_B^\alpha = (\mathcal{J}_{BA})^\alpha_\beta \, dz_A^\beta \qquad \text{and} \qquad \frac{\partial}{\partial z_B^\alpha} = (\mathcal{J}_{BA}^{-1})^\beta_\alpha \frac{\partial}{\partial z_A^\beta}.$$

and, conjugating everything,

$$d\bar{z}_B^\alpha = \overline{(\mathcal{J}_{BA})^\alpha_\beta} \, d\bar{z}_A^\beta \qquad \text{and} \qquad \frac{\partial}{\partial \bar{z}_B^\alpha} = \overline{(\mathcal{J}_{BA}^{-1})^\beta_\alpha} \frac{\partial}{\partial \bar{z}_A^\beta}.$$

Recalling that the change of coordinates are holomorphic, we have $\partial z^\alpha_B / \partial \bar z^\beta_A = 0$, that is

$$\frac{\partial x^\alpha_B}{\partial x^\beta_A} = \frac{\partial x^{\alpha+n}_B}{\partial x^{\beta+n}_A} = \operatorname{Re}\frac{\partial z^\alpha_B}{\partial z^\beta_A} = \operatorname{Re}[(\mathcal{J}_{BA})^\alpha_\beta],$$
$$\frac{\partial x^{\alpha+n}_B}{\partial x^\beta_A} = -\frac{\partial x^\alpha_B}{\partial x^{\beta+n}_A} = \operatorname{Im}\frac{\partial z^\alpha_B}{\partial z^\beta_A} = \operatorname{Im}[(\mathcal{J}_{BA})^\alpha_\beta],$$

and

$$dx^a_B = \frac{\partial x^a_B}{\partial x^b_A}\,dx^b_A = \begin{cases} \left(\operatorname{Re}\dfrac{\partial z^a_B}{\partial z^\beta_A}\right) dx^\beta_A - \left(\operatorname{Im}\dfrac{\partial z^a_B}{\partial z^\beta_A}\right) dx^{\beta+n}_A & \text{if } 1 \le a \le n, \\ \left(\operatorname{Im}\dfrac{\partial z^{a-n}_B}{\partial z^\beta_A}\right) dx^\beta_A + \left(\operatorname{Re}\dfrac{\partial z^{a-n}_B}{\partial z^\beta_A}\right) dx^{\beta+n}_A & \text{if } n \le a-n \le n, \end{cases}$$

that is

$$\left(\frac{\partial x^a_B}{\partial x^b_A}\right) = \left(\begin{array}{c|c} \operatorname{Re}\mathcal{J}_{BA} & -\operatorname{Im}\mathcal{J}_{BA} \\ \hline \operatorname{Im}\mathcal{J}_{BA} & \operatorname{Re}\mathcal{J}_{BA} \end{array}\right).$$

Now we briefly examine $T^{1,0}\tilde M$. We know that $\{\partial_\alpha, \dot\partial_\beta\}$ is a local frame for $T^{1,0}\tilde M$; let $\{dz^\alpha, dv^\beta\}$ be its dual coframe. Then again we find

$$\begin{aligned} dz^\alpha_B &= \frac{\partial z^\alpha_B}{\partial z^\beta_A}\,dz^\beta_A = (\mathcal{J}_{BA})^\alpha_\beta\,dz^\beta_A, \\ dv^\alpha_B &= \frac{\partial z^\alpha_B}{\partial z^\beta_A}\,dv^\beta_A + \frac{\partial^2 z^\alpha_B}{\partial z^\beta_A \partial z^\gamma_A}\,v^\gamma_A\,dz^\beta_A = (\mathcal{J}_{BA})^\alpha_\beta\,dv^\beta_A + (H_{BA})^\alpha_{\beta\gamma}\,v^\gamma_A\,dz^\beta_A, \end{aligned} \tag{2.1.6}$$

and

$$\begin{aligned} (\partial_\alpha)_B &= \frac{\partial z^\beta_A}{\partial z^\alpha_B}(\partial_\beta)_A - \frac{\partial z^\delta_A}{\partial z^\alpha_B}\frac{\partial^2 z^\alpha_B}{\partial z^\beta_A \partial z^\gamma_A}\,v^\gamma_A\,\frac{\partial z^\beta_A}{\partial z^\alpha_B}(\dot\partial_\delta)_A \\ &= (\mathcal{J}^{-1}_{BA})^\beta_\alpha(\partial_\beta)_A - (\mathcal{J}^{-1}_{BA})^\beta_\alpha(H_{BA})^\mu_{\beta\gamma}(\mathcal{J}^{-1}_{BA})^\delta_\mu\,v^\gamma_A\,(\dot\partial_\delta)_A, \\ (\dot\partial_\alpha)_B &= \frac{\partial z^\beta_A}{\partial z^\alpha_B}(\dot\partial_\beta)_A = (\mathcal{J}^{-1}_{BA})^\beta_\alpha(\dot\partial_\beta)_A. \end{aligned} \tag{2.1.7}$$

2.1.3. Horizontal and vertical bundles

The next step is to define the vertical bundles.

Definition 2.1.1: The *real vertical bundle* is

$$\mathcal{V}_\mathbb{R} = \ker d\pi \subset T_\mathbb{R}\tilde M,$$

while the *complexified vertical bundle* is

$$\mathcal{V}_\mathbb{C} = \mathcal{V}_\mathbb{R} \otimes_\mathbb{R} \mathbb{C} = \ker d\pi \subset T_\mathbb{C}\tilde M.$$

Since the projection π is holomorphic, $d\pi$ commutes with J. So $\mathcal{V}_{\mathbb{C}}$ splits as $\mathcal{V}_{\mathbb{C}} = \mathcal{V}^{1,0} \oplus \mathcal{V}^{0,1}$, and we may define the *(complex) vertical bundle*

$$\mathcal{V} = \mathcal{V}^{1,0} = \ker d\pi \subset T^{1,0}\tilde{M}.$$

Clearly, $\tilde{\pi}: \mathcal{V} \to \tilde{M}$ is a holomorphic vector bundle of rank n, whose local frame is given by $\{\dot{\partial}_1, \ldots, \dot{\partial}_n\}$. Analogously, a local frame for $\mathcal{V}_{\mathbb{R}}$ is $\{\dot{\partial}_1^o, \ldots, \dot{\partial}_{2n}^o\}$.

Take $u \in \tilde{M}$. Then the natural isomorphism $\iota_u: (T_{\mathbb{R}}M)_{\pi(u)} \to (\mathcal{V}_{\mathbb{R}})_u$ commutes with J; hence the $(\mathcal{V}_{\mathbb{C}})_u$-valued extension to $(T_{\mathbb{C}}M)_{\pi(u)}$ sends $T^{1,0}_{\pi(u)}M$ onto $\mathcal{V}_u$, and commutes with J and the isomorphisms ^{o} and $_o$. In particular, we can define the canonical isomorphisms $\iota_v: T^{1,0}_{\pi(v)}M \to \mathcal{V}_v$ by setting

$$\iota_v(w) = \big(\iota_{v^o}(w^o)\big)_o.$$

Definition 2.1.2: The *real radial vertical vector field* $\iota^o: \tilde{M} \to \mathcal{V}_{\mathbb{R}}$ and the *(complex) radial vertical vector field* $\iota: \tilde{M} \to \mathcal{V}$ are defined as usual:

$$\iota^o(u) = \iota_u(u) \qquad \text{and} \qquad \iota(v) = \iota_v(v),$$

for all $u \in T_{\mathbb{R}}M$ and $v \in T^{1,0}M$. They of course satisfy

$$\forall v \in T^{1,0}M \qquad\qquad \iota^o(v^o) = \big(\iota(v)\big)^o.$$

Now take $\xi \in \mathcal{X}(T_{\mathbb{R}}M)$. Then we know that $d\xi: T_{\mathbb{R}}M \to T_{\mathbb{R}}(T_{\mathbb{R}}M)$ is given by

$$d\xi(u) = u^a\partial_a^o + u^a\frac{\partial\xi^b}{\partial x^a}\dot{\partial}_b^o.$$

Analogously, we can take $\xi \in \mathcal{X}(T^{1,0}M)$, getting $d\xi: T_{\mathbb{C}}M \to T_{\mathbb{C}}(T^{1,0}M)$; if

$$v = v^\alpha\frac{\partial}{\partial z^\alpha} + v^{\bar{\alpha}}\frac{\partial}{\partial \bar{z}^\alpha} \in T_{\mathbb{C}}M,$$

then

$$d\xi(v) = v^\alpha\left[\partial_\alpha + \frac{\partial\xi^\beta}{\partial z^\alpha}\dot{\partial}_\beta + \frac{\partial\overline{\xi^\beta}}{\partial z^\alpha}\dot{\partial}_{\bar{\beta}}\right] + v^{\bar{\alpha}}\left[\partial_{\bar{\alpha}} + \frac{\partial\xi^\beta}{\partial \bar{z}^\alpha}\dot{\partial}_\beta + \frac{\partial\overline{\xi^\beta}}{\partial \bar{z}^\alpha}\dot{\partial}_{\bar{\beta}}\right]. \tag{2.1.8}$$

In particular,

$$\forall u \in T_{\mathbb{R}}M \qquad\qquad d\xi(u) = d\xi^o(u), \tag{2.1.9}$$

because $u \in T_{\mathbb{R}}M$ iff $u^{\bar{\alpha}} = \overline{u^\alpha}$ for $\alpha = 1, \ldots, n$. Moreover,

LEMMA 2.1.1: *Let M be a complex manifold, and take $p \in M$ and $\xi \in \mathcal{X}(T^{1,0}M)$. Then the following assertions are equivalent:*

(i) *ξ is holomorphic at p, that is $\bar\partial\xi_p = 0$;*
(ii) *$d\xi(T^{1,0}_pM) \subset T^{1,0}_{\xi(p)}(T^{1,0}M)$;*
(iii) *$d\xi_p \circ J = J \circ d\xi_p$ on $T_{\mathbb{C}}M$;*
(iv) *$d\xi^o_p \circ J = J \circ d\xi^o_p$ on $T_{\mathbb{R}}M$.*

Proof: (ii)$\Longrightarrow$(i)$\Longrightarrow$(iii): it follows from (2.1.8).

(iii)$\Longrightarrow$(iv). Take $u \in (T_{\mathbb{R}}M)_p$; then $Ju \in (T_{\mathbb{R}}M)_p$ and (2.1.9) yields

$$d\xi^o_p(Ju) = d\xi_p(Ju) = J\big(d\xi_p(u)\big) = J\big(d\xi^o_p(u)\big).$$

(iv)$\Longrightarrow$(ii). Take $v \in T^{1,0}_pM$, and set $u = v^o$. Then (2.1.9) yields

$$d\xi_p(v) = \frac{1}{2}[d\xi_p(u) - i\,d\xi_p(Ju)] = \frac{1}{2}[d\xi^o_p(u) - i\,d\xi^o_p(Ju)].$$

Applying J we then get

$$\begin{aligned} J\big(d\xi_p(v)\big) &= \frac{1}{2}\left[J\big(d\xi^o_p(u)\big) - iJ\big(d\xi^o_p(Ju)\big)\right] = \frac{i}{2}[d\xi^o_p(u) - i\,d\xi^o_p(Ju)] \\ &= \frac{i}{2}\left[d\xi_p(u) - i\,d\xi_p(Ju)\right] = i\,d\xi_p(v). \end{aligned}$$

□

We shall denote by $\mathcal{X}_{\mathcal{O}}(\tilde M)$ the space of holomorphic sections of $\tilde M$, where when $\tilde M$ is considered as a subset of $T_{\mathbb{R}}M$ "holomorphic" means that the condition given by Lemma 2.1.1.(iv) is satisfied. More generally, if $p\colon E \to B$ is a complex vector bundle on a complex manifold, we shall denote by $\mathcal{X}_{\mathcal{O}}(E)$ the space of holomorphic sections of E.

DEFINITION 2.1.3: Let $p^{1,0}\colon T_{\mathbb{C}}(T^{1,0}M) \to T^{1,0}(T^{1,0}M)$ be the natural projection; note that $p^{1,0}$ restricted to $T_{\mathbb{R}}(T^{1,0}M)$ coincides with the isomorphism ${}_o$. Then setting

$$d^{1,0} = p^{1,0} \circ d$$

we have

$$d^{1,0}\xi(v) = v^\alpha\left[\partial_\alpha + \frac{\partial\xi^\beta}{\partial z^\alpha}\dot\partial_\beta\right] + v^{\bar\alpha}\frac{\partial\xi^\beta}{\partial\bar z^\alpha}\dot\partial_\beta,$$

for all $\xi \in \mathcal{X}(T^{1,0}M)$ and all $v \in T_{\mathbb{C}}M$. It is easy to check that

$$\forall v \in T^{1,0}M \qquad d^{1,0}\xi(v^o) = \big(d\xi^o(v^o)\big)_o, \tag{2.1.10}$$

and so $d^{1,0}$ is the correct analogue on $T^{1,0}M$ of d on $T_{\mathbb{R}}M$. Furthermore, $d^{1,0}\xi_p$ commutes with J iff ξ is holomorphic at p.

Now let $\tilde{D}: \mathcal{X}(T_{\mathbb{R}}M) \to \mathcal{X}(T^*_{\mathbb{R}}M \otimes T_{\mathbb{R}}M)$ be a (real) non-linear connection on M. The usual isomorphisms give rise to a map $\tilde{D}_o: \mathcal{X}(T^{1,0}M) \to \mathcal{X}\big((T^{1,0}M)^*_{\mathbb{R}} \otimes T^{1,0}M\big)$ — where $(T^{1,0}M)^*_{\mathbb{R}}$ is the space of $\mathbb{R}$-linear functionals on $T^{1,0}M$ — by setting

$$(\tilde{D}_o\xi)(v) = \big(\tilde{D}\xi^o(v^o)\big)_o.$$

Clearly, $\tilde{D}_o$ satisfies

$$\tilde{D}_o\xi'_p - \tilde{D}_o\xi_p = \iota_v^{-1} \circ (d^{1,0}\xi'_p - d^{1,0}\xi_p)$$

for all $p \in M$ and $\xi, \xi' \in \mathcal{X}(T^{1,0}M)$ such that $\xi(p) = \xi'(p) = v$. At this point we need the following

Lemma 2.1.2: *Let V be a complex vector space. Then every $\mathbb{R}$-linear functional $\omega \in V^*_{\mathbb{R}}$ is the sum of a $\mathbb{C}$-linear functional $\omega' \in V^*$ and a $\mathbb{C}$-antilinear functional $\omega'' \in \overline{V}^*$; moreover, this decomposition is unique. In other words, $V^*_{\mathbb{R}} = V^* \oplus \overline{V}^*$.*

Proof: Assume $\omega = \omega' + \omega''$, with $\omega' \in V^*$ and $\omega'' \in \overline{V}^*$. Then

$$\omega(v) = \omega'(v) + \omega''(v), \qquad \omega(iv) = i\omega'(v) - i\omega''(v),$$

and hence

$$\omega'(v) = \frac{\omega(v) - i\omega(iv)}{2}, \qquad \omega''(v) = \frac{\omega(v) + i\omega(iv)}{2}.$$

On the other hand, it is easy to check that ω' and ω'' so defined are as required, and we are done. □

In our case, this implies that we may write $\tilde{D}_o = \tilde{D}' + \tilde{D}''$, with

$$\tilde{D}'\xi \in \mathcal{X}\big((T^{1,0}M)^* \otimes T^{1,0}M\big) \qquad \text{and} \qquad \tilde{D}''\xi \in \mathcal{X}\big((\overline{T^{1,0}M})^* \otimes T^{1,0}M\big).$$

Recalling the canonical isomorphisms, we can think of $\tilde{D}'\xi$ as belonging to the space $\mathcal{X}(\bigwedge^{1,0}M \otimes T^{1,0}M)$, and of $\tilde{D}''\xi$ as belonging to $\mathcal{X}(\bigwedge^{0,1}M \otimes T^{1,0}M)$. We are then ready for the following definition:

Definition 2.1.4: A *complexified non-linear connection* is a map

$$\tilde{D}: \mathcal{X}(T^{1,0}M) \to \mathcal{X}(T^*_{\mathbb{C}}M \otimes T^{1,0}M)$$

such that

$$\tilde{D}o \equiv 0$$

and

$$\tilde{D}\xi'_p - \tilde{D}\xi_p = \iota_v^{-1} \circ (d^{1,0}\xi'_p - d^{1,0}\xi_p) \tag{2.1.11}$$

for all $p \in M$ and $\xi, \xi' \in \mathcal{X}(T^{1,0}M)$ with $\xi(p) = \xi'(p) = v$.

It is easy to check that if we define $\tilde{D}^o\colon \mathcal{X}(T_{\mathbb{R}}M) \to \mathcal{X}(T^*_{\mathbb{R}}M \otimes T_{\mathbb{R}}M)$ by setting

$$\tilde{D}^o\xi(u) = \left(\tilde{D}\xi_o(u_o)\right)^o$$

we get a (real) non-linear connection, thanks to (2.1.10).

If $\tilde{D}$ is a complexified non-linear connection, we can uniquely write

$$\tilde{D} = \tilde{D}' + \tilde{D}'',$$

with

$$\tilde{D}'\colon \mathcal{X}(T^{1,0}M) \to \mathcal{X}(\textstyle\bigwedge^{1,0}M \otimes T^{1,0}M) \quad \text{and} \quad \tilde{D}''\colon \mathcal{X}(T^{1,0}M) \to \mathcal{X}(\textstyle\bigwedge^{0,1}M \otimes T^{1,0}M).$$

DEFINITION 2.1.5: We shall say that a complexified non-linear connection $\tilde{D}$ is a *complex non-linear connection* if

$$\forall \xi \in \mathcal{X}_{\mathcal{O}}(T^{1,0}M) \qquad \tilde{D}''\xi \equiv 0. \tag{2.1.12}$$

Since $\tilde{D}'\xi \circ J = J \circ \tilde{D}'\xi$ and $\tilde{D}''\xi \circ J = -J \circ \tilde{D}''\xi$ for any $\xi \in \mathcal{X}(T^{1,0}M)$, (2.1.12) is equivalent to

$$\forall \xi \in \mathcal{X}_{\mathcal{O}}(T^{1,0}M) \qquad \tilde{D}\xi \circ J = J \circ \tilde{D}\xi. \tag{2.1.13}$$

Note that (2.1.13) is supposed to hold only for holomorphic vector fields. However, the locality of $\tilde{D}$ immediately implies that

$$\tilde{D}\xi_p \circ J = J \circ \tilde{D}\xi_p$$

for all $\xi \in \mathcal{X}(T^{1,0}M)$ which are holomorphic in a neighborhood of $p \in M$.

As in the real case, we can associate to complex non-linear connections both complex horizontal bundles and complex horizontal maps.

DEFINITION 2.1.6: A *complex horizontal bundle* is a complex subbundle $\mathcal{H}_{\mathbb{C}} \subset T_{\mathbb{C}}\tilde{M}$ which is J-invariant, conjugation invariant (that is $\overline{\mathcal{H}_{\mathbb{C}}} = \mathcal{H}_{\mathbb{C}}$) and such that

$$T_{\mathbb{C}}\tilde{M} = \mathcal{H}_{\mathbb{C}} \oplus \mathcal{V}_{\mathbb{C}}.$$

Since $\mathcal{H}_{\mathbb{C}}$ is J-invariant, we can write $\mathcal{H}_{\mathbb{C}} = \mathcal{H}^{1,0} \oplus \mathcal{H}^{0,1}$, where $\mathcal{H}^{1,0} = \mathcal{H}_{\mathbb{C}} \cap T^{1,0}\tilde{M}$; moreover, being $\mathcal{H}_{\mathbb{C}}$ conjugation invariant, $\mathcal{H}^{0,1} = \overline{\mathcal{H}^{1,0}}$. This means that a complex horizontal bundle is completely determined by its $(1,0)$-part $\mathcal{H}^{1,0}$; later on, we shall often simply write $\mathcal{H}$ instead of $\mathcal{H}^{1,0}$.

DEFINITION 2.1.7: A *complex horizontal map* is a complex linear bundle map $\Theta\colon \mathcal{V}_{\mathbb{C}} \to T_{\mathbb{C}}\tilde{M}$ commuting with J and the conjugation and such that

$$(d\pi \circ \Theta)_v\big|_{\mathcal{V}^{1,0}} = \iota_v^{-1}\big|_{\mathcal{V}^{1,0}}$$

for all $v \in \tilde{M}$. Since $\mathcal{V}_{\mathbb{C}} = \mathcal{V}^{1,0} \oplus \mathcal{V}^{0,1}$, $\mathcal{V}^{0,1} = \overline{\mathcal{V}^{1,0}}$ and Θ commutes with both J and the conjugation, it is clear that $\Theta(\mathcal{V}^{1,0}) \subset T^{1,0}\tilde{M}$ and Θ is completely determined by its behavior on $\mathcal{V}^{1,0}$.

It is easy to recover in this case too the correspondences among complex non-linear connections, horizontal bundles and horizontal maps. Let $\mathcal{H}_{\mathbb{C}}$ be a complex horizontal bundle, and $\kappa\colon T_{\mathbb{C}}\tilde{M} \to \mathcal{V}_{\mathbb{C}}$ the associated vertical projection. Since $\mathcal{H}_{\mathbb{C}}$ is J-invariant and conjugation invariant, κ commutes with J and the conjugation; in particular, κ sends $T^{1,0}\tilde{M}$ onto $\mathcal{V}^{1,0}$. We define a complex non-linear connection $\tilde{D}_{\mathcal{H}_{\mathbb{C}}}$ on M by setting

$$\tilde{D}_{\mathcal{H}_{\mathbb{C}}}\xi_p = \iota^{-1}_{\xi(p)} \circ \kappa_{\xi(p)} \circ d^{1,0}\xi_p$$

for any $p \in M$ and $\xi \in \mathcal{X}(T^{1,0}M)$. Since $d^{1,0}\xi'_p - d^{1,0}\xi_p$ sends $(T_{\mathbb{C}}M)_p$ into $\mathcal{V}^{1,0}_v$ as soon as $\xi(p) = \xi'(p) = v$, it is clear that $\tilde{D}_{\mathcal{H}_{\mathbb{C}}}$ is a complexified non-linear connection. Furthermore, if ξ is holomorphic then both κ and $d^{1,0}\xi$ commute with J, and so $\tilde{D}_{\mathcal{H}_{\mathbb{C}}}$ is complex.

Now let $\tilde{D}$ be a complex non-linear connection. Take $v \in \tilde{M}_p$ and $\xi \in \mathcal{X}(T^{1,0}M)$ such that $\xi(p) = v$. Then define $\Theta^{\tilde{D}}_v\colon \mathcal{V}^{1,0}_v \to T^{1,0}_v\tilde{M}$ by

$$\Theta^{\tilde{D}}_v = d^{1,0}\xi_p \circ \iota_v^{-1} - \iota_v \circ \tilde{D}\xi_p \circ \iota_v^{-1}.$$

This definition does not depend on ξ but only on v, by (2.1.11). If we extend $\Theta^{\tilde{D}}$ to $\mathcal{V}^{0,1}$ asking that Θ commutes with conjugation, and then to $\mathcal{V}_{\mathbb{C}}$ by complex linearity, we obtain a complex horizontal map. Indeed, $\Theta^{\tilde{D}}$ commutes with conjugation by definition. Next, since we may take ξ to be holomorphic in a neighborhood of p, it commutes with J. And so, since $d\pi \circ d^{1,0}\xi_p = \mathrm{id}$ on $T^{1,0}M$, we have shown that $\Theta^{\tilde{D}}$ actually is a complex horizontal map.

Finally, if Θ is a complex horizontal map, it is clear that $\mathcal{H}^{\Theta}_{\mathbb{C}} = \Theta(\mathcal{V}_{\mathbb{C}})$ is a complex horizontal bundle. The arguments used in the real case can be then repeated almost word by word to show that we have defined a one-to-one correspondence among complex horizontal bundles, complex non-linear connections and complex horizontal maps.

We end this subsection describing what we mean by homogeneity in this context.

Definition 2.1.8: We shall say that a complex non-linear connection $\tilde{D}$ is *homogeneous* iff

$$\tilde{D}(\mu_\zeta \circ \xi) = \mu_\zeta \circ \tilde{D}\xi$$

for all $\zeta \in \mathbb{C}^*$ and $\xi \in \mathcal{X}(T^{1,0}M)$, where $\mu_\zeta\colon T^{1,0}M \to T^{1,0}M$ is the multiplication by ζ.

Definition 2.1.9: A complex horizontal bundle $\mathcal{H}_{\mathbb{C}}$ is *homogeneous* iff

$$(\mathcal{H}_{\mathbb{C}})_{\mu_\zeta(v)} = d(\mu_\zeta)_v((\mathcal{H}_{\mathbb{C}})_v)$$

for all $\zeta \in \mathbb{C}^*$ and $v \in \tilde{M}$.

Definition 2.1.10: A complex horizontal map Θ is *homogeneous* iff

$$d(\mu_\zeta)_v \circ \Theta_v \circ \iota_v = \Theta_{\mu_\zeta(v)} \circ \iota_{\mu_\zeta(v)}$$

for all $\zeta \in \mathbb{C}^*$ and $v \in \tilde{M}$.

Again, it is easy to prove exactly as in the real case that the previously defined correspondences preserve homogeneity.

2.1.4. Local coordinates

Our next aim is to find the expression in local coordinates of complex non-linear connections. Let $\tilde{D}$ be a generic complexified non-linear connection, with associated covariant differentiation $\tilde{\nabla}$. Take $\xi \in \mathcal{X}(T^{1,0}M)$, $p \in M$ and set $v = \xi(p)$. Then

$$d^{1,0}\xi\left(\frac{\partial}{\partial z^\alpha}\right) = \partial_\alpha + \frac{\partial \xi^\beta}{\partial z^\alpha}\dot{\partial}_\beta, \qquad d^{1,0}\xi\left(\frac{\partial}{\partial \bar{z}^\alpha}\right) = \frac{\partial \xi^\beta}{\partial \bar{z}^\alpha}\dot{\partial}_\beta.$$

By definition, $\iota_v(\tilde{\nabla}_{\partial/\partial z^\alpha}\xi) - d^{1,0}\xi_p(\partial/\partial z^\alpha)$ and $\iota_v(\tilde{\nabla}_{\partial/\partial \bar{z}^\alpha}\xi) - d^{1,0}\xi_p(\partial/\partial \bar{z}^\alpha)$ must depend only on v. Therefore there are $\Gamma^\beta_\alpha(v)$, $\Gamma^\beta_{\bar{\alpha}}(v) \in \mathbb{C}$ such that

$$\iota_v(\tilde{\nabla}_{\partial/\partial z^\alpha}\xi) - d^{1,0}\xi_p(\partial/\partial z^\alpha) = -\partial_\alpha + \Gamma^\beta_\alpha(v)\dot{\partial}_\beta,$$
$$\iota_v(\tilde{\nabla}_{\partial/\partial \bar{z}^\alpha}\xi) - d^{1,0}\xi_p(\partial/\partial \bar{z}^\alpha) = \Gamma^\beta_{\bar{\alpha}}(v)\dot{\partial}_\beta.$$

Thus

$$\tilde{\nabla}_w\xi(p) = \left\{ w^\alpha\left(\frac{\partial \xi^\beta}{\partial z^\alpha}(p) + \Gamma^\beta_\alpha(\xi(p))\right) + w^{\bar{\alpha}}\left(\frac{\partial \xi^\beta}{\partial \bar{z}^\alpha} + \Gamma^\beta_{\bar{\alpha}}(\xi(p))\right)\right\} \left.\frac{\partial}{\partial z^\beta}\right|_p$$

for all $w = w^\alpha(\partial/\partial z^\alpha) + w^{\bar{\alpha}}(\partial/\partial \bar{z}^\alpha) \in (T_{\mathbb{C}}M)_p$. In particular, $\tilde{D}$ is complex iff

$$\Gamma^\beta_{\bar{\alpha}} \equiv 0.$$

The Γ^β_α's are the *Christoffel symbols* of the complex non-linear connection $\tilde{D}$.

Assuming now $\tilde{D}$ complex, it is easy to show that in local coordinates $\Theta^{\tilde{D}}$ is given by

$$\delta_\alpha = \Theta^{\tilde{D}}(\dot{\partial}_\alpha) = \partial_\alpha - \Gamma^\beta_\alpha\dot{\partial}_\beta,$$
$$\delta_{\bar{\alpha}} = \Theta^{\tilde{D}}(\dot{\partial}_{\bar{\alpha}}) = \partial_{\bar{\alpha}} - \overline{\Gamma^\beta_\alpha}\dot{\partial}_{\bar{\beta}}.$$

Clearly, $\{\delta_1, \dots, \delta_n\}$ is a local frame for $\mathcal{H}^{1,0}$, and $\{\delta_1, \dots, \delta_n, \dot{\partial}_1, \dots, \dot{\partial}_n\}$ a local frame for $T^{1,0}\tilde{M}$. We shall denote by $\{dz^1, \dots, dz^n, \psi^1, \dots, \psi^n\}$ the dual coframe; clearly,

$$\psi^\alpha = dv^\alpha + \Gamma^\alpha_\beta\, dz^\beta.$$

DEFINITION 2.1.11: We also set

$$\chi_v = \Theta_v \circ \iota_v : T_v^{1,0}M \to \mathcal{H}_v^{1,0};$$

then the *(complex) radial horizontal vector field* $\chi \in \mathcal{X}(\mathcal{H}^{1,0})$ is given by

$$\chi = \Theta \circ \iota.$$

Clearly, $\chi(\partial/\partial z^\alpha) = \delta_\alpha$.

Finally, let $\mathcal{H}_{\mathbf{C}}$ be a subbundle of $T_{\mathbf{C}}\tilde{M}$ such that $T_{\mathbf{C}}\tilde{M} = \mathcal{H}_{\mathbf{C}} \oplus \mathcal{V}_{\mathbf{C}}$. Let $\{\delta_\alpha, \delta_{\bar\alpha}\}$ be a local frame for $\mathcal{H}_{\mathbf{C}}$ such that $d\pi(\delta_\alpha) = \partial/\partial z^\alpha$ and $d\pi(\delta_{\bar\alpha}) = \partial/\partial \bar{z}^\alpha$. Then we should have

$$\delta_\alpha = \partial_\alpha - \Gamma_\alpha^\beta \dot\partial_\beta - \Gamma_\alpha^{\bar\beta} \dot\partial_{\bar\beta},$$
$$\delta_{\bar\alpha} = \partial_{\bar\alpha} - \Gamma_{\bar\alpha}^\beta \dot\partial_\beta - \Gamma_{\bar\alpha}^{\bar\beta} \dot\partial_{\bar\beta},$$

for suitable functions Γ_α^β, $\Gamma_\alpha^{\bar\beta}$, $\Gamma_{\bar\alpha}^\beta$ and $\Gamma_{\bar\alpha}^{\bar\beta}$. Therefore $\overline{\mathcal{H}_{\mathbf{C}}} = \mathcal{H}_{\mathbf{C}}$ iff $\overline{\delta_\alpha} = \delta_{\bar\alpha}$, that is iff

$$\Gamma_{\bar\alpha}^{\bar\beta} = \overline{\Gamma_\alpha^\beta} \qquad \text{and} \qquad \Gamma_\alpha^{\bar\beta} = \overline{\Gamma_{\bar\alpha}^\beta}. \tag{2.1.14}$$

Analogously, $\mathcal{H}_{\mathbf{C}}$ is J-invariant iff $J\delta_\alpha = i\delta_\alpha$, that is iff

$$\Gamma_\alpha^{\bar\beta} \equiv 0 \equiv \Gamma_{\bar\alpha}^\beta.$$

We end this subsection with two remarks proved exactly as in the real case. First of all, a complex horizontal bundle $\mathcal{H}_{\mathbf{C}}$ is homogeneous iff its Christoffel symbols are, that is iff

$$\Gamma_\alpha^\beta(\mu_\zeta(v)) = \zeta \Gamma_\alpha^\beta(v)$$

for all $\zeta \in \mathbf{C}^*$ and $v \in \tilde{M}$. Secondly, and lastly, the behavior of the Christoffel symbols under change of coordinates is

$$(\Gamma_\alpha^\beta)_B = (\mathcal{J}_{BA})_\gamma^\beta (\Gamma_\delta^\gamma)_A (\mathcal{J}_{BA}^{-1})_\alpha^\delta - (H_{BA})_{\gamma\delta}^\beta (\mathcal{J}_{BA}^{-1})_\alpha^\gamma v_A^\delta.$$

2.2. Complex vertical connections

2.2.1. Definitions

DEFINITION 2.2.1: A *complexified vertical connection* on a complex manifold M is a linear connection on the complexified vertical bundle $\mathcal{V}_{\mathbb{C}}$, that is a linear map $D: \mathcal{X}(\mathcal{V}_{\mathbb{C}}) \to \mathcal{X}(T^*_{\mathbb{C}}\tilde{M} \otimes \mathcal{V}_{\mathbb{C}})$ such that

$$D(fV) = f\,DV + df \otimes V$$

for every $f \in C^\infty(\tilde{M})$ and $V \in \mathcal{X}(\mathcal{V}_{\mathbb{C}})$.

This definition makes almost no mention of the complex structure of $\tilde{M}$, and thus it is too general for our aims. So we make three assumptions. First of all, we assume that D commutes with the conjugation, that is that

$$\forall V \in \mathcal{X}(\mathcal{V}_{\mathbb{C}}) \qquad D(\overline{V}) = \overline{DV}. \tag{2.2.1}$$

In particular, since $\mathcal{V}_{\mathbb{C}} = \mathcal{V}^{1,0} \oplus \mathcal{V}^{0,1}$ and $\mathcal{V}^{0,1} = \overline{\mathcal{V}^{1,0}}$, this implies that D is completely determined by its behavior on $\mathcal{X}(\mathcal{V}^{1,0})$.

Second, we shall assume that D restricted to $\mathcal{V}^{1,0}$ commutes with J, that is that

$$\forall V \in \mathcal{X}(\mathcal{V}^{1,0}) \qquad D(JV) = J \circ DV. \tag{2.2.2}$$

As a consequence, $DV \in \mathcal{X}(T^*_{\mathbb{C}}\tilde{M} \otimes \mathcal{V}^{1,0})$ for all $V \in \mathcal{X}(\mathcal{V}^{1,0})$; indeed, (2.2.2) implies

$$J \circ DV = D(JV) = D(iV) = iDV.$$

Finally, we shall assume that for *holomorphic* sections V of $\mathcal{V}^{1,0}$ the 1-form DV commutes with J, that is

$$\forall V \in \mathcal{X}_{\mathcal{O}}(\mathcal{V}^{1,0}) \qquad (DV) \circ J = J \circ (DV), \tag{2.2.3}$$

where, as before, $\mathcal{X}_{\mathcal{O}}$ stands for holomorphic sections. Since $T^*_{\mathbb{C}}\tilde{M} = \bigwedge^{1,0}\tilde{M} \oplus \bigwedge^{0,1}\tilde{M}$, we may decompose the connection D as a sum

$$D = D' + D'',$$

with $D': \mathcal{X}(\mathcal{V}^{1,0}) \to \mathcal{X}(\bigwedge^{1,0}\tilde{M} \otimes \mathcal{V}^{1,0})$ and $D'': \mathcal{X}(\mathcal{V}^{1,0}) \to \mathcal{X}(\bigwedge^{0,1}\tilde{M} \otimes \mathcal{V}^{1,0})$. Then (2.2.3) is equivalent to

$$\forall V \in \mathcal{X}_{\mathcal{O}}(\mathcal{V}^{1,0}) \qquad D''V \equiv 0.$$

DEFINITION 2.2.2: A *complex vertical connection* is a complexified vertical connection D satisfying (2.2.1), (2.2.2) and (2.2.3). In particular, we shall consider D as defined over $\mathcal{X}(\mathcal{V}^{1,0})$ and with values in $\mathcal{X}(T^*_{\mathbb{C}}\tilde{M} \otimes \mathcal{V}^{1,0})$, extended, when necessary, to the whole $\mathcal{V}_{\mathbb{C}}$ by (2.2.1) and complex linearity.

Let $D: \mathcal{X}(\mathcal{V}_{\mathbb{C}}) \to \mathcal{X}(T^*_{\mathbb{C}}\tilde{M} \otimes \mathcal{V}_{\mathbb{C}})$ be a complexified vertical connection. In local coordinates, writing $V = V^\alpha \dot\partial_\alpha + V^{\bar\alpha}\dot\partial_{\bar\alpha}$, we get

$$DV = dV^\alpha \otimes \dot\partial_\alpha + dV^{\bar\alpha} \otimes \dot\partial_{\bar\alpha} + V^\alpha D\dot\partial_\alpha + V^{\bar\alpha} D\dot\partial_{\bar\alpha},$$

and

$$D\dot\partial_\alpha = \omega^\beta_\alpha \otimes \dot\partial_\beta + \omega^{\bar\beta}_\alpha \otimes \dot\partial_{\bar\beta} \qquad \text{and} \qquad D\dot\partial_{\bar\alpha} = \omega^\beta_{\bar\alpha} \otimes \dot\partial_\beta + \omega^{\bar\beta}_{\bar\alpha} \otimes \dot\partial_{\bar\beta},$$

where ω^β_α, $\omega^{\bar\beta}_\alpha$, $\omega^\beta_{\bar\alpha}$ and $\omega^{\bar\beta}_{\bar\alpha}$ are locally defined 1-forms. Then (2.2.1) is equivalent to

$$\omega^{\bar\beta}_{\bar\alpha} = \overline{\omega^\beta_\alpha} \qquad \text{and} \qquad \omega^\beta_{\bar\alpha} = \overline{\omega^{\bar\beta}_\alpha};$$

(2.2.2) on the other hand is equivalent to

$$\omega^{\bar\beta}_\alpha \equiv 0 \equiv \omega^\beta_{\bar\alpha}.$$

Finally, (2.2.3) is equivalent to requiring that every ω^β_α is a (1,0)-form, that is that

$$\omega^\beta_\alpha = \tilde\Gamma^\beta_{\alpha;\mu}\, dz^\mu + \tilde\Gamma^\beta_{\alpha\gamma}\, dv^\gamma,$$

for suitable coefficients $\tilde\Gamma^\beta_{\alpha;\mu}$ and $\tilde\Gamma^\beta_{\alpha\gamma}$.

Summing up, locally a complex vertical connection is given by

$$DV = \left[dV^\beta + V^\alpha \omega^\beta_\alpha\right] \otimes \dot\partial_\beta$$

for all $V \in \mathcal{X}(\mathcal{V}^{1,0})$, where the ω^β_α's are local (1,0)-forms. In particular,

$$\begin{aligned} \nabla_X V &= [X(V^\alpha) + \omega^\alpha_\beta(X) V^\beta]\dot\partial_\alpha, & \nabla_X \overline{V} &= \overline{\nabla_{\overline{X}} V}, \\ \nabla_{\overline{X}} V &= \overline{X}(V^\alpha)\dot\partial_\alpha, & \nabla_{\overline{X}} \overline{V} &= \overline{\nabla_X V}, \end{aligned} \tag{2.2.4}$$

for all $X \in T^{1,0}\tilde{M}$.

2.2.2. Good vertical connections

From now on, we shall write $\mathcal{V}$ in place of $\mathcal{V}^{1,0}$. The idea is that, exactly as in the real case, we can associate a complex non-linear connection to some complex vertical connections.

Let D be a complex vertical connection, and denote again by $\Lambda: T^{1,0}\tilde{M} \to \mathcal{V}$ the bundle map defined by

$$\Lambda(X) = \nabla_X \iota.$$

In local coordinates,

$$\Lambda(X) = [\dot X^\alpha + \omega^\alpha_\beta(X) v^\beta]\dot\partial_\alpha.$$

Definition 2.2.3: We say that the complex vertical connection D is *good* if the bundle map $\Lambda|_{\mathcal{V}}: \mathcal{V} \to \mathcal{V}$ is a bundle isomorphism.

Exactly as in the real case, we see that D is good iff $\mathcal{H}_{\mathbb{C}} = \mathcal{H} \oplus \overline{\mathcal{H}}$ is a complex horizontal bundle, where $\mathcal{H} = \ker \Lambda \subset T^{1,0}\tilde{M}$. In local coordinates, D is good iff the matrix

$$L^\alpha_\beta = \delta^\alpha_\beta + \tilde{\Gamma}^\alpha_{\gamma\beta} v^\gamma$$

is invertible. If $\big((L^{-1})^\mu_\nu\big)$ denotes the inverse matrix, a local frame for $\mathcal{H}$ is given by

$$\delta_\alpha = \partial_\alpha - (L^{-1})^\mu_\nu \tilde{\Gamma}^\nu_{\gamma;\alpha} v^\gamma \dot{\partial}_\mu = \partial_\alpha - \Gamma^\mu_\alpha \dot{\partial}_\mu,$$

where

$$\Gamma^\mu_\alpha = (L^{-1})^\mu_\nu \tilde{\Gamma}^\nu_{\gamma;\alpha} v^\gamma$$

are the Christoffel symbols of the complex non-linear connection associated to $\mathcal{H}$.

So let $D\colon \mathcal{X}(\mathcal{V}) \to \mathcal{X}(T^*_{\mathbb{C}}\tilde{M} \otimes \mathcal{V})$ be a good complex vertical connection. We shall denote by $\{dz^\mu, \psi^\alpha\}$ the local coframe dual to $\{\delta_\mu, \dot{\partial}_\alpha\}$, where

$$\psi^\alpha = dv^\alpha + \Gamma^\alpha_\mu \, dz^\mu.$$

Correspondingly, we may decompose

$$\omega^\alpha_\beta = \Gamma^\alpha_{\beta;\mu} \, dz^\mu + \Gamma^\alpha_{\beta\gamma} \psi^\gamma,$$

where

$$\Gamma^\alpha_{\beta\gamma} = \tilde{\Gamma}^\alpha_{\beta\gamma} \qquad \text{and} \qquad \Gamma^\alpha_{\beta;\mu} = \tilde{\Gamma}^\alpha_{\beta;\mu} - \tilde{\Gamma}^\alpha_{\beta\gamma} \Gamma^\beta_\mu.$$

In particular, we again have

$$\Gamma^\alpha_{\beta;\mu} v^\beta = \Gamma^\alpha_\mu.$$

Furthermore, writing $X = X^\mu \delta_\mu + \dot{X}^\gamma \dot{\partial}_\gamma$, we have

$$\nabla_X V = \left\{ X^\mu \left[\delta_\mu(V^\alpha) + \Gamma^\alpha_{\beta;\mu} V^\beta \right] + \dot{X}^\gamma \left[\dot{\partial}_\gamma(V^\alpha) + \Gamma^\alpha_{\beta\gamma} V^\beta \right] \right\} \dot{\partial}_\alpha,$$
$$\nabla_{\overline{X}} V = \left\{ \overline{X^\mu} \delta_{\bar{\mu}}(V^\alpha) + \overline{\dot{X}^\gamma} \dot{\partial}_{\bar{\gamma}}(V^\alpha) \right\} \dot{\partial}_\alpha.$$

The behavior of ω^α_β, $\Gamma^\alpha_{\beta\gamma}$ and $\Gamma^\alpha_{\beta;\mu}$ under change of coordinates is exactly as expected:

$$\begin{aligned} (\omega^\beta_\alpha)_B &= (\mathcal{J}^{-1}_{BA})^\gamma_\alpha (\omega^\delta_\gamma)_A (\mathcal{J}_{BA})^\beta_\delta - (\mathcal{J}^{-1}_{BA})^\gamma_\alpha (H_{BA})^\beta_{\gamma\mu} \, dz^\mu; \\ (\Gamma^\beta_{\alpha;\mu})_B &= (\mathcal{J}^{-1}_{BA})^\gamma_\alpha (\mathcal{J}^{-1}_{BA})^\nu_\mu (\Gamma^\delta_{\gamma;\nu})_A (\mathcal{J}_{BA})^\beta_\delta - (\mathcal{J}^{-1}_{BA})^\gamma_\alpha (\mathcal{J}^{-1}_{BA})^\nu_\mu (H_{BA})^\beta_{\gamma\nu}; \\ (\Gamma^\beta_{\alpha\gamma})_B &= (\mathcal{J}^{-1}_{BA})^\rho_\alpha (\mathcal{J}^{-1}_{BA})^\sigma_\gamma (\Gamma^\tau_{\rho\sigma})_A (\mathcal{J}_{BA})^\beta_\tau. \end{aligned} \tag{2.2.5}$$

Let $\Theta\colon \mathcal{V} \to \mathcal{H}$ be the complex horizontal map associated to the complex horizontal bundle $\mathcal{H}_{\mathbb{C}}$. In local coordinates, $\Theta(\dot{\partial}_\alpha) = \delta_\alpha$. We may then define a complex linear connection on $\mathcal{H}$ by setting

$$\nabla_X H = \Theta \left[\nabla_X (\Theta^{-1} H) \right]$$

for any $X \in T_{\mathbb{C}}\tilde{M}$ and $H \in \mathcal{X}(\mathcal{H})$. By linearity, this yields a complex linear connection on $T_{\mathbb{C}}\tilde{M}$, still denoted by D.

2.2.3. Connections on complex tensor bundles

DEFINITION 2.2.4: Let $D: \mathcal{X}(\mathcal{V}) \to \mathcal{X}(T^*_{\mathbb{C}}\tilde{M} \otimes \mathcal{V})$ be a good complex vertical connection, and let D also denote the associated complex linear connection on $T_{\mathbb{C}}\tilde{M}$. Then we can define a complex linear connection on $T^*_{\mathbb{C}}\tilde{M}$, which we shall denote by $\nabla: \mathcal{X}(T^*_{\mathbb{C}}\tilde{M}) \to \mathcal{X}(T^*_{\mathbb{C}}\tilde{M} \otimes T^*_{\mathbb{C}}\tilde{M})$, by requiring

$$(\nabla\varphi)(X) + \varphi(DX) = d\big(\varphi(X)\big),$$

for all $\varphi \in \mathcal{X}(T^*_{\mathbb{C}}\tilde{M})$ and $X \in \mathcal{X}(T_{\mathbb{C}}\tilde{M})$.

It is easy to check that ∇ commutes with the conjugation, that is

$$\nabla\overline{\varphi} = \overline{\nabla\varphi},$$

and with J^* on $\bigwedge^{1,0}\tilde{M}$. In particular, it preserves the type, that is

$$\varphi \in A^{1,0}(\tilde{M}) \Longrightarrow \nabla\varphi \in \mathcal{X}(T^*_{\mathbb{C}}\tilde{M} \otimes \textstyle\bigwedge^{1,0}\tilde{M}).$$

In local coordinates, if $\varphi = \varphi_\mu\, dz^\mu + \dot{\varphi}_\alpha \psi^\alpha$, then

$$\nabla\varphi = [d\varphi_\mu - \varphi_\nu \omega^\nu_\mu] \otimes dz^\mu + [d\dot{\varphi}_\alpha - \dot{\varphi}_\beta \omega^\beta_\alpha] \otimes \psi^\alpha.$$

Clearly, we can extend ∇ to a complex linear connection

$$\nabla: \mathcal{X}(T^{(r,s)}_{\mathbb{C}}\tilde{M}) \to \mathcal{X}(T^*_{\mathbb{C}}\tilde{M} \otimes T^{(r,s)}_{\mathbb{C}}\tilde{M})$$

commuting with the conjugation and preserving the types (i.e., the behavior with respect to J and J^*), where, of course, $T^{(r,s)}_{\mathbb{C}}\tilde{M} = T^{(r,s)}\tilde{M} \otimes_{\mathbb{R}} \mathbb{C}$. Then, still following the real example, we introduce an exterior differential for $T^{(r,s)}_{\mathbb{C}}\tilde{M}$-valued (p,q)-forms.

DEFINITION 2.2.5: If $\varphi = \varphi^\alpha \otimes e_\alpha$ is a $T^{(r,s)}_{\mathbb{C}}\tilde{M}$-valued (p,q)-form, where $\{e_\alpha\}$ is a local frame for $T^{(r,s)}_{\mathbb{C}}\tilde{M}$ composed by tensor products of $\{\delta_\mu, \dot{\partial}_\alpha\}$, $\{dz^\mu, \psi^\alpha\}$ and their conjugates, we set

$$D\varphi = d\varphi^\alpha \otimes e_\alpha + (-1)^{p+q} \varphi^\alpha \wedge \nabla e_\alpha.$$

It is easy to check, exactly as in the real case, that $D\varphi$ is well-defined. Furthermore, we can decompose D as $D = D' + D''$, according to the type. In local coordinates,

$$\begin{aligned} D'\varphi &= \partial\varphi^\alpha \otimes e_\alpha + (-1)^{p+q}\varphi^\alpha \wedge \nabla e_\alpha, \\ D''\varphi &= \bar{\partial}\varphi^\alpha \otimes e_\alpha, \end{aligned} \tag{2.2.6}$$

where we used the fact that the connection forms of a complex linear connection are of type (1,0).

2.2.4. The torsions

DEFINITION 2.2.6: The tangent bundle $T^{1,0}M$ (and hence $\tilde{M}$) is naturally equipped with a (1,0)-form, the *canonical form*

$$\eta = dz^\mu \otimes \partial_\mu + dv^\alpha \otimes \dot{\partial}_\alpha \in \mathcal{X}(\textstyle\bigwedge^{1,0}\tilde{M} \otimes T^{1,0}\tilde{M}).$$

It is easy to check that η is a well-defined $T^{1,0}\tilde{M}$-valued global (1,0)-form on $\tilde{M}$.

Now let $D\colon \mathcal{X}(\mathcal{V}) \to \mathcal{X}(T^*_{\mathbb{C}}\tilde{M} \otimes \mathcal{V})$ be a good complex vertical connection, and let D also denote the associated complex linear connection on $\tilde{M}$. Then it is easy to see that

$$\eta = dz^\mu \otimes \delta_\mu + \psi^\alpha \otimes \dot{\partial}_\alpha;$$

in particular,

$$\eta \circ \iota = \iota \qquad \text{and} \qquad \eta \circ \chi = \chi.$$

We have introduced in the previous subsection an exterior differential D; so it is very natural to consider

$$D\eta = (-dz^\nu \wedge \omega^\mu_\nu) \otimes \delta_\mu + (d\psi^\alpha - \psi^\beta \wedge \omega^\alpha_\beta) \otimes \dot{\partial}_\alpha.$$

Now $D\eta$ is a $T^{1,0}\tilde{M}$-valued 2-form obtained by differentiating a (1,0)-form; thus it is the sum of a $T^{1,0}\tilde{M}$-valued (2,0)-form $D'\eta$ and a $T^{1,0}\tilde{M}$-valued (1,1)-form $D''\eta$.

DEFINITION 2.2.7: We shall call $\theta = D'\eta$ the $(2,0)$-*torsion* of the connection, and $\tau = D''\eta$ the $(1,1)$-*torsion* of the connection.

Locally, we may write

$$\theta = \theta^\mu \otimes \delta_\mu + \dot{\theta}^\alpha \otimes \dot{\partial}_\alpha \qquad \text{and} \qquad \tau = \tau^\alpha \otimes \dot{\partial}_\alpha,$$

where

$$\begin{aligned}
\theta^\mu &= -dz^\nu \wedge \omega^\mu_\nu = \tfrac{1}{2}[\Gamma^\mu_{\nu;\sigma} - \Gamma^\mu_{\sigma;\nu}]\, dz^\sigma \wedge dz^\nu + \Gamma^\mu_{\nu\gamma}\psi^\gamma \wedge dz^\nu;\\
\dot{\theta}^\alpha &= \partial\psi^\alpha - \psi^\beta \wedge \omega^\alpha_\beta\\
&= \tfrac{1}{2}[\delta_\mu(\Gamma^\alpha_\nu) - \delta_\nu(\Gamma^\alpha_\mu)]\, dz^\mu \wedge dz^\nu + [\dot{\partial}_\beta(\Gamma^\alpha_\mu) - \Gamma^\alpha_{\beta;\mu}]\, \psi^\beta \wedge dz^\mu \qquad (2.2.7)\\
&\quad + \tfrac{1}{2}[\Gamma^\alpha_{\beta\gamma} - \Gamma^\alpha_{\gamma\beta}]\, \psi^\beta \wedge \psi^\gamma;\\
\tau^\alpha &= \bar{\partial}\psi^\alpha = -\delta_{\bar{\nu}}(\Gamma^\alpha_\mu)\, dz^\mu \wedge d\bar{z}^\nu - \dot{\partial}_{\bar{\beta}}(\Gamma^\alpha_\mu)\, dz^\mu \wedge \overline{\psi^\beta}.
\end{aligned}$$

We shall see in the next section that one of the prominent features of the Chern-Finsler connection is that $\dot{\theta}^\alpha \equiv 0$, a fact without references in the real setting. Anyway, an immediate corollary is

COROLLARY 2.2.1: *The* $(1,1)$-*torsion* τ *vanishes iff the frame* $\{\delta_\mu, \dot{\partial}_\alpha\}$ *is holomorhic.*

Proof: Indeed the frame $\{\delta_\mu, \dot{\partial}_\alpha\}$ is holomorphic iff its dual coframe $\{dz^\mu, \psi^\alpha\}$ is, which happens iff the forms ψ^α are holomorphic, that is iff $\tau^\alpha = \bar{\partial}\psi^\alpha = 0$ for $\alpha = 1, \ldots, n$. □

The torsion(s) and the covariant derivative enjoy the usual relationship:

PROPOSITION 2.2.2: *Let $D: \mathcal{X}(T_{\mathbb{C}}\tilde{M}) \to \mathcal{X}(T^*_{\mathbb{C}}\tilde{M} \otimes T_{\mathbb{C}}\tilde{M})$ be the complex linear connection induced by a good complex vertical connection. Then for any $X, Y \in \mathcal{X}(T^{1,0}\tilde{M})$ we have*

$$\begin{aligned} \nabla_X Y - \nabla_Y X &= [X,Y] + \theta(X,Y); \\ \nabla_X \overline{Y} - \nabla_{\overline{Y}} X &= [X,\overline{Y}] + \tau(X,\overline{Y}) + \bar{\tau}(X,\overline{Y}). \end{aligned}$$

Proof: Write $X = X^\mu \delta_\mu + \dot{X}^\alpha \dot{\partial}_\alpha$ and $Y = Y^\nu \delta_\nu + \dot{Y}^\beta \dot{\partial}_\beta$. Then

$$\begin{aligned} \nabla_X Y &= \big[X(Y^\nu) + \omega^\nu_\mu(X) Y^\mu\big]\delta_\nu + \big[X(\dot{Y}^\beta) + \omega^\beta_\alpha(X)\dot{Y}^\alpha\big]\dot{\partial}_\beta, \\ \nabla_Y X &= \big[Y(X^\nu) + \omega^\nu_\mu(Y) X^\mu\big]\delta_\nu + \big[Y(\dot{X}^\beta) + \omega^\beta_\alpha(Y)\dot{X}^\alpha\big]\dot{\partial}_\beta, \\ \theta(X,Y) &= [\omega^\nu_\mu(X)Y^\mu - \omega^\nu_\mu(Y)X^\mu]\delta_\nu + [\omega^\beta_\alpha(X)\dot{Y}^\alpha - \omega^\beta_\alpha(Y)\dot{X}^\alpha]\dot{\partial}_\beta + \partial\psi^\beta(X,Y)\dot{\partial}_\beta, \\ [X,Y] &= \big(X(Y^\nu) - Y(X^\nu)\big)\delta_\nu + \psi^\beta([X,Y])\dot{\partial}_\beta, \\ \partial\psi^\beta(X,Y) &= X(\dot{Y}^\beta) - Y(\dot{X}^\beta) - \psi^\beta([X,Y]), \end{aligned}$$

where we used the fact that $[\delta_\mu, \delta_\nu], [\delta_\mu, \dot{\partial}_\alpha] \in \mathcal{V}^{1,0}$, and the first formula is proved. For the second one,

$$\begin{aligned} \nabla_X \overline{Y} &= X(\overline{Y^\nu})\delta_{\bar{\nu}} + X(\overline{\dot{Y}^\beta})\dot{\partial}_{\bar{\beta}}, \qquad & \nabla_{\overline{Y}} X &= \overline{Y}(X^\mu)\delta_\mu + \overline{Y}(\dot{X}^\alpha)\dot{\partial}_\alpha, \\ \tau(X,\overline{Y}) &= \bar{\partial}\psi^\beta(X,\overline{Y})\dot{\partial}_\beta, \qquad & \bar{\tau}(X,\overline{Y}) &= \partial\overline{\psi^\beta}(X,\overline{Y})\dot{\partial}_{\bar{\beta}}, \\ \bar{\partial}\psi^\beta(X,\overline{Y}) &= -\overline{Y}(\dot{X}^\beta) - \psi^\beta([X,\overline{Y}]), \qquad & \partial\overline{\psi^\beta}(X,\overline{Y}) &= X(\overline{\dot{Y}^\beta}) - \overline{\psi^\beta}([X,\overline{Y}]), \end{aligned}$$

$$[X,\overline{Y}] = X(\overline{Y^\nu})\delta_{\bar{\nu}} - \overline{Y}(X^\mu)\delta_\mu + \psi^\beta([X,\overline{Y}])\dot{\partial}_\beta + \overline{\psi^\beta}([X,\overline{Y}])\dot{\partial}_{\bar{\beta}},$$

where we used the fact that $[\delta_\mu, \delta_{\bar{\nu}}], [\delta_\mu, \dot{\partial}_{\bar{\beta}}] \in \mathcal{V}_{\mathbb{C}}$, and we are done. □

2.2.5. The curvature

DEFINITION 2.2.8: The *curvature tensor* $R: \mathcal{X}(T^{1,0}\tilde{M}) \to \mathcal{X}(\bigwedge^2(T^*_{\mathbb{C}}\tilde{M}) \otimes T^{1,0}\tilde{M})$ of a good complex vertical connection D is given by $R = D \circ D$, that is

$$\forall X \in \mathcal{X}(T^{1,0}\tilde{M}) \qquad R_X = D(DX).$$

Analogously the *curvature operator* $\Omega \in \mathcal{X}(\bigwedge^2(T^*_{\mathbb{C}}\tilde{M}) \otimes \bigwedge^{1,0}\tilde{M} \otimes T^{1,0}\tilde{M})$ of D is defined by

$$\Omega(X,Y)Z = R_Z(X,Y).$$

Locally, Ω is given by

$$\Omega = \Omega^\alpha_\beta \otimes [dz^\beta \otimes \delta_\alpha + \psi^\beta \otimes \dot{\partial}_\alpha],$$

where

$$\Omega^\alpha_\beta = d\omega^\alpha_\beta - \omega^\gamma_\beta \wedge \omega^\alpha_\gamma.$$

As in the real case, everything is globally well-defined.

Decomposing Ω into types, we get

$$\Omega = \Omega' + \Omega'',$$

where Ω' is a (2,0)-form and Ω'' a (1,1)-form. Locally,

$$(\Omega')^\alpha_\beta = \partial\omega^\alpha_\beta - \omega^\gamma_\beta \wedge \omega^\alpha_\gamma, \qquad (\Omega'')^\alpha_\beta = \bar\partial\omega^\alpha_\beta.$$

Ω has no (0,2)-components because the connection forms are (1,0)-forms. As we shall see in the next section, Ω' vanishes identically when we consider the curvature operator of the Chern-Finsler connection, leaving Ω'' as the only meaningful part of the curvature.

The relation between curvature and covariant derivatives is the usual one:

PROPOSITION 2.2.3: *Let $D: \mathcal{X}(T^{1,0}\tilde M) \to \mathcal{X}(T^*_{\mathbb{C}}\tilde M \otimes T^{1,0}\tilde M)$ be the complex linear connection on $\tilde M$ induced by a good complex vertical connection. Then for any $X, Y \in \mathcal{X}(T^{1,0}\tilde M)$ we have*

$$\nabla_X\nabla_Y - \nabla_Y\nabla_X = \nabla_{[X,Y]} + \Omega'(X,Y);$$
$$\nabla_X\nabla_{\overline{Y}} - \nabla_{\overline{Y}}\nabla_X = \nabla_{[X,\overline{Y}]} + \Omega''(X,\overline{Y});$$
$$\nabla_{\overline{X}}\nabla_{\overline{Y}} - \nabla_{\overline{Y}}\nabla_{\overline{X}} = \nabla_{[\overline{X},\overline{Y}]}.$$

Proof: Write $X = X^\mu\delta_\mu + \dot X^\alpha\dot\partial_\alpha$, $Y = Y^\nu\delta_\nu + \dot Y^\beta\dot\partial_\beta$, and take a vertical vector field $V = V^\alpha\dot\partial_\alpha$. Then

$$\nabla_Y V = [Y(V^\alpha) + \omega^\alpha_\beta(Y)V^\beta]\dot\partial_\alpha, \qquad \nabla_X V = [X(V^\alpha) + \omega^\alpha_\beta(X)V^\beta]\dot\partial_\alpha,$$
$$\nabla_X\nabla_Y V = \Big[X\big(Y(V^\alpha)\big) + X\big(\omega^\alpha_\beta(Y)\big)V^\beta + \omega^\alpha_\beta(Y)X(V^\beta)$$
$$+\omega^\alpha_\beta(X)Y(V^\beta) + \omega^\alpha_\gamma(X)\omega^\gamma_\beta(Y)V^\beta\Big]\dot\partial_\alpha,$$
$$\nabla_Y\nabla_X V = \Big[Y\big(X(V^\alpha)\big) + Y\big(\omega^\alpha_\beta(X)\big)V^\beta + \omega^\alpha_\beta(X)Y(V^\beta)$$
$$+\omega^\alpha_\beta(Y)X(V^\beta) + \omega^\alpha_\gamma(Y)\omega^\gamma_\beta(X)V^\beta\Big]\dot\partial_\alpha,$$
$$\nabla_{[X,Y]}V = \left[X\big(Y(V^\alpha)\big) - Y\big(X(V^\alpha)\big) + \omega^\alpha_\beta([X,Y])V^\beta\right]\dot\partial_\alpha,$$
$$\Omega'(X,Y)V = \Big[\partial\omega^\alpha_\beta(X,Y) - \omega^\gamma_\beta(X)\omega^\alpha_\gamma(Y) + \omega^\gamma_\beta(Y)\omega^\alpha_\gamma(X)\Big]V^\beta\dot\partial_\alpha,$$
$$\partial\omega^\alpha_\beta(X,Y) = X\big(\omega^\alpha_\beta(Y)\big) - Y\big(\omega^\alpha_\beta(X)\big) - \omega^\alpha_\beta([X,Y]),$$

and so the first formula (applied to a vertical vector field) is verified. For the second one,

$$\nabla_{\overline{Y}}V = \overline{Y}(V^\alpha)\dot\partial_\alpha, \qquad \nabla_X V = [X(V^\alpha) + \omega^\alpha_\beta(X)V^\beta]\dot\partial_\alpha,$$
$$\nabla_X\nabla_{\overline{Y}}V = \left[X\big(\overline{Y}(V^\alpha)\big) + \omega^\alpha_\beta(X)\overline{Y}(V^\beta)\right]\dot\partial_\alpha,$$
$$\nabla_{\overline{Y}}\nabla_X V = \left[\overline{Y}\big(X(V^\alpha)\big) + \overline{Y}\big(\omega^\alpha_\beta(X)\big)V^\beta + \omega^\alpha_\beta(X)\overline{Y}(V^\beta)\right]\dot\partial_\alpha,$$
$$\nabla_{[X,\overline{Y}]}V = \left[X\big(\overline{Y}(V^\alpha)\big) - \overline{Y}\big(X(V^\alpha)\big) + \omega^\alpha_\beta([X,\overline{Y}])V^\beta\right]\dot\partial_\alpha,$$

$$\Omega''(X,\overline{Y})V = \big(\bar\partial\omega^\alpha_\beta(X,\overline{Y})\big)V^\beta\dot\partial_\alpha = \left[-\overline{Y}\big(\omega^\alpha_\beta(X)\big) - \omega^\alpha_\beta([X,\overline{Y}])\right] V^\beta\dot\partial_\alpha,$$

and we are done. Finally,

$$\nabla_{\overline{X}}\nabla_{\overline{Y}}V = \overline{X}\big(\overline{Y}(V^\alpha)\big)\dot\partial_\alpha, \qquad \nabla_{\overline{Y}}\nabla_{\overline{X}}V = \overline{Y}\big(\overline{X}(V^\alpha)\big)\dot\partial_\alpha,$$
$$\nabla_{[\overline{X},\overline{Y}]}V = \left[\overline{X}\big(\overline{Y}(V^\alpha)\big) - \overline{Y}\big(\overline{X}(V^\alpha)\big) + \omega^\alpha_\beta([\overline{X},\overline{Y}])V^\beta\right]\dot\partial_\alpha,$$

and, since $[\overline{X},\overline{Y}] \in T^{0,1}\tilde{M}$ so that $\omega^\alpha_\beta([\overline{X},\overline{Y}]) = 0$, we are again done. The same computations prove the formulas for horizontal vector fields. □

We end this section recovering the Bianchi identities in this setting.

PROPOSITION 2.2.4: *Let $D\colon \mathcal{X}(T^{1,0}\tilde{M}) \to \mathcal{X}(T^*_{\mathbb{C}}\tilde{M} \otimes T^{1,0}\tilde{M})$ be the complex linear connection on $\tilde{M}$ induced by a good complex vertical connection. Then*

$$D'\theta = \eta \wedge \Omega', \qquad D''\tau = 0,$$
$$D''\theta + D'\tau = \eta \wedge \Omega'',$$
$$D'\Omega' = 0, \qquad D''\Omega'' = 0,$$
$$D''\Omega' + D'\Omega'' = 0.$$

Proof: It suffices to compute. First of all,

$$\partial\theta^\mu + \theta^\nu \wedge \omega^\mu_\nu = dz^\nu \wedge \partial\omega^\mu_\nu - dz^\nu \wedge \omega^\sigma_\nu \wedge \omega^\mu_\sigma = dz^\nu \wedge (\Omega')^\mu_\nu;$$
$$\partial\dot\theta^\alpha + \dot\theta^\beta \wedge \omega^\alpha_\beta = -\partial\psi^\beta \wedge \omega^\alpha_\beta + \omega^\beta \wedge \partial\omega^\alpha_\beta + \partial\psi^\beta \wedge \omega^\alpha_\beta - \psi^\beta \wedge \omega^\gamma_\beta \wedge \omega^\alpha_\gamma$$
$$= \psi^\beta \wedge (\Omega')^\alpha_\beta,$$

and so $D'\theta = \eta \wedge \Omega'$. Next

$$\bar\partial\theta^\mu = dz^\nu \wedge \bar\partial\omega^\mu_\nu = dz^\nu \wedge (\Omega'')^\mu_\nu,$$
$$\bar\partial\dot\theta^\alpha = \bar\partial\partial\psi^\alpha - \bar\partial\psi^\beta \wedge \omega^\alpha_\beta + \psi^\beta \wedge \bar\partial\omega^\alpha_\beta$$
$$= \bar\partial\partial\psi^\alpha - \bar\partial\psi^\beta \wedge \omega^\alpha_\beta + \psi^\beta \wedge (\Omega'')^\alpha_\beta,$$
$$\partial\tau^\alpha + \tau^\beta \wedge \omega^\alpha_\beta = \partial\bar\partial\psi^\alpha + \bar\partial\psi^\beta \wedge \omega^\alpha_\beta = -\bar\partial\partial\psi^\alpha + \bar\partial\psi^\beta \wedge \omega^\alpha_\beta,$$

and so $D''\theta + D'\tau = \eta \wedge \Omega''$. Now

$$\partial(\Omega')^\alpha_\beta - (\Omega')^\alpha_\gamma \wedge \omega^\gamma_\beta + (\Omega')^\gamma_\beta \wedge \omega^\alpha_\gamma$$
$$= -\partial\omega^\gamma_\beta \wedge \omega^\alpha_\gamma + \omega^\gamma_\beta \wedge \partial\omega^\alpha_\gamma - \omega^\gamma_\beta \wedge \partial\omega^\alpha_\gamma + \omega^\gamma_\beta \wedge \omega^\delta_\gamma \wedge \omega^\alpha_\delta$$
$$+ \partial\omega^\gamma_\beta \wedge \omega^\alpha_\gamma - \omega^\gamma_\beta \wedge \omega^\delta_\gamma \wedge \omega^\alpha_\delta = 0,$$

and $D'\Omega' = 0$. On the other hand,

$$\bar\partial(\Omega')^\alpha_\beta = \bar\partial\partial\omega^\alpha_\beta - \bar\partial\omega^\gamma_\beta \wedge \omega^\alpha_\gamma + \omega^\gamma_\beta \wedge \bar\partial\omega^\alpha_\gamma,$$
$$\partial(\Omega'')^\alpha_\beta - \omega^\gamma_\beta \wedge (\Omega'')^\alpha_\gamma + (\Omega'')^\gamma_\beta \wedge \omega^\alpha_\gamma = \partial\bar\partial\omega^\alpha_\beta - \omega^\gamma_\beta \wedge \bar\partial\omega^\alpha_\gamma + \bar\partial\omega^\gamma_\beta \wedge \omega^\alpha_\gamma,$$

and $D''\Omega' + D'\Omega'' = 0$. Finally, $D''\tau = 0$ and $D''\Omega'' = 0$ are trivial. □

2.3. The Chern-Finsler connection

2.3.1. Complex Finsler metrics

DEFINITION 2.3.1: A *complex Finsler metric* on a complex manifold M is a continuous function $F: T^{1,0}M \to \mathbb{R}^+$ satisfying

(i) $G = F^2$ is smooth on $\tilde{M}$;
(ii) $F(v) > 0$ for all $v \in \tilde{M}$;
(iii) $F\big(\mu_\zeta(v)\big) = |\zeta| F(v)$ for all $v \in T^{1,0}M$ and $\zeta \in \mathbb{C}$.

We shall systematically denote by G the function $G = F^2$. Note that it is important to ask for the smoothness of G only on $\tilde{M}$: as we shall show momentarily, G is smooth on the whole of $T^{1,0}M$ iff F is the norm associated to a Hermitian metric. If this is the case, we shall say that F *comes from* a Hermitian metric.

Again, the easiest example of complex Finsler metrics not coming from a Hermitian metric is given by a Minkowski metric.

DEFINITION 2.3.2: A *complex Minkowski space* is $\mathbb{C}^n$ endowed with a complex Finsler metric $F: \mathbb{C}^n \times \mathbb{C}^n \cong T^{1,0}\mathbb{C}^n \to \mathbb{R}^+$ given by

$$\forall p \in \mathbb{C}^n \; \forall v \in T^{1,0}_p\mathbb{C}^n \cong \mathbb{C}^n \qquad F(p; v) = \|v\|,$$

where $\|\cdot\|: \mathbb{C}^n \to \mathbb{R}^+$ is a complex norm (with strongly convex unit ball) on $\mathbb{C}^n$. If $\|\cdot\|$ is not the norm associated to a Hermitian product, then F does not come from a Hermitian metric.

In the past twenty years, a number of (not necessarily smooth) complex Finsler metrics have become a very useful tool in geometric function theory of holomorphic mappings — and indeed this is the origin of our interest in complex Finsler geometry. The two most important such metrics are indubitably the Kobayashi and Carathéodory metrics, which we presently define.

DEFINITION 2.3.3: Let M be a complex manifold; then the *Kobayashi metric* $F_K: T^{1,0}M \to \mathbb{R}^+$ of M is given by

$$F_K(p; v) = \inf\big\{ |\xi| \bigm| \exists \varphi \in \mathrm{Hol}(\Delta, M) : \varphi(0) = p, \quad \varphi_0(\xi) = v \big\},$$

for all $(p; v) \in T^{1,0}M$, where Δ is the unit disk in $\mathbb{C}$ and $\mathrm{Hol}(\Delta, M)$ is the space of holomorphic maps from Δ into M.

DEFINITION 2.3.4: Let M be a complex manifold; then the *Carathéodory metric* $F_C: T^{1,0}M \to \mathbb{R}^+$ of M is given by

$$F_C(p; v) = \sup\big\{ |df_p(v)| \bigm| f \in \mathrm{Hol}(M, \Delta) \quad f(p) = 0 \big\},$$

for all $(p; v) \in T^{1,0}M$.

In general, the Kobayashi metric is only upper semicontinuous on $\tilde{M}$, and the Carathéodory metric is only continuous. However, Lempert in his fundamental paper [L] proved that if $D \subset\subset \mathbb{C}^n$ is a strongly convex bounded domain of $\mathbb{C}^n$ then the Kobayashi metric (and the Carathéodory metric too, for they coincide there) is a *smooth* complex Finsler metric, furthermore enjoying a number of very interesting properties (existence of complex geodesics, of Monge-Ampère potentials, and so on). Most of our chapter 3 is devoted to recover Lempert's results under only differential geometric assumptions on the (smooth) complex Finsler metric, instead of under a priori assumptions on the geometry of the manifold. Our assumptions will imply that the given complex Finsler metric necessarily is the Kobayashi metric of the manifold; therefore the unique part of Lempert's results actually depending on the strong convexity of the domain is the proof of the smoothness of the Kobayashi metric. For more informations on Kobayashi and Carathéodory metrics see, e.g., [A] and references therein.

To start working on general smooth complex Finsler metrics, we need a few notations and general formulas. We shall denote by indexes like α, $\bar{\beta}$ and so on the derivatives with respect to the v-coordinates; for instance,

$$G_{\alpha\bar{\beta}} = \frac{\partial^2 G}{\partial v^\alpha \partial \bar{v}^\beta}.$$

On the other hand, the derivatives with respect to the z-coordinates will be denoted by indexes after a semicolon; for instance,

$$G_{;\mu\nu} = \frac{\partial^2 G}{\partial z^\mu \partial z^\nu} \qquad \text{or} \qquad G_{\alpha;\bar{\nu}} = \frac{\partial^2 G}{\partial \bar{z}^\nu \partial v^\alpha}.$$

For our aims, we ought to focus on a smaller class of Finsler metrics.

Definition 2.3.5: A complex Finsler metric F will be said *strongly pseudoconvex* if the Levi matrix $(G_{\alpha\bar{\beta}})$ is positive definite on $\tilde{M}$.

This is equivalent to requiring that all the *F-indicatrices*

$$I_F(p) = \{v \in T_p^{1,0}M \mid F(v) < 1\}$$

are strongly pseudoconvexes. As we shall see in 2.3.2, this hypothesis will allow us to define a Hermitian metric on the vertical bundle.

The main (actually, almost the unique) property of the function G is its (1,1)-homogeneity: we have

$$G(p;\zeta v) = \zeta\bar{\zeta}\, G(p;v) \tag{2.3.1}$$

for all $(p;v) \in T^{1,0}M$ and $\zeta \in \mathbb{C}$. We now collect a number of formulas we shall use later on which are consequences of (2.3.1). First of all, differentiating with respect to v^α and $\overline{v^\beta}$ we get

$$\begin{aligned} G_\alpha(p;\zeta v) &= \bar{\zeta} G_\alpha(p;v), \\ G_{\alpha\bar{\beta}}(p;\zeta v) &= G_{\alpha\bar{\beta}}(p;v), \\ G_{\alpha\beta}(p;\zeta v) &= (\bar{\zeta}/\zeta) G_{\alpha\beta}(p;v). \end{aligned} \tag{2.3.2}$$

Thus differentiating with respect to ζ or $\bar{\zeta}$ and then setting $\zeta = 1$ we get

$$G_{\alpha\bar{\beta}}\overline{v^\beta} = G_\alpha, \qquad G_{\alpha\beta}v^\beta = 0, \tag{2.3.3}$$

and

$$G_{\alpha\beta\gamma}v^\gamma = -G_{\alpha\beta}, \qquad G_{\alpha\beta\bar{\gamma}}\overline{v^\gamma} = G_{\alpha\beta}, \qquad G_{\alpha\bar{\beta}\gamma}v^\gamma = 0, \tag{2.3.4}$$

where everything is evaluated at $(p; v)$.

On the other hand, differentiating directly (2.3.1) with respect to ζ or $\bar{\zeta}$ and putting eventually $\zeta = 1$ we get

$$G_\alpha v^\alpha = G, \qquad G_{\alpha\beta}v^\alpha v^\beta = 0, \qquad G_{\alpha\bar{\beta}}v^\alpha\overline{v^\beta} = G. \tag{2.3.5}$$

It is clear that we may get other formulas applying any differential operator acting only on the z-coordinates, or just by conjugation. For instance, we get

$$G_{\bar{\alpha};\mu}\overline{v^\alpha} = G_{;\mu}, \tag{2.3.6}$$

and so on.

By the way, this is enough to prove the claim about the smoothness of G:

LEMMA 2.3.1: *Let F be a complex Finsler metric on a complex manifold M. Then $G = F^2$ is smooth on $T^{1,0}M$ iff F comes from a Hermitian metric on M.*

Proof: One direction is clear. Conversely, assume G smooth (C^2 is enough) on $T^{1,0}M$. Take $(p; v) \in \tilde{M}$; then (2.3.5) yields

$$G(p; v) = G_{\alpha\bar{\beta}}(p; v)v^\alpha\overline{v^\beta}.$$

Take $t > 0$; by (2.3.2) $G_{\alpha\bar{\beta}}(p; tv) = G_{\alpha\bar{\beta}}(p; v)$. Hence

$$\forall t > 0 \qquad G(p; v) = G_{\alpha\bar{\beta}}(p; tv)v^\alpha\overline{v^\beta}.$$

Letting t go to zero, by smoothness we obtain

$$G(p; v) = G_{\alpha\bar{\beta}}(p; o_p)v^\alpha\overline{v^\beta}.$$

It is easy to check that setting $g_{\alpha\bar{\beta}}(p) = G_{\alpha\bar{\beta}}(p; o_p)$ one gets a Hermitian metric on M, whose associated norm is exactly F. □

Assuming from now on (unless explicitly noted otherwise) F strongly pseudoconvex, we get another bunch of formulas which we shall need later.

As usual in Hermitian geometry, we shall denote by $(G^{\bar{\beta}\alpha})$ the inverse matrix of $(G_{\alpha\bar{\beta}})$. First of all, applying $G^{\bar{\beta}\alpha}$ to the first equation in (2.3.3) we get

$$G^{\bar{\beta}\alpha}G_\alpha = \overline{v^\beta}, \tag{2.3.7}$$

and thus, applying (2.3.6),

$$G_{\bar{\beta};\mu}G^{\bar{\beta}\alpha}G_\alpha = G_{;\mu}. \tag{2.3.8}$$

Recalling that $(G^{\bar\beta\alpha})$ is the inverse matrix of $(G_{\alpha\bar\beta})$, we may also compute derivatives of $G^{\bar\beta\alpha}$:

$$DG^{\bar\beta\alpha} = -G^{\bar\nu\alpha}G^{\bar\beta\mu}(DG_{\mu\bar\nu}), \tag{2.3.9}$$

where D denotes any first order linear differential operator. As a consequence of (2.3.4) and (2.3.9) we get for instance

$$G^{\bar\beta\alpha}_{\bar\sigma}\,\overline{v^\sigma} = -G^{\bar\nu\alpha}G^{\bar\beta\mu}G_{\mu\bar\nu\bar\sigma}\,\overline{v^\sigma} = 0, \tag{2.3.10}$$

and recalling also (2.3.7) we obtain

$$G_{\bar\beta}G^{\bar\beta\alpha}_{\gamma} = -G_{\bar\beta}G^{\bar\beta\mu}G^{\bar\nu\alpha}G_{\mu\bar\nu\gamma} = -G^{\bar\nu\alpha}G_{\mu\bar\nu\gamma}v^\mu = 0. \tag{2.3.11}$$

2.3.2. The Chern-Finsler connection

Now we may start to work. To any Hermitian metric is associated a unique complex linear connection such that the metric tensor is parallel: the Chern connection. The main goal of this section is to define the analogue for strongly pseudoconvex Finsler metrics.

The first observation is that F defines a Hermitian metric on the vertical bundle $\mathcal{V}$. Indeed, if $v \in \tilde M$ and W_1, $W_2 \in \mathcal{V}_v$, with $W_j = W_j^\alpha \dot\partial_\alpha$, we set

$$\langle W_1, W_2\rangle_v = G_{\alpha\bar\beta}(v)W_1^\alpha\overline{W_2^\beta}.$$

Being F strongly pseudoconvex, $\langle\,,\rangle$ is a Hermitian metric as soon as it is well-defined. But in fact

$$(G_{\alpha\bar\beta})_B = (\mathcal{J}_{BA}^{-1})^\gamma_\alpha\overline{(\mathcal{J}_{BA}^{-1})^\delta_\beta}(G_{\gamma\bar\delta})_A,$$
$$(W_j^\alpha)_B = (\mathcal{J}_{BA})^\alpha_\gamma(W_j^\gamma)_A,$$

and everything works. Note that the third equation in (2.3.5) says that

$$G = \langle\iota,\iota\rangle;$$

so ι is an isometric embedding of $\tilde M$ into $\mathcal{V}$.

The main theorem of this subsection states that there is a unique good complex vertical connection D making this Hermitian structure parallel:

THEOREM 2.3.2: *Let $F\colon T^{1,0}M \to \mathbb{R}^+$ be a strongly pseudoconvex Finsler metric on a complex manifold M, and let $\langle\,,\rangle$ denote the Hermitian structure on $\mathcal{V}$ induced by F. Then there is a unique complex vertical connection $D\colon \mathcal{X}(\mathcal{V}) \to \mathcal{X}(T^*_{\mathbb{C}}\tilde M \otimes \mathcal{V})$ such that*

$$X\langle V, W\rangle = \langle\nabla_X V, W\rangle + \langle V, \nabla_{\overline{X}}W\rangle \tag{2.3.12}$$

for all $X \in T^{1,0}\tilde M$ and V, $W \in \mathcal{X}(\mathcal{V})$. Furthermore, this connection is good.

Proof: Assume such a connection exists; we shall recover the connection forms ω^α_β, showing its uniqueness. The tensor

$$\mathcal{G} = G_{\alpha\bar\beta}\,\psi^\alpha \otimes \overline{\psi^\beta}$$

(where here $\{\psi^\alpha\}$ is just the dual coframe of $\{\dot\partial_\alpha\}$) is clearly globally defined. Since the covariant derivative ∇ commutes with contractions, (2.3.12) is equivalent to $\nabla\mathcal{G} = 0$, that is to

$$dG_{\alpha\bar\beta} - G_{\gamma\bar\beta}\,\omega^\gamma_\alpha - G_{\alpha\bar\gamma}\,\overline{\omega^\gamma_\beta} = 0.$$

Since the ω^γ_α's are (1,0)-forms, this is equivalent to $\partial G_{\alpha\bar\beta} = G_{\gamma\bar\beta}\,\omega^\gamma_\alpha$, that is to

$$\omega^\alpha_\beta = G^{\bar\tau\alpha}\,\partial G_{\beta\bar\tau}, \tag{2.3.13}$$

and the uniqueness is proved. In particular,

$$\tilde\Gamma^\alpha_{\beta\gamma} = G^{\bar\tau\alpha} G_{\beta\bar\tau\gamma};$$

so $\tilde\Gamma^\alpha_{\beta\gamma} v^\beta = 0$ by (2.3.4), and the connection is good.

Conversely, take $V \in \mathcal{X}(\mathcal{V})$, $v \in \tilde M$ and $X \in T^{1,0}_v\tilde M$. To define $\nabla_X V \in \mathcal{V}_v$ it suffices to know its product with any other element of $\mathcal{V}_v$; so we set

$$\langle \nabla_X V, W\rangle = X\langle V, W\rangle$$

for any section $W \in \mathcal{X}(\mathcal{V})$ which is holomorphic in a neighborhood of v, so that $X\langle V, W\rangle(v)$ depends only on the value of W at v, and not on the derivatives of W. Once defined $\nabla_X V$, we define $\nabla_{\overline X} V$ by

$$\langle \nabla_{\overline X} V, W\rangle = \overline X\langle V, W\rangle - \langle V, \nabla_X W\rangle,$$

where this time W is any section of $\mathcal{V}$. Then it is easy to check that the map $D\colon \mathcal{X}(\mathcal{V}) \to \mathcal{X}(T^*_{\mathbb{C}}\tilde M \otimes \mathcal{V})$ so defined is a complex vertical connection satisfying (2.3.12), and we are done. □

DEFINITION 2.3.6: The unique good complex vertical connection D whose existence is asserted in the previous theorem is the *Chern-Finsler connection* associated to the strongly pseudoconvex Finsler metric F.

The connection matrix is given by

$$\omega^\alpha_\beta = G^{\bar\tau\alpha}\partial G_{\beta\bar\tau} = G^{\bar\tau\alpha}(G_{\beta\bar\tau;\mu}\,dz^\mu + G_{\beta\bar\tau\gamma}\,dv^\gamma),$$

that is

$$\tilde\Gamma^\alpha_{\beta\gamma} = G^{\bar\tau\alpha} G_{\beta\bar\tau\gamma} \qquad \text{and} \qquad \tilde\Gamma^\alpha_{\beta;\mu} = G^{\bar\tau\alpha} G_{\beta\bar\tau;\mu}.$$

In particular,

$$\Gamma^\alpha_{\mu} = \tilde\Gamma^\alpha_{\beta;\mu} v^\beta = G^{\bar\tau\alpha} G_{\bar\tau;\mu}.$$

Denoting by $\mathcal{H} \subset T^{1,0}\tilde{M}$ the complex horizontal bundle associated to the Chern-Finsler connection, the natural local frame $\{\delta_1, \ldots, \delta_n\}$ for $\mathcal{H}$ is given by

$$\delta_\mu = \partial_\mu - \Gamma^\alpha_\mu \dot{\partial}_\alpha = \partial_\mu - (G^{\bar{\tau}\alpha} G_{\bar{\tau};\mu}) \dot{\partial}_\alpha.$$

From now on we shall work only with the frame $\{\delta_\mu, \dot{\partial}_\alpha\}$ and its dual coframe $\{dz^\mu, \psi^\alpha\}$, where

$$\psi^\alpha = dv^\alpha + \Gamma^\alpha_\mu \, dz^\mu = dv^\alpha + G^{\bar{\tau}\alpha} G_{\bar{\tau};\mu} \, dz^\mu.$$

Writing

$$\omega^\alpha_\beta = \Gamma^\alpha_{\beta;\mu} \, dz^\mu + \Gamma^\alpha_{\beta\gamma} \psi^\gamma$$

we get

$$\begin{aligned} \Gamma^\alpha_{\beta\gamma} &= G^{\bar{\tau}\alpha} G_{\beta\bar{\tau}\gamma} = \Gamma^\alpha_{\gamma\beta}, \\ \Gamma^\alpha_{\beta;\mu} &= G^{\bar{\tau}\alpha} \delta_\mu (G_{\beta\bar{\tau}}) = G^{\bar{\tau}\alpha}(G_{\beta\bar{\tau};\mu} - G_{\beta\bar{\tau}\gamma} \Gamma^\gamma_\mu). \end{aligned} \tag{2.3.14}$$

Note that

$$\Gamma^\alpha_{\beta;\mu} = \dot{\partial}_\beta (\Gamma^\alpha_\mu); \tag{2.3.15}$$

for this reason *from now on we shall write* $\Gamma^\alpha_{;\mu}$ *instead of* Γ^α_μ. Furthermore,

$$\Gamma^{\bar{\alpha}}_{;\bar{\mu}} = \overline{\Gamma^\alpha_{;\mu}},$$

more or less by definition — cf. (2.1.14).

Being D a good complex vertical connection, it extends to a complex linear connection on $\tilde{M}$. Using the complex horizontal map $\Theta\colon \mathcal{V} \to \mathcal{H}$ we can transfer the Hermitian structure $\langle\, , \rangle$ on $\mathcal{H}$ just by setting

$$\forall H, K \in \mathcal{H}_v \qquad \langle H, K \rangle_v = \langle \Theta^{-1}(H), \Theta^{-1}(K) \rangle_v,$$

and then we can define a Hermitian structure on $T^{1,0}\tilde{M}$ by requiring $\mathcal{H}$ to be orthogonal to $\mathcal{V}$. It is easy to check that these definitions are compatible enough so to get

$$X\langle Y, Z \rangle = \langle \nabla_X Y, Z \rangle + \langle Y, \nabla_{\overline{X}} Z \rangle$$

for any $X \in T^{1,0}\tilde{M}$ and Y, $Z \in \mathcal{X}(T^{1,0}\tilde{M})$.

2.3.3. Some computations

The Chern-Finsler connection enjoys a number of interesting features, making computations much simpler than for a generic good complex vertical connection. For instance, we have

LEMMA 2.3.3: *Let D be the Chern-Finsler connection associated to a strongly pseudoconvex Finsler metric F, and let $\{\delta_1, \ldots, \delta_n\}$ be the corresponding local horizontal frame. Then*

$$\begin{aligned} &[\delta_\mu, \delta_\nu] = 0, && [\delta_\mu, \dot{\partial}_\alpha] = \Gamma^\sigma_{\alpha;\mu} \dot{\partial}_\sigma, && [\dot{\partial}_\alpha, \dot{\partial}_\beta] = 0, \\ &[\delta_\mu, \delta_{\bar{\nu}}] = \delta_{\bar{\nu}}(\Gamma^\sigma_{;\mu}) \dot{\partial}_\sigma - \delta_\mu(\Gamma^{\bar{\tau}}_{;\bar{\nu}}) \dot{\partial}_{\bar{\tau}}, && [\delta_\mu, \dot{\partial}_{\bar{\alpha}}] = \Gamma^\sigma_{\bar{\alpha};\mu} \dot{\partial}_\sigma, && [\dot{\partial}_\alpha, \dot{\partial}_{\bar{\beta}}] = 0, \end{aligned}$$

for all $1 \leq \alpha,\ \beta,\ \mu,\ \nu \leq n$, *where* $\Gamma^{\sigma}_{\bar\alpha;\mu} = \dot\partial_{\bar\alpha}(\Gamma^{\sigma}_{;\mu})$.

Proof: It suffices to compute. First of all,

$$[\delta_\mu, \delta_\nu] = (\Gamma^{\alpha}_{;\mu\nu} - \Gamma^{\alpha}_{;\nu\mu} + \Gamma^{\alpha}_{\sigma;\nu}\Gamma^{\sigma}_{;\mu} - \Gamma^{\alpha}_{\sigma;\mu}\Gamma^{\sigma}_{;\nu})\dot\partial_\alpha,$$

where $\Gamma^{\alpha}_{;\mu\nu} = \partial_\nu(\Gamma^{\alpha}_{;\mu})$ and so on. Now,

$$\Gamma^{\alpha}_{;\mu\nu} = G^{\bar\tau\alpha}(G_{\bar\tau;\mu\nu} - G_{\sigma\bar\tau;\nu}\Gamma^{\sigma}_{;\mu}), \qquad \Gamma^{\alpha}_{\sigma;\nu}\Gamma^{\sigma}_{;\mu} = G^{\bar\tau\alpha}(G_{\sigma\bar\tau;\nu}\Gamma^{\sigma}_{;\mu} - G_{\sigma\bar\tau\rho}\Gamma^{\rho}_{;\nu}\Gamma^{\sigma}_{;\mu}),$$
$$\Gamma^{\alpha}_{;\nu\mu} = G^{\bar\tau\alpha}(G_{\bar\tau;\nu\mu} - G_{\sigma\bar\tau;\mu}\Gamma^{\sigma}_{;\nu}), \qquad \Gamma^{\alpha}_{\sigma;\mu}\Gamma^{\sigma}_{;\nu} = G^{\bar\tau\alpha}(G_{\sigma\bar\tau;\mu}\Gamma^{\sigma}_{;\nu} - G_{\sigma\bar\tau\rho}\Gamma^{\rho}_{;\mu}\Gamma^{\sigma}_{;\nu}),$$

and so $[\delta_\mu, \delta_\nu] = 0$. Note that we have actually proved that

$$\delta_\nu(\Gamma^{\alpha}_{;\mu}) = \delta_\mu(\Gamma^{\alpha}_{;\nu}). \tag{2.3.16}$$

Next,

$$[\delta_\mu, \dot\partial_\alpha] = [\partial_\mu - \Gamma^{\sigma}_{;\mu}\dot\partial_\sigma, \dot\partial_\alpha] = \dot\partial_\alpha(\Gamma^{\sigma}_{;\mu})\dot\partial_\sigma = \Gamma^{\sigma}_{\alpha;\mu}\dot\partial_\sigma.$$

Now,

$$\begin{aligned}[\delta_\mu, \delta_{\bar\nu}] &= [\delta_\mu - \Gamma^{\alpha}_{;\mu}\dot\partial_\alpha, \partial_{\bar\nu} - \Gamma^{\bar\beta}_{;\bar\nu}\dot\partial_{\bar\beta}] = \Gamma^{\alpha}_{;\mu\bar\nu}\dot\partial_\alpha - \Gamma^{\bar\beta}_{;\bar\nu\mu}\dot\partial_{\bar\beta} - \Gamma^{\bar\beta}_{;\bar\nu}\Gamma^{\alpha}_{\bar\beta;\mu}\dot\partial_\alpha + \Gamma^{\alpha}_{;\mu}\Gamma^{\bar\beta}_{\alpha;\bar\nu}\dot\partial_{\bar\beta}\\ &= \delta_{\bar\nu}(\Gamma^{\alpha}_{;\mu})\dot\partial_\alpha - \delta_\mu(\Gamma^{\bar\beta}_{;\bar\nu})\dot\partial_{\bar\beta},\end{aligned}$$

where $\Gamma^{\alpha}_{;\mu\bar\nu} = \partial_{\bar\nu}(\Gamma^{\alpha}_{;\mu})$, and so on.

Furthermore,

$$[\delta_\mu, \dot\partial_{\bar\alpha}] = [\partial_\mu - \Gamma^{\sigma}_{;\mu}\dot\partial_\sigma, \dot\partial_{\bar\alpha}] = \dot\partial_{\bar\alpha}(\Gamma^{\sigma}_{;\mu})\dot\partial_\sigma = \Gamma^{\sigma}_{\bar\alpha;\mu}\dot\partial_\sigma.$$

The remaining assertions are trivial. □

LEMMA 2.3.4: *Let D be the Chern-Finsler connection associated to a strongly pseudoconvex Finsler metric F, and let* $\{\delta_1, \dots, \delta_n\}$ *be the corresponding local horizontal frame. Then*

$$\delta_\mu(G) = \delta_{\bar\mu}(G) = \delta_{\bar\mu}(G_\alpha) = 0$$

for all $1 \leq \alpha,\ \mu,\ \nu \leq n$.

Proof: Indeed,

$$\delta_\mu(G) = G_{;\mu} - \Gamma^{\sigma}_{;\mu}G_\sigma = G_{;\mu} - G^{\bar\tau\sigma}G_{\bar\tau;\mu}G_\sigma = G_{;\mu} - G_{\bar\tau;\mu}\overline{v^\nu} = 0.$$

Next $\delta_{\bar\mu}(G) = \overline{\delta_\mu(G)}$, because G is real-valued. Finally,

$$\delta_{\bar\mu}(G_\alpha) = G_{\alpha;\bar\mu} - \Gamma^{\bar\tau}_{;\bar\mu}G_{\alpha\bar\tau} = G_{\alpha;\bar\mu} - G_{\alpha;\bar\mu} = 0.$$

□

Another feature of the Chern-Finsler connection is that the radial horizontal vector field χ induces an embedding of $\tilde M$ into $\mathcal{H}$ which respects the Lie algebra structure.

Definition 2.3.7: A (never vanishing) vector field $\xi \in \mathcal{X}(\tilde{M})$ may be lifted in two different ways to vector fields in $T^{1,0}\tilde{M}$: via the *horizontal lift*

$$\xi^H(v) = \chi_v\Big(\xi\big(\pi(v)\big)\Big),$$

and via the *vertical lift*

$$\xi^V(v) = \iota_v\Big(\xi\big(\pi(v)\big)\Big).$$

A consequence of Lemma 2.3.3 is that the horizontal lift is a Lie algebra homomorphism:

Proposition 2.3.5: *Let D be the Chern-Finsler connection associated to a strongly pseudoconvex Finsler metric F on a complex manifold M. Then:*

(i) $[\mathcal{X}(\mathcal{H}), \mathcal{X}(\mathcal{H})] \subset \mathcal{X}(\mathcal{H})$ *and* $[\mathcal{X}(\mathcal{V}), \mathcal{X}(\mathcal{V})] \subset \mathcal{X}(\mathcal{V})$;

(ii) *if* ξ_1, $\xi_2 \in \mathcal{X}(\tilde{M})$ *then* $[\xi_1^H, \xi_2^H] = [\xi_1, \xi_2]^H$, $[\xi_1^V, \xi_2^V] = 0$ *and* $[\xi_1^H, \xi_2^V] \in \mathcal{X}(\mathcal{V})$.

Proof: (i) Take H_1, $H_2 \in \mathcal{X}(\mathcal{H})$. Locally, $H_j = H_j^\mu \delta_\mu$; hence

$$[H_1, H_2] = (H_1^\nu \delta_\nu(H_2^\mu) - H_2^\nu \delta_\nu(H_1^\mu))\delta_\mu \tag{2.3.17}$$

(where we used Lemma 2.3.3) is horizontal. Analogously, if V_1, $V_2 \in \mathcal{X}(\mathcal{V})$ with $V_j = V_j^\alpha \dot{\partial}_\alpha$, we get

$$[V_1, V_2] = (V_1^\beta \dot{\partial}_\beta(V_2^\alpha) - V_2^\beta \dot{\partial}_\beta(V_1^\alpha))\dot{\partial}_\alpha, \tag{2.3.18}$$

which is vertical.

(ii) Locally, $\xi_j = \xi_j^\mu(\partial/\partial z^\mu)$ and $\xi_j^H = (\xi_j^\mu \circ \pi)\delta_\mu$; so (2.3.17) yields

$$[\xi_1^H, \xi_2^H] = ((\xi_1^\nu \circ \pi)\delta_\nu(\xi_2^\mu \circ \pi) - (\xi_2^\nu \circ \pi)\delta_\nu(\xi_1^\mu \circ \pi))\delta_\mu.$$

Now $\delta_\nu(\xi_j^\mu \circ \pi) = (\partial \xi_j^\mu / \partial z^\nu) \circ \pi$; therefore

$$[\xi_1^H, \xi_2^H] = \left[\left(\xi_1^\nu \frac{\partial \xi_2^\mu}{\partial z^\nu} - \xi_2^\nu \frac{\partial \xi_1^\mu}{\partial z^\nu}\right) \circ \pi\right] \delta_\mu = [\xi_1, \xi_2]^H.$$

On the other hand, $\xi_j^V = (\xi_j^\alpha \circ \pi)\dot{\partial}_\alpha$ and $\dot{\partial}_\beta(\xi_j^\alpha \circ \pi) = 0$ yield

$$[\xi_1^V, \xi_2^V] = 0.$$

Finally,

$$[\xi_1^H, \xi_2^V] = \left[\left(\xi_1^\mu \frac{\partial \xi_2^\alpha}{\partial z^\mu}\right) \circ \pi + ((\xi_1^\mu \xi_2^\beta) \circ \pi)\Gamma_{\beta;\mu}^\alpha\right] \dot{\partial}_\alpha,$$

again by Lemma 2.3.3. □

Note that, as a consequence of (ii), the obvious map of $\mathcal{X}(\mathcal{V})$ into $\mathcal{X}(\mathcal{H})$ induced by the complex horizontal map $\Theta: \mathcal{V} \to \mathcal{X}$ is *not* an isomorphism of Lie algebras; it suffices to remark that $\Theta(\xi^V) = \xi^H$ for all $\xi \in \mathcal{X}(\tilde{M})$.

2.3.4. Torsions and curvatures

Also the torsions and the curvatures of the Chern-Finsler connection enjoy a number of nice properties. Some of them are collected in the next couple of results.

PROPOSITION 2.3.6: *Let D be the Chern-Finsler connection associated to a strongly pseudoconvex Finsler metric F on a complex manifold M. Then*

(i) $\dot{\theta}^\alpha = 0$ *for* $\alpha = 1, \ldots, n$;

(ii) $\Omega' \equiv 0$.

Proof: (i) It follows from (2.3.16), (2.3.14), (2.3.15) and (2.2.7).

(ii) By definition,

$$\omega^\beta_\alpha = G^{\bar{\tau}\beta} \partial G_{\alpha\bar{\tau}}.$$

So

$$\begin{aligned}\partial\omega^\beta_\alpha &= \partial G^{\bar{\tau}\beta} \wedge \partial G_{\alpha\bar{\tau}} = -G^{\bar{\tau}\mu} G^{\bar{\nu}\beta} \partial G_{\mu\bar{\nu}} \wedge \partial G_{\alpha\bar{\tau}} \\ &= (G^{\bar{\tau}\mu} \partial G_{\alpha\bar{\tau}}) \wedge (G^{\bar{\nu}\beta} \partial G_{\mu\bar{\nu}}) = \omega^\mu_\alpha \wedge \omega^\beta_\mu.\end{aligned}$$

□

If we write $\eta = \eta^{\mathcal{H}} + \eta^{\mathcal{V}}$, where

$$\eta^{\mathcal{H}} = dz^\mu \otimes \delta_\mu \qquad \text{and} \qquad \eta^{\mathcal{V}} = \psi^\alpha \otimes \dot{\partial}_\alpha,$$

then Proposition 2.3.6.(i) amounts to say

$$\theta = D\eta^{\mathcal{H}} = D'\eta^{\mathcal{H}} \qquad \text{and} \qquad \tau = D\eta^{\mathcal{V}} = D''\eta^{\mathcal{V}}.$$

For future references, we record here the local expression of $\theta = \theta^\sigma \otimes \delta_\sigma$ and $\tau = \tau^\alpha \otimes \dot{\partial}_\alpha$ for the Chern-Finsler connection:

$$\begin{aligned}\theta^\sigma &= \tfrac{1}{2}[\Gamma^\sigma_{\nu;\mu} - \Gamma^\sigma_{\mu;\nu}]\, dz^\mu \wedge dz^\nu + \Gamma^\sigma_{\nu\gamma} \psi^\gamma \wedge dz^\nu; \\ \tau^\alpha &= -\delta_{\bar{\nu}}(\Gamma^\alpha_{;\mu})\, dz^\mu \wedge d\bar{z}^\nu - \Gamma^\alpha_{\bar{\beta};\mu}\, dz^\mu \wedge \overline{\psi^\beta}.\end{aligned} \tag{2.3.19}$$

As a consequence of Proposition 2.3.6.(ii), $\Omega = \Omega''$ for the Chern-Finsler connection; in particular,

$$\nabla_X \nabla_Y - \nabla_Y \nabla_X = \nabla_{[X,Y]}, \tag{2.3.20}$$

for all $X, Y \in T^{1,0}\tilde{M}$. From now on *we shall always write Ω instead of Ω''.*

The Bianchi identities too assume a simpler form:

COROLLARY 2.3.7: *Let D be the Chern-Finsler connection associated to a strongly pseudoconvex Finsler metric F on a complex manifold M. Then*

$$\begin{aligned} D'\theta &= 0, & D''\theta &= \eta^{\mathcal{H}} \wedge \Omega, \\ D'\tau &= \eta^{\mathcal{V}} \wedge \Omega, & D''\tau &= 0, \\ D'\Omega &= 0, & D''\Omega &= 0. \end{aligned}$$

Proof: It follows from Propositions 2.2.4 and 2.3.6. □

In local coordinates, the curvature operator is given by

$$\Omega^\alpha_\beta = R^\alpha_{\beta;\mu\bar\nu}\, dz^\mu \wedge d\bar z^\nu + R^\alpha_{\beta\delta;\bar\nu}\, \psi^\delta \wedge d\bar z^\nu + R^\alpha_{\beta\bar\gamma;\mu}\, dz^\mu \wedge \overline{\psi^\gamma} + R^\alpha_{\beta\delta\bar\gamma}\, \psi^\delta \wedge \overline{\psi^\gamma},$$

where

$$\begin{aligned} R^\alpha_{\beta;\mu\bar\nu} &= -\delta_{\bar\nu}(\Gamma^\alpha_{\beta;\mu}) - \Gamma^\alpha_{\beta\sigma}\delta_{\bar\nu}(\Gamma^\sigma_{;\mu}),\\ R^\alpha_{\beta\delta;\bar\nu} &= -\delta_{\bar\nu}(\Gamma^\alpha_{\beta\delta}) = R^\alpha_{\delta\beta;\bar\nu},\\ R^\alpha_{\beta\bar\gamma;\mu} &= -\dot\partial_{\bar\gamma}(\Gamma^\alpha_{\beta;\mu}) - \Gamma^\alpha_{\beta\sigma}\Gamma^\sigma_{\bar\gamma;\mu},\\ R^\alpha_{\beta\delta\bar\gamma} &= -\dot\partial_{\bar\gamma}(\Gamma^\alpha_{\beta\delta}) = R^\alpha_{\delta\beta\bar\gamma}. \end{aligned} \tag{2.3.21}$$

In particular, since

$$\begin{aligned} (D'\tau)^\alpha = (\eta^{\mathcal V}\wedge\Omega)^\alpha = \psi^\sigma \wedge \Omega^\alpha_\sigma &= R^\alpha_{\sigma;\mu\bar\nu}\, \psi^\sigma \wedge dz^\mu \wedge d\bar z^\nu + R^\alpha_{\sigma\delta;\bar\nu}\, \psi^\sigma \wedge \psi^\delta \wedge d\bar z^\nu\\ &\quad + R^\alpha_{\sigma\bar\gamma;\mu}\, \psi^\sigma \wedge dz^\mu \wedge \overline{\psi^\gamma} + R^\alpha_{\sigma\delta\bar\gamma}\, \psi^\sigma \wedge \psi^\delta \wedge \overline{\psi^\gamma}\\ &= -R^\alpha_{\sigma;\mu\bar\nu}\, dz^\mu \wedge \psi^\sigma \wedge d\bar z^\nu - R^\alpha_{\sigma\bar\gamma;\mu}\, dz^\mu \wedge \psi^\sigma \wedge \overline{\psi^\gamma}, \end{aligned}$$

the vanishing of τ implies the vanishing of most of the curvature; cf. Corollary 2.2.1.

Another consequence of (2.3.21) is an unexpected relation between Ω and τ:

LEMMA 2.3.8: *Let D be the Chern-Finsler connection associated to a strongly pseudoconvex Finsler metric F on a complex manifold M. Then*

$$\tau = \Omega(\cdot\,,\cdot)\iota.$$

Proof: Recalling (2.3.21), (2.3.3), (2.3.4) and

$$\Gamma^\alpha_{\beta;\mu} v^\beta = \Gamma^\alpha_{;\mu}, \qquad \Gamma^\alpha_{\beta\gamma} v^\beta = 0,$$

we have

$$\begin{aligned} R^\alpha_{\beta;\mu\bar\nu} v^\beta &= -\delta_{\bar\nu}(\Gamma^\alpha_{;\mu}),\\ R^\alpha_{\beta\delta;\bar\nu} v^\beta &= 0,\\ R^\alpha_{\beta\bar\gamma;\mu} v^\beta &= -\Gamma^\alpha_{\bar\gamma;\mu},\\ R^\alpha_{\beta\delta\bar\gamma} v^\beta &= 0, \end{aligned}$$

and the assertion follows from (2.3.19). □

2.3.5. Kähler Finsler metrics

In the Hermitian case, the vanishing of the torsion is equivalent to the metric being Kähler. In our case the situation is a bit subtler, for instance because our torsion θ has a horizontal part and a mixed part.

DEFINITION 2.3.8: A form $\gamma \in A(\tilde M)$ is *horizontal* if it vanishes contracted with any $V \in \mathcal X(\mathcal V)$. The decomposition $T^{1,0}\tilde M = \mathcal H \oplus \mathcal V$ induces a projection p_H^* of $A(\tilde M)$ onto the horizontal forms; the *horizontal part* of a form γ is then $p_H^*(\gamma)$.

There is a corresponding projection on the vertical forms, of course, but we shall not need it now because the vertical part of both torsions θ and τ is zero. For this reason, the form $\theta - p_H^*(\theta)$ will be called the *mixed part* of θ. In local coordinates,

$$p_H^*(\theta) = (\Gamma^{\sigma}_{\nu;\mu}\, dz^{\mu} \wedge dz^{\nu}) \otimes \delta_{\sigma} \qquad \text{and} \qquad \theta - p_H^*(\theta) = (\Gamma^{\sigma}_{\nu\gamma}\, \psi^{\gamma} \wedge dz^{\nu}) \otimes \delta_{\sigma}.$$

The next proposition discusses the meaning of the vanishing of the (2,0)-torsion θ or of one of its parts.

PROPOSITION 2.3.9: *Let F be a strongly pseudoconvex Finsler metric on a complex manifold M. Then:*

(i) *the mixed part of the $(2,0)$-torsion θ vanishes iff F comes from a Hermitian metric;*

(ii) *θ itself vanishes iff F comes from a Hermitian Kähler metric.*

Proof: (i) The mixed part of the torsion vanishes iff $G_{\beta\bar{\mu}\gamma} = 0$ for all β, μ and γ. Conjugating, this is equivalent to having $\dot{\partial}_{\gamma}(G_{\beta\bar{\mu}}) = 0 = \dot{\partial}_{\bar{\gamma}}(G_{\beta\bar{\mu}})$, that is $G_{\beta\bar{\mu}}(v)$ depends only on $\pi(v)$ — and this happens iff F comes from a Hermitian metric.

(ii) It follows from (i) and the fact that when F comes from a Hermitian metric $g = (g_{\alpha\bar{\beta}})$ one has

$$\Gamma^{\alpha}_{\beta;\mu} = g^{\bar{\tau}\alpha} \frac{\partial g_{\beta\bar{\tau}}}{\partial z^{\mu}}.$$

□

The meaning of the next definition is then clear:

DEFINITION 2.3.9: We say that a strongly pseudoconvex Finsler metric F is *strongly Kähler* if the horizontal part of the (2,0)-torsion vanishes, that is iff

$$\forall H, K \in \mathcal{H} \qquad\qquad \theta(H,K) = 0.$$

It turns out that this definition is too stringent. The point is that, as we shall see in the next two sections, from a geometrical point of view only the vanishing of suitable contractions of θ is relevant.

DEFINITION 2.3.10: We shall say that F is *Kähler* if

$$\forall H \in \mathcal{H} \qquad\qquad \theta(H,\chi) = 0,$$

and that F is *weakly Kähler* if

$$\forall H \in \mathcal{H} \qquad\qquad \langle \theta(H,\chi), \chi \rangle = 0.$$

In local coordinates, F is strongly Kähler iff

$$\Gamma^{\alpha}_{\mu;\nu} = \Gamma^{\alpha}_{\nu;\mu};$$

it is Kähler iff

$$\Gamma^{\alpha}_{\mu;\nu} v^{\mu} = \Gamma^{\alpha}_{\nu;\mu} v^{\mu};$$

it is weakly Kähler iff

$$G_{\alpha}[\Gamma^{\alpha}_{\mu;\nu} - \Gamma^{\alpha}_{\nu;\mu}] v^{\mu} = 0,$$

that is iff

$$0 = [G_{\mu;\nu} - G_{\nu;\mu} + G_{\nu\sigma}\Gamma^{\sigma}_{;\mu}] v^{\mu} = [G_{\mu\bar{\tau};\nu} - G_{\nu\bar{\tau};\mu} + G_{\nu\sigma\bar{\tau}}\Gamma^{\sigma}_{;\mu}] v^{\mu}\overline{v^{\tau}}. \tag{2.3.22}$$

(and so this is the condition we called Kähler in [AP2]). In particular, if F comes from a Hermitian metric then these three conditions are all equivalent to the usual Kähler condition, because $G_{\nu\sigma\bar{\tau}} \equiv 0$ for a Finsler metric coming from a Hermitian metric.

There are other characterizations of strongly Kähler Finsler metrics.

DEFINITION 2.3.11: To a complex Finsler metric F we may associate the *fundamental form*

$$\Phi = iG_{\alpha\bar{\beta}}\, dz^{\alpha} \wedge \overline{dz^{\beta}},$$

which is a well-defined real (1,1)-form on $\tilde{M}$, thanks to (2.1.6) and (2.1.7).

Then the strong Kähler condition is equivalent to the vanishing of the horizontal part of $d\Phi$. To express it more clearly, set

$$d_H = p_H^* \circ d, \qquad \partial_H = p_H^* \circ \partial \qquad \text{and} \qquad \bar{\partial}_H = p_H^* \circ \bar{\partial},$$

so that again $d_H = \partial_H + \bar{\partial}_H$.

THEOREM 2.3.10: *Let F be a strongly pseudoconvex Finsler metric on a complex manifold M. Then the following assertions are equivalent:*

(i) *F is a strongly Kähler Finsler metric;*
(ii) *$\nabla_H K - \nabla_K H = [H, K]$ for all H, $K \in \mathcal{X}(\mathcal{H})$;*
(iii) *$d_H\Phi = 0$;*
(iv) *$\partial_H\Phi = 0$;*
(v) *for any $p_0 \in M$ there is a neighborhood U of p_0 in M and a real-valued smooth function $\phi \in C^{\infty}(\pi^{-1}(U))$ such that $\Phi = i\partial_H\bar{\partial}_H\phi$ on $\pi^{-1}(U)$.*

Proof: (i) $\iff$ (ii): Proposition 2.2.2.

(iii) $\iff$ (iv) holds simply because Φ is a real (1,1)-form.

(iv) $\iff$ (i). Indeed, (2.3.14) yields

$$\partial\Phi(X, Y, \overline{Z}) = i\langle\theta(X, Y), Z\rangle$$

for all X, Y, $Z \in T^{1,0}\tilde{M}$; hence $\partial_H\Phi$ vanishes iff $p_H^* \circ \theta$ vanishes, that is iff F is strongly Kähler.

(v) $\iff$ (iii) One direction is clear. For the converse, let γ be any horizontal form. In local coordinates, defined on a coordinate neighborhood of the form $\pi^{-1}(U)$, one has

$$\gamma|_{(p;v)} = \gamma_{AB}(p;v)\, dz^A \wedge \overline{dz^B},$$

for suitable multi-indeces A and $\bar{B}$. On U we may then consider the family of forms

$$\gamma_v|_p = \gamma_{A\bar{B}}(p;v)\, dz^A \wedge \overline{dz^B} \in A(U),$$

where here $\{dz^j\}$ is the dual frame of $\{\partial/\partial z^j\}$; in other words, we are considering the v-coordinates just as parameters.

The gist is that the following formula holds:

$$(d_H\gamma)_v = d(\gamma_v).$$

Then we may now apply the Dolbeault and Serre theorems (with parameters) to Φ_v in a possibly smaller neighborhood of p_0 — still denoted by U — to get a function $\phi_v \in C^\infty(U, \mathbb{R})$ depending smoothly on v such that $\Phi_v = i\partial\bar{\partial}\phi_v$. Then setting

$$\phi(p;v) = \phi_v(p)$$

we get $\Phi = i\partial_H\bar{\partial}_H\phi$, as required. □

From this point of view, a strongly pseudoconvex Finsler metric is Kähler iff

$$d_H\Phi(\cdot,\chi,\cdot) \equiv 0,$$

and it is weakly Kähler iff

$$d_H\Phi(\cdot,\chi,\bar{\chi}) \equiv 0.$$

2.4. Variations of the length integral

2.4.1. The setting

Let $F\colon T^{1,0}M \to \mathbb{R}^+$ be a strongly pseudoconvex Finsler metric on a complex manifold M. To F we may associate a function $F^o\colon T_{\mathbb{R}}M \to \mathbb{R}^+$ just by setting

$$\forall u \in T_{\mathbb{R}}M \qquad F^o(u) = F(u_o).$$

F^o satisfies all the properties defining a real Finsler metric but perhaps the last one: the indicatrices are not necessarily strongly convex. Nevertheless, we may use it to measure the length of curves, and so to define geodesics; and one of the main results of this section is a theorem ensuring the local existence and uniqueness of geodesics for weakly Kähler Finsler metrics — a striking by-product of the complex structure. Of course, it is conceivable (cf. [P2]) that the indicatrices of F^o could be necessarily strongly convex if F is weakly Kähler; nevertheless, the main point here is that it is possible to develop a theory of geodesics in weakly Kähler Finsler manifolds without assuming the real convexity.

Let us fix the notations needed to study variations of the length integral in this setting. Let $\sigma\colon [a,b] \to M$ be a regular curve; we define $\dot\sigma\colon [a,b] \to \tilde M$ by setting

$$\dot\sigma(t) = \frac{d\sigma^\alpha}{dt}(t)\left.\frac{\partial}{\partial z^\alpha}\right|_{\sigma(t)}.$$

Then the length of σ is given by

$$L(\sigma) = \int_a^b F\big(\dot\sigma(t)\big)\,dt,$$

exactly as in the real case.

Let $\Sigma\colon (-\varepsilon,\varepsilon)\times[a,b] \to M$ be a regular variation of a given regular curve $\sigma_0\colon [a,b] \to M$. We shall again set $\ell_\Sigma(s) = L(\sigma_s)$. Let $p\colon \Sigma^*(T^{1,0}M) \to (-\varepsilon,\varepsilon)\times[a,b]$ be the pull-back bundle, and $\gamma\colon \Sigma^*(T^{1,0}M) \to T^{1,0}M$ be the bundle map such that the diagram

$$\begin{array}{ccc} \Sigma^*(T^{1,0}M) & \xrightarrow{\ \gamma\ } & T^{1,0}M \\ {\scriptstyle p}\big\downarrow & & \big\downarrow{\scriptstyle \pi} \\ (-\varepsilon,\varepsilon)\times[a,b] & \xrightarrow{\ \Sigma\ } & M \end{array}$$

commutes. An element $\xi \in \mathcal{X}\big(\Sigma^*(T^{1,0}M)\big)$ can be written locally as

$$\xi(s,t) = v^\alpha(s,t)\left.\frac{\partial}{\partial z^\alpha}\right|_{(s,t)} = \left(u^a(s,t)\left.\frac{\partial}{\partial x^a}\right|_{(s,t)}\right)_o,$$

where we used the notations introduced in subsection 2.1.1. A local frame on $T_{\mathbb{R}}\big(\Sigma^*(T^{1,0}M)\big)$ is given by $\{\partial_s, \partial_t, \dot\partial_a^o\}$, where

$$\partial_s = \partial/\partial s, \qquad \partial_t = \partial/\partial t \qquad \text{and} \qquad \dot\partial_a^o = \partial/\partial u^a,$$

or by $\{\partial_s, \partial_t, \dot\partial_\alpha, i\dot\partial_\alpha\}$, where $\dot\partial_\alpha = \partial/\partial v^\alpha$.

Two particularly important sections of $\Sigma^*(T^{1,0}M)$ are

$$T = \gamma^{-1}\left(d^{1,0}\Sigma\left(\frac{\partial}{\partial t}\right)\right) = \frac{\partial\Sigma^\alpha}{\partial t}\frac{\partial}{\partial z^\alpha},$$

and

$$U = \gamma^{-1}\left(d^{1,0}\Sigma\left(\frac{\partial}{\partial s}\right)\right) = \frac{\partial\Sigma^\alpha}{\partial s}\frac{\partial}{\partial z^\alpha};$$

the section U is still called the *transversal vector* of the variation Σ.

Again, setting $\Sigma^*\tilde M = \gamma^{-1}(\tilde M)$, we have $T \in \mathcal{X}(\Sigma^*\tilde M)$ and

$$T(s,t) = \gamma^{-1}\big(\dot\sigma_s(t)\big).$$

Now we pull-back $T^{1,0}\tilde{M}$ over $\Sigma^*\tilde{M}$ by using γ, obtaining the commutative diagram

$$\begin{array}{ccc} \gamma^*(T^{1,0}\tilde{M}) & \xrightarrow{\tilde{\gamma}} & T^{1,0}\tilde{M} \\ \downarrow & & \downarrow \\ \Sigma^*\tilde{M} & \xrightarrow{\gamma} & \tilde{M} \\ \downarrow & & \downarrow \\ (-\varepsilon,\varepsilon)\times[a,b] & \xrightarrow{\Sigma} & M \end{array} ;$$

note that $\gamma^*(T^{1,0}\tilde{M})$ is a complex vector bundle over a real manifold. The bundle map $\tilde{\gamma}$ induçes a Hermitian structure on $\gamma^*(T^{1,0}\tilde{M})$ by

$$\forall X, Y \in \gamma^*(T^{1,0}\tilde{M})_v \langle X, Y\rangle_v = \langle\tilde{\gamma}(X), \tilde{\gamma}(Y)\rangle_{\gamma(v)}.$$

Analogously, the Chern connection D gives rise to a complex linear connection

$$D^*\colon \mathcal{X}(\gamma^*(T^{1,0}\tilde{M})) \to \mathcal{X}(T^*_{\mathbb{C}}(\Sigma^*\tilde{M}) \otimes \gamma^*(T^{1,0}\tilde{M})),$$

where $T^*_{\mathbb{C}}(\Sigma^*\tilde{M}) = T^*_{\mathbb{R}}(\Sigma^*\tilde{M}) \otimes \mathbb{C}$, by setting

$$\nabla^*_X Y = \tilde{\gamma}^{-1}\left(\nabla_{d^{1,0}\gamma(X)}\tilde{\gamma}(Y)\right),$$
$$\nabla^*_{\overline{X}} Y = \tilde{\gamma}^{-1}\left(\nabla_{\overline{d^{1,0}\gamma(X)}}\tilde{\gamma}(Y)\right),$$

for all $X \in T_{\mathbb{R}}(\Sigma^*\tilde{M})$ and $Y \in \mathcal{X}(\gamma^*(T^{1,0}\tilde{M}))$. We remark here that if we write

$$X = X^s\partial_s + X^t\partial_t + \dot{X}^a_o\dot{\partial}^o_a \in T_{\mathbb{R}}(\Sigma^*\tilde{M})$$

then

$$d^{1,0}\gamma(X) = \left[X^s\frac{\partial\Sigma^\alpha}{\partial s} + X^t\frac{\partial\Sigma^\alpha}{\partial t}\right]\partial_\alpha + \dot{X}^\alpha\dot{\partial}_\alpha = \left(d\gamma(X)\right)_o,$$

where $\dot{X}^\alpha = \dot{X}^\alpha_o + i\dot{X}^{\alpha+n}_o$. Furthermore we have

$$\begin{aligned} X\langle Y, Z\rangle &= X\left(\langle\tilde{\gamma}(Y), \tilde{\gamma}(Z)\rangle_\gamma\right) = d\gamma(X)\left(\langle\tilde{\gamma}(Y), \tilde{\gamma}(Z)\rangle\right) \\ &= \left(d^{1,0}\gamma(X) + \overline{d^{1,0}\gamma(X)}\right)\left(\langle\tilde{\gamma}(Y), \tilde{\gamma}(Z)\rangle\right) \\ &= \langle\nabla^*_X Y, Z\rangle + \langle Y, \nabla^*_{\overline{X}} Z\rangle + \langle\nabla^*_{\overline{X}} Y, Z\rangle + \langle Y, \nabla^*_X Z\rangle, \end{aligned} \tag{2.4.1}$$

for all $X \in T_{\mathbb{R}}(\Sigma^*\tilde{M})$ and $Y, Z \in \mathcal{X}(\gamma^*(T^{1,0}\tilde{M}))$.

We may also decompose $T_{\mathbb{R}}(\Sigma^*\tilde{M}) = \mathcal{H}^* \oplus \mathcal{V}^*$, where as usual a local frame for $\mathcal{V}^*$ is given by $\{\dot{\partial}_\alpha, i\dot{\partial}_\alpha\}$, and a local frame for $\mathcal{H}^*$ is given by

$$\delta_t = \partial_t - (\Gamma^\mu_{;\alpha}\circ\gamma)\frac{\partial\Sigma^\alpha}{\partial t}\dot{\partial}_\mu, \qquad \delta_s = \partial_s - (\Gamma^\mu_{;\alpha}\circ\gamma)\frac{\partial\Sigma^\alpha}{\partial s}\dot{\partial}_\mu.$$

Therefore, setting $T^H = d^{1,0}\gamma(\delta_t)$ and $U^H = d^{1,0}\gamma(\delta_s)$, we have

$$T^H(v) = \frac{\partial\Sigma^\mu}{\partial t}(s,t)\delta_\mu|_{\gamma(v)} = \chi_{\gamma(v)}\left(\dot{\sigma}_s(t)\right) \in \mathcal{H}_{\gamma(v)}$$

and

$$U^H(v) = \frac{\partial \Sigma^\mu}{\partial s}(s,t)\delta_\mu|_{\gamma(v)} = \chi_{\gamma(v)}\big(\gamma(U(s,t))\big) \in \mathcal{H}_{\gamma(v)},$$

for all $v \in \Sigma^*\tilde{M}_{(s,t)}$. In particular,

$$T^H(\gamma^{-1}(\dot\sigma_s)) = \chi(\dot\sigma_s). \tag{2.4.2}$$

Again, we have a bundle map $\Xi: T_{\mathbb{R}}(\Sigma^*\tilde{M}) \to \gamma^*(T^{1,0}\tilde{M})$ such that the diagram

$$\begin{array}{ccc} T_{\mathbb{R}}(\Sigma^*\tilde{M}) & \xrightarrow{\ \Xi\ } & \gamma^*(T^{1,0}\tilde{M}) \\ & {\scriptstyle d^{1,0}\gamma}\searrow & \downarrow{\scriptstyle\tilde\gamma} \\ & & T^{1,0}\tilde{M} \end{array}$$

commutes. Using Ξ we may prove three final formulas:

$$\begin{aligned}\tilde\gamma(\nabla^*_X\Xi(Y) - \nabla^*_Y\Xi(X)) &= \nabla_{d^{1,0}\gamma(X)}\, d^{1,0}\gamma(Y) - \nabla_{d^{1,0}\gamma(Y)}\, d^{1,0}\gamma(X) \\ &= [d^{1,0}\gamma(X), d^{1,0}\gamma(Y)] + \theta(d^{1,0}\gamma(X), d^{1,0}\gamma(Y)),\end{aligned} \tag{2.4.3}$$

for all $X, Y \in \mathcal{X}(T_{\mathbb{R}}(\Sigma^*\tilde{M}))$;

$$\begin{aligned}\tilde\gamma\circ(\nabla^*_X\nabla^*_Y - \nabla^*_Y\nabla^*_X) &= (\nabla_{d^{1,0}\gamma(X)}\nabla_{d^{1,0}\gamma(Y)} - \nabla_{d^{1,0}\gamma(Y)}\nabla_{d^{1,0}\gamma(X)})\circ\tilde\gamma \\ &= \nabla_{[d^{1,0}\gamma(X), d^{1,0}\gamma(Y)]}\circ\tilde\gamma,\end{aligned} \tag{2.4.4}$$

and

$$\tilde\gamma\circ(\nabla^*_X\nabla^*_{\bar Y} - \nabla^*_{\bar Y}\nabla^*_X) = (\nabla^*_{[d^{1,0}\gamma(X),\overline{d^{1,0}\gamma(Y)}]} + \Omega(d^{1,0}\gamma(X), \overline{d^{1,0}\gamma(Y)})\circ\tilde\gamma, \tag{2.4.5}$$

for all $X, Y \in T_{\mathbb{R}}(\Sigma^*\tilde{M})$.

2.4.2. The first variation formula

We are now able to prove the first variation formula for weakly Kähler Finsler metrics:

THEOREM 2.4.1: *Let $F: T^{1,0}M \to \mathbb{R}^+$ be a weakly Kähler Finsler metric on a complex manifold M. Take a regular curve $\sigma_0: [a,b] \to M$ with $F(\dot\sigma_0) \equiv c_0 > 0$, and a regular variation $\Sigma: (-\varepsilon, \varepsilon) \times [a,b] \to M$ of σ_0. Then*

$$\frac{d\ell_\Sigma}{ds}(0) = \frac{1}{c_0}\left\{\mathrm{Re}\langle U^H, T^H\rangle_{\dot\sigma_0}\Big|_a^b - \mathrm{Re}\int_a^b \langle U^H, \nabla_{T^H+\overline{T^H}}T^H\rangle_{\dot\sigma_0}\,dt\right\}.$$

In particular, if Σ is a fixed variation, that is $\Sigma(\cdot,a) \equiv \sigma_0(a)$ and $\Sigma(\cdot,b) \equiv \sigma_0(b)$, we have

$$\frac{d\ell_\Sigma}{ds}(0) = -\frac{1}{c_0}\,\mathrm{Re}\int_a^b \langle U^H, \nabla_{T^H+\overline{T^H}}T^H\rangle_{\dot\sigma_0}\,dt. \tag{2.4.6}$$

Proof: By definition,

$$\ell_\Sigma(s) = \int_a^b \left(G(\dot\sigma_s)\right)^{1/2} dt;$$

therefore

$$\frac{d\ell_\Sigma}{ds} = \frac{1}{2c_s}\int_a^b \frac{\partial}{\partial s}[G(\dot\sigma_s)]\, dt = \frac{1}{2c_s}\int_a^b \frac{\partial}{\partial s}\langle \Xi(\delta_t), \Xi(\delta_t)\rangle_T\, dt,$$

where $c_s \equiv F(\dot\sigma_s)$ and we used

$$G(\dot\sigma_s) = \langle \chi(\dot\sigma_s), \chi(\dot\sigma_s)\rangle_{\dot\sigma_s} = \langle \Xi(\delta_t), \Xi(\delta_t)\rangle_T,$$

by (2.4.2). Now, using (2.4.1) and (2.4.3), we get

$$\begin{aligned}
\frac{1}{2}\frac{\partial}{\partial s}\langle \Xi(\delta_t), \Xi(\delta_t)\rangle_T &= \frac{1}{2}\delta_s\langle \Xi(\delta_t), \Xi(\delta_t)\rangle_T \\
&= \frac{1}{2}\Big\{\langle \nabla^*_{\delta_s}\Xi(\delta_t), \Xi(\delta_t)\rangle_T + \langle \Xi(\delta_t), \nabla^*_{\overline{\delta_s}}\Xi(\delta_t)\rangle_T \\
&\qquad + \langle \nabla^*_{\overline{\delta_s}}\Xi(\delta_t), \Xi(\delta_t)\rangle_T + \langle \Xi(\delta_t), \nabla^*_{\delta_s}\Xi(\delta_t)\rangle_T\Big\} \\
&= \mathrm{Re}\Big\{\langle \nabla^*_{\delta_s}\Xi(\delta_t), \Xi(\delta_t)\rangle_T + \langle \nabla^*_{\overline{\delta_s}}\Xi(\delta_t), \Xi(\delta_t)\rangle_T\Big\} \\
&= \mathrm{Re}\Big\{\langle \nabla^*_{\delta_t}\Xi(\delta_s), \Xi(\delta_t)\rangle_T + \langle [U^H, T^H] + \nabla_{\overline{U^H}}T^H, T^H\rangle_{\dot\sigma_s} \\
&\qquad + \langle \theta(U^H, T^H), T^H\rangle_{\dot\sigma_s}\Big\}.
\end{aligned}$$

Since F is weakly Kähler, (2.4.2) yields

$$\langle \theta(U^H, T^H), T^H\rangle_{\dot\sigma_s} = 0.$$

Furthermore,

$$\begin{aligned}
[U^H, T^H] &= \left\{\frac{\partial\Sigma^\nu}{\partial s}\delta_\nu\left(\frac{\partial\Sigma^\mu}{\partial t}\right) - \frac{\partial\Sigma^\nu}{\partial t}\delta_\nu\left(\frac{\partial\Sigma^\mu}{\partial s}\right)\right\}\delta_\mu, \\
\nabla_{\overline{U^H}}T^H &= \frac{\partial\overline{\Sigma^\nu}}{\partial s}\delta_{\bar\nu}\left(\frac{\partial\Sigma^\mu}{\partial t}\right)\delta_\mu;
\end{aligned}$$

since

$$\begin{aligned}
\frac{\partial\Sigma^\nu}{\partial s}\delta_\nu\left(\frac{\partial\Sigma^\mu}{\partial t}\right) + \frac{\partial\overline{\Sigma^\nu}}{\partial s}\delta_{\bar\nu}\left(\frac{\partial\Sigma^\mu}{\partial t}\right) &= \frac{\partial^2\Sigma^\mu}{\partial s\partial t} = \frac{\partial^2\Sigma^\mu}{\partial t\partial s} \\
&= \frac{\partial\Sigma^\nu}{\partial t}\delta_\nu\left(\frac{\partial\Sigma^\mu}{\partial s}\right) + \frac{\partial\overline{\Sigma^\nu}}{\partial t}\delta_{\bar\nu}\left(\frac{\partial\Sigma^\mu}{\partial s}\right),
\end{aligned} \tag{2.4.7}$$

we get

$$[U^H, T^H] + \nabla_{\overline{U^H}}T^H = \frac{\partial\overline{\Sigma^\nu}}{\partial t}\delta_{\bar\nu}\left(\frac{\partial\Sigma^\mu}{\partial s}\right)\delta_\mu = \nabla_{\overline{T^H}}U^H. \tag{2.4.8}$$

Then

$$
\begin{aligned}
\frac{1}{2}\frac{\partial}{\partial s}\langle \Xi(\delta_t), \Xi(\delta_t)\rangle_T &= \operatorname{Re}\left\{ \langle \nabla^*_{\delta_t} \Xi(\delta_s), \Xi(\delta_t)\rangle_T + \langle \nabla^*_{\overline{\delta_t}} \Xi(\delta_s), \Xi(\delta_t)\rangle_T \right\} \\
&= \operatorname{Re}\left\{ \delta_t \langle \Xi(\delta_s), \Xi(\delta_t)\rangle_T - \langle \Xi(\delta_s), \nabla^*_{\delta_t + \overline{\delta_t}} \Xi(\delta_t)\rangle_T \right\} \qquad (2.4.9) \\
&= \operatorname{Re}\left\{ \frac{\partial}{\partial t}\langle U^H, T^H\rangle_{\dot\sigma_s} - \langle U^H, \nabla_{T^H + \overline{T^H}} T^H\rangle_{\dot\sigma_s} \right\},
\end{aligned}
$$

and the assertion follows. □

As a corollary we get the equation of geodesics:

COROLLARY 2.4.2: *Let $F\colon T^{1,0}M \to \mathbb{R}^+$ be a weakly Kähler Finsler metric on a complex manifold M, and let $\sigma\colon [a,b] \to M$ be a regular curve with $F(\dot\sigma) \equiv c_0 > 0$. Then σ is a geodesic for F iff*

$$\nabla_{T^H + \overline{T^H}} T^H \equiv 0, \tag{2.4.10}$$

where $T^H(v) = \chi_v(\dot\sigma(t)) \in \mathcal{H}_v$ for all $v \in \tilde M_{\sigma(t)}$.

Proof: It follows immediately from (2.4.6). □

COROLLARY 2.4.3: *Let $F\colon T^{1,0}M \to \mathbb{R}^+$ be a weakly Kähler Finsler metric on a complex manifold M. Then for any $p \in M$ and $v \in \tilde M_p$ there exists a unique geodesic $\sigma\colon (-\varepsilon, \varepsilon) \to M$ such that $\sigma(0) = p$ and $\dot\sigma(0) = v$.*

Proof: In local coordinates we have

$$\nabla_{T^H + \overline{T^H}} T^H = \left[(\dot\sigma^\mu \delta_\mu + \overline{\dot\sigma^\mu}\delta_{\bar\mu})(\dot\sigma^\alpha) + \Gamma^\alpha_{\nu;\mu}(\dot\sigma)\dot\sigma^\mu \dot\sigma^\nu \right] \delta_\alpha = [\ddot\sigma^\alpha + \Gamma^\alpha_{;\mu}(\dot\sigma)\dot\sigma^\mu]\delta_\alpha.$$

So (2.4.10) is a quasi-linear ODE system, and the assertion follows. □

So the standard ODE arguments apply in this case too, and we may recover for weakly Kähler Finsler metrics the usual theory of geodesics, as developed for instance in subsection 1.6.2.

In general, looking at the proof of Theorem 2.4.1 we see that the equation of geodesics for a strongly pseudoconvex Finsler metric F not necessarily weakly Kähler is

$$\nabla_{T^H + \overline{T^H}} T^H - \theta^*(T^H, \overline{T^H}) = 0, \tag{2.4.11}$$

where θ^* is defined by

$$\forall H, K, L \in \mathcal{H}_v \qquad \langle \theta(H,K), L\rangle_v = \langle H, \theta^*(L, \overline{K})\rangle_v.$$

In local coordinates,

$$\theta^* = G^{\bar\nu\alpha} G_{\beta\bar\gamma}(\Gamma^{\bar\gamma}_{\bar\mu;\bar\nu} - \Gamma^{\bar\gamma}_{\bar\nu;\bar\mu})\, dz^\beta \wedge d\bar z^\mu \otimes \delta_\alpha.$$

2.4.3. The second variation formula

Our next goal is the second variation formula, which holds for Kähler Finsler metrics. To express it correctly, we need two further ingredients.

DEFINITION 2.4.1: The *horizontal (1,1)-torsion* $\tau^{\mathcal{H}}$ is defined simply by

$$\tau^{\mathcal{H}}(X,\overline{Y}) = \Theta\big(\tau(X,\overline{Y})\big) = \Omega(X,\overline{Y})\chi.$$

DEFINITION 2.4.2: The *symmetric product* $\langle\!\langle\, ,\rangle\!\rangle\colon \mathcal{H}\times\mathcal{H}\to\mathbb{C}$ is locally given by

$$\forall H,K\in\mathcal{H}_v \qquad \langle\!\langle H,K\rangle\!\rangle_v = G_{\alpha\beta}(v)\,H^\alpha K^\beta.$$

It is clearly globally well-defined, and it satisfies

$$\forall H\in\mathcal{H} \qquad \langle\!\langle H,\chi\rangle\!\rangle = 0.$$

Needless to say, one can define a symmetric product on $\mathcal{V}$ in the same way.

Then the second variation formula for Kähler Finsler metrics is:

THEOREM 2.4.4: *Let $F\colon T^{1,0}M \to \mathbb{R}^+$ be a Kähler Finsler metric on a complex manifold M. Take a geodesic $\sigma_0\colon[a,b]\to M$ with $F(\dot\sigma_0)\equiv 1$, and a regular variation $\Sigma\colon(-\varepsilon,\varepsilon)\times[a,b]\to M$ of σ_0. Then*

$$\begin{aligned}\frac{d^2\ell_\Sigma}{ds^2}(0) &= \operatorname{Re}\langle\nabla_{U^H+\overline{U^H}}U^H,T^H\rangle_{\dot\sigma_0}\Big|_a^b\\ &\quad+\int_a^b\Big\{\|\nabla_{T^H+\overline{T^H}}U^H\|^2_{\dot\sigma_0}-\Big|\frac{\partial}{\partial t}\operatorname{Re}\langle U^H,T^H\rangle_{\dot\sigma_0}\Big|^2\\ &\qquad-\operatorname{Re}\Big[\langle\Omega(T^H,\overline{U^H})U^H,T^H\rangle_{\dot\sigma_0}-\langle\Omega(U^H,\overline{T^H})U^H,T^H\rangle_{\dot\sigma_0}\\ &\qquad+\langle\!\langle\tau^{\mathcal{H}}(U^H,\overline{T^H}),U^H\rangle\!\rangle_{\dot\sigma_0}-\langle\!\langle\tau^{\mathcal{H}}(T^H,\overline{U^H}),U^H\rangle\!\rangle_{\dot\sigma_0}\Big]\Big\}\,dt.\end{aligned}$$

In particular, if Σ is a fixed variation such that $\operatorname{Re}\langle U^H,T^H\rangle_{\dot\sigma_0}\equiv 0$ we have

$$\begin{aligned}\frac{d^2\ell_\Sigma}{ds^2}(0) &= \int_a^b\Big\{\|\nabla_{T^H+\overline{T^H}}U^H\|^2_{\dot\sigma_0}\\ &\qquad-\operatorname{Re}\Big[\langle\Omega(T^H,\overline{U^H})U^H,T^H\rangle_{\dot\sigma_0}-\langle\Omega(U^H,\overline{T^H})U^H,T^H\rangle_{\dot\sigma_0}\\ &\qquad+\langle\!\langle\tau^{\mathcal{H}}(U^H,\overline{T^H}),U^H\rangle\!\rangle_{\dot\sigma_0}-\langle\!\langle\tau^{\mathcal{H}}(T^H,\overline{U^H}),U^H\rangle\!\rangle_{\dot\sigma_0}\Big]\Big\}\,dt.\end{aligned}$$

Proof: During the proof of the first variation formula — in (2.4.9) — we saw that

$$\frac{d\ell_\Sigma}{ds}(s) = \operatorname{Re}\int_a^b\frac{\langle\nabla^*_{\delta_t}\Xi(\delta_s),\Xi(\delta_t)\rangle_T+\langle\nabla^*_{\overline{\delta_t}}\Xi(\delta_s),\Xi(\delta_t)\rangle_T}{\big(\langle\Xi(\delta_t),\Xi(\delta_t)\rangle_T\big)^{1/2}}\,dt.$$

So we need to compute

$$\frac{\partial}{\partial s}\left[\frac{\langle\nabla^*_{\delta_t}\Xi(\delta_s),\Xi(\delta_t)\rangle_T+\langle\nabla^*_{\overline{\delta_t}}\Xi(\delta_s),\Xi(\delta_t)\rangle_T}{\left(\langle\Xi(\delta_t),\Xi(\delta_t)\rangle_T\right)^{1/2}}\right]$$
$$=\frac{\delta_s\langle\nabla^*_{\delta_t}\Xi(\delta_s),\Xi(\delta_t)\rangle_T+\delta_s\langle\nabla^*_{\overline{\delta_t}}\Xi(\delta_s),\Xi(\delta_t)\rangle_T}{\left(\langle\Xi(\delta_t),\Xi(\delta_t)\rangle_T\right)^{1/2}}$$
$$-\frac{1}{2}\frac{\langle\nabla^*_{\delta_t}\Xi(\delta_s),\Xi(\delta_t)\rangle_T+\langle\nabla^*_{\overline{\delta_t}}\Xi(\delta_s),\Xi(\delta_t)\rangle_T}{\left(\langle\Xi(\delta_t),\Xi(\delta_t)\rangle_T\right)^{3/2}}\delta_s\langle\Xi(\delta_t),\Xi(\delta_t)\rangle_T. \tag{2.4.12}$$

Since, when $s=0$, the denominator of the first term on the right-hand side is equal to 1, and the denominator of the second term is equal to 2, we do not need to consider them any further. Let us call (I) the numerator of the first term, and (II) the numerator of the second term. First of all, (2.4.9) yields

$$\frac{1}{2}\operatorname{Re}(\mathrm{II})=\left|\operatorname{Re}\left[\frac{\partial}{\partial t}\langle U^H,T^H\rangle_{\dot\sigma_s}-\langle U^H,\nabla_{T^H+\overline{T^H}}T^H\rangle_{\dot\sigma_s}\right]\right|^2;$$

in particular, for $s=0$ we get

$$\frac{1}{2}\operatorname{Re}(\mathrm{II})(0)=\left|\frac{\partial}{\partial t}\operatorname{Re}\langle U^H,T^H\rangle_{\dot\sigma_0}\right|^2, \tag{2.4.13}$$

because σ_0 is a geodesic.

The computation of (I) is quite longer. First of all, using (2.4.3), (2.4.4) and (2.4.5) we get

$$\begin{aligned}
(\mathrm{I})=&\langle\nabla^*_{\delta_s}\nabla^*_{\delta_t}\Xi(\delta_s),\Xi(\delta_t)\rangle_T+\langle\nabla^*_{\overline{\delta_s}}\nabla^*_{\delta_t}\Xi(\delta_s),\Xi(\delta_t)\rangle_T\\
&+\langle\nabla^*_{\delta_t}\Xi(\delta_s),\nabla^*_{\delta_s}\Xi(\delta_t)\rangle_T+\langle\nabla^*_{\delta_t}\Xi(\delta_s),\nabla^*_{\overline{\delta_s}}\Xi(\delta_t)\rangle_T\\
&+\langle\nabla^*_{\delta_s}\nabla^*_{\overline{\delta_t}}\Xi(\delta_s),\Xi(\delta_t)\rangle_T+\langle\nabla^*_{\overline{\delta_s}}\nabla^*_{\overline{\delta_t}}\Xi(\delta_s),\Xi(\delta_t)\rangle_T\\
&+\langle\nabla^*_{\overline{\delta_t}}\Xi(\delta_s),\nabla^*_{\delta_s}\Xi(\delta_t)\rangle_T+\langle\nabla^*_{\overline{\delta_t}}\Xi(\delta_s),\nabla^*_{\overline{\delta_s}}\Xi(\delta_t)\rangle_T\\
=&\langle\nabla^*_{\delta_t}\nabla^*_{\delta_s}\Xi(\delta_s),\Xi(\delta_t)\rangle_T-\langle\nabla_{[T^H,U^H]}U^H,T^H\rangle_{\dot\sigma_s}\\
&+\langle\nabla^*_{\delta_t}\nabla^*_{\overline{\delta_s}}\Xi(\delta_s),\Xi(\delta_t)\rangle_T-\langle\nabla_{[T^H,\overline{U^H}]}U^H,T^H\rangle_{\dot\sigma_s}-\langle\Omega(T^H,\overline{U^H})U^H,T^H\rangle_{\dot\sigma_s}\\
&+\langle\nabla^*_{\delta_t}\Xi(\delta_s),\nabla^*_{\delta_t}\Xi(\delta_s)\rangle_T+\langle\nabla_{T^H}U^H,[U^H,T^H]+\nabla_{\overline{U^H}}T^H\rangle_{\dot\sigma_s}\\
&+\langle\nabla_{T^H}U^H,\theta(U^H,T^H)\rangle_{\dot\sigma_s}\\
&+\langle\nabla^*_{\overline{\delta_t}}\nabla^*_{\delta_s}\Xi(\delta_s),\Xi(\delta_t)\rangle_T-\langle\nabla_{[\overline{T^H},U^H]}U^H,T^H\rangle_{\dot\sigma_s}+\langle\Omega(U^H,\overline{T^H})U^H,T^H\rangle_{\dot\sigma_s}\\
&+\langle\nabla^*_{\overline{\delta_t}}\nabla^*_{\overline{\delta_s}}\Xi(\delta_s),\Xi(\delta_t)\rangle_T-\langle\nabla_{[\overline{T^H},\overline{U^H}]}U^H,T^H\rangle_{\dot\sigma_s}\\
&+\langle\nabla^*_{\overline{\delta_t}}\Xi(\delta_s),\nabla^*_{\delta_t}\Xi(\delta_s)\rangle_T+\langle\nabla_{\overline{T^H}}U^H,[U^H,T^H]+\nabla_{\overline{U^H}}T^H\rangle_{\dot\sigma_s}\\
&+\langle\nabla_{\overline{T^H}}U^H,\theta(U^H,T^H)\rangle_{\dot\sigma_s}.
\end{aligned}$$

Recalling (2.4.8), (2.4.2) and that F is Kähler we get

$$\begin{aligned}(\mathrm{I}) = &\langle \nabla^*_{\delta_t} \nabla^*_{\delta_s} \Xi(\delta_s), \Xi(\delta_t)\rangle_T + \langle \nabla^*_{\delta_t} \nabla^*_{\overline{\delta_s}} \Xi(\delta_s), \Xi(\delta_t)\rangle_T \\ &+ \langle \nabla^*_{\overline{\delta_t}} \nabla^*_{\delta_s} \Xi(\delta_s), \Xi(\delta_t)\rangle_T + \langle \nabla^*_{\overline{\delta_t}} \nabla^*_{\overline{\delta_s}} \Xi(\delta_s), \Xi(\delta_t)\rangle_T \\ &- \langle \Omega(T^H, \overline{U^H})U^H, T^H\rangle_{\dot\sigma_s} + \langle \Omega(U^H, \overline{T^H})U^H, T^H\rangle_{\dot\sigma_s} \\ &- \langle \nabla_{[T^H,U^H]+[T^H,\overline{U^H}]+[\overline{T^H},U^H]+[\overline{T^H},\overline{U^H}]} U^H, T^H\rangle_{\dot\sigma_s} \\ &+ \langle \nabla^*_{\delta_t} \Xi(\delta_s), \nabla^*_{\delta_t} \Xi(\delta_s)\rangle_T + \langle \nabla^*_{\overline{\delta_t}} \Xi(\delta_s), \nabla^*_{\delta_t} \Xi(\delta_s)\rangle_T \\ &+ \langle \nabla^*_{\delta_t} \Xi(\delta_s), \nabla^*_{\overline{\delta_t}} \Xi(\delta_s)\rangle_T + \langle \nabla^*_{\overline{\delta_t}} \Xi(\delta_s), \nabla^*_{\overline{\delta_t}} \Xi(\delta_s)\rangle_T.\end{aligned}$$

Now Lemma 2.3.3 yields

$$\begin{aligned}[T^H, U^H] &= \left[\frac{\partial\Sigma^\mu}{\partial t}\delta_\mu\left(\frac{\partial\Sigma^\nu}{\partial s}\right) - \frac{\partial\Sigma^\mu}{\partial s}\delta_\mu\left(\frac{\partial\Sigma^\nu}{\partial t}\right)\right]\delta_\nu, \\ [T^H, \overline{U^H}] &= \frac{\partial\Sigma^\mu}{\partial t}\frac{\partial\overline{\Sigma^\nu}}{\partial s}\left[\delta_{\bar\nu}(\Gamma^\alpha_{;\mu})\dot\partial_\alpha - \delta_\mu(\Gamma^{\bar\beta}_{;\bar\nu})\dot\partial_{\bar\beta}\right] \\ &\quad + \frac{\partial\Sigma^\mu}{\partial t}\delta_\mu\left(\frac{\partial\overline{\Sigma^\nu}}{\partial s}\right)\delta_{\bar\nu} - \frac{\partial\overline{\Sigma^\nu}}{\partial s}\delta_{\bar\nu}\left(\frac{\partial\Sigma^\mu}{\partial t}\right)\delta_\mu, \\ [\overline{T^H}, U^H] &= \frac{\partial\overline{\Sigma^\mu}}{\partial t}\frac{\partial\Sigma^\nu}{\partial s}\left[\delta_\nu(\Gamma^{\bar\beta}_{;\bar\mu})\dot\partial_{\bar\beta} - \delta_{\bar\mu}(\Gamma^\alpha_{;\nu})\dot\partial_\alpha\right] \\ &\quad + \frac{\partial\overline{\Sigma^\mu}}{\partial t}\delta_{\bar\mu}\left(\frac{\partial\Sigma^\nu}{\partial s}\right)\delta_\nu - \frac{\partial\Sigma^\nu}{\partial s}\delta_\nu\left(\frac{\partial\overline{\Sigma^\mu}}{\partial t}\right)\delta_{\bar\mu}, \\ [\overline{T^H}, \overline{U^H}] &= \left[\frac{\partial\overline{\Sigma^\mu}}{\partial t}\delta_{\bar\mu}\left(\frac{\partial\overline{\Sigma^\nu}}{\partial s}\right) - \frac{\partial\overline{\Sigma^\mu}}{\partial s}\delta_{\bar\mu}\left(\frac{\partial\overline{\Sigma^\nu}}{\partial t}\right)\right]\delta_{\bar\nu},\end{aligned}$$

and so (2.4.7) and (2.3.19) yield

$$\begin{aligned}&[T^H, U^H] + [T^H, \overline{U^H}] + [\overline{T^H}, U^H] + [\overline{T^H}, \overline{U^H}] \\ &\qquad = \tau(U^H, \overline{T^H}) - \tau(T^H, \overline{U^H}) + \overline{\tau(U^H, \overline{T^H})} - \overline{\tau(T^H, \overline{U^H})}.\end{aligned}$$

Furthermore, if $V \in \mathcal{V}$ we have

$$\begin{aligned}\langle \nabla_V U^H, T^H\rangle_{\dot\sigma_s} &= G_{\alpha\bar\beta}(\dot\sigma_s)V^\gamma\left[\dot\partial_\gamma\left(\frac{\partial\Sigma^\alpha}{\partial s}\right) + \Gamma^\alpha_{\delta\gamma}(\dot\sigma_s)\frac{\partial\Sigma^\delta}{\partial s}\right]\overline{\dot\sigma^\beta_s} \\ &= G_\alpha(\dot\sigma_s)\Gamma^\alpha_{\delta\gamma}(\dot\sigma_s)\frac{\partial\Sigma^\delta}{\partial s}V^\gamma = G_{\delta\gamma}(\dot\sigma_s)\frac{\partial\Sigma^\delta}{\partial s}V^\gamma \\ &= \langle\!\langle \Theta(V), U^H\rangle\!\rangle_{\dot\sigma_s},\end{aligned}$$

and

$$\langle \nabla_{\overline{V}} U^H, T^H\rangle_{\dot\sigma_s} = G_{\alpha\bar\beta}(\dot\sigma_s)\overline{V^\gamma}\left[\dot\partial_{\bar\gamma}\left(\frac{\partial\Sigma^\alpha}{\partial s}\right)\right]\overline{\dot\sigma^\beta_s} = 0.$$

Therefore

$$\begin{aligned}&\langle \nabla_{[T^H,U^H]+[T^H,\overline{U^H}]+[\overline{T^H},U^H]+[\overline{T^H},\overline{U^H}]} U^H, T^H\rangle_{\dot\sigma_s} \\ &\qquad = \langle\!\langle \tau^{\mathcal{H}}(U^H, \overline{T^H}), U^H\rangle\!\rangle_{\dot\sigma_s} - \langle\!\langle \tau^{\mathcal{H}}(T^H, \overline{U^H}), U^H\rangle\!\rangle_{\dot\sigma_s},\end{aligned}$$

and thus

$$
\begin{aligned}
(\mathrm{I}) = \delta_t \langle \nabla^*_{\delta_s+\overline{\delta_s}} \Xi(\delta_s), \Xi(\delta_t)\rangle_T - \langle \nabla^*_{\delta_s+\overline{\delta_s}} \Xi(\delta_s), \nabla^*_{\delta_t+\overline{\delta_t}} \Xi(\delta_t)\rangle_T \\
- \langle \Omega(T^H, \overline{U^H})U^H, T^H\rangle_{\dot\sigma_s} + \langle \Omega(U^H, \overline{T^H})U^H, T^H\rangle_{\dot\sigma_s} \\
- \langle\!\langle \tau^{\mathcal{H}}(U^H, \overline{T^H}), U^H\rangle\!\rangle_{\dot\sigma_s} + \langle\!\langle \tau^{\mathcal{H}}(T^H, \overline{U^H}), U^H\rangle\!\rangle_{\dot\sigma_s} + \|\nabla^*_{\delta_t+\overline{\delta_t}} \Xi(\delta_s)\|^2_T.
\end{aligned}
\tag{2.4.14}
$$

Recalling that for $s = 0$ we have $\nabla^*_{\delta_t+\overline{\delta_t}} \Xi(\delta_t) \equiv 0$ because σ_0 is a geodesic, (2.4.12), (2.4.13) and (2.4.14) evaluated at $s = 0$ yield the assertion. □

In particular, as we shall discuss in chapter 3, Theorem 2.4.4 will allow us to express the Morse index form in complex terms.

2.5. Holomorphic curvature

2.5.1. Symmetry properties of the horizontal curvature

The aim of this section is to study in some details the curvature operator of a strongly pseudoconvex Finsler metric. So let $F: T^{1,0}M \to \mathbb{R}^+$ be a strongly pseudoconvex Finsler metric on a complex manifold M, and let D denote the associated Chern-Finsler connection. As we have seen in the previous section, from a geometrical point of view the most interesting part of the curvature is the horizontal one, that is

$$\langle \Omega(H, \overline{K})L, M\rangle_v$$

where H, K, L, $M \in \mathcal{H}_v$ and $v \in \tilde{M}$. In local coordinates,

$$\langle \Omega(H, \overline{K})L, M\rangle_v = G_{\sigma\bar\beta} R^\sigma_{\alpha;\mu\bar\nu} H^\mu \overline{K^\nu} L^\alpha \overline{M^\beta}.$$

Of course, it is important to know its symmetries:

PROPOSITION 2.5.1: *Take $v \in \tilde{M}$ and H, K, L, $M \in \mathcal{H}_v$. Then*

$$\langle \Omega(\overline{K}, H)L, M\rangle_v = -\langle \Omega(H, \overline{K})L, M\rangle_v; \tag{2.5.1}$$

$$\langle \Omega(K, \overline{H})M, L\rangle_v = \overline{\langle \Omega(H, \overline{K})L, M\rangle_v}. \tag{2.5.2}$$

Furthermore, if $\bar\partial_H \theta = 0$ we also have

$$\langle \Omega(L, \overline{K})H, M\rangle_v = \langle \Omega(H, \overline{K})L, M\rangle_v = \langle \Omega(H, \overline{M})L, K\rangle_v. \tag{2.5.3}$$

Proof: (2.5.1) follows immediately from the observation $\Omega^\alpha_\beta(\overline{K}, H) = -\Omega^\alpha_\beta(H, \overline{K})$. To prove (2.5.2), we start from

$$\Omega^\alpha_\beta = \bar\partial \omega^\alpha_\beta = \bar\partial(G^{\bar\tau\alpha} \partial G_{\beta\bar\tau}) = -G^{\bar\tau\mu} G^{\bar\nu\alpha} \bar\partial G_{\mu\bar\nu} \wedge \partial G_{\beta\bar\tau} + G^{\bar\tau\alpha} \bar\partial\partial G_{\beta\bar\tau};$$

in particular,

$$G_{\alpha\bar\gamma}\Omega^{\alpha}_{\beta} = -G^{\bar\tau\mu}\bar\partial G_{\mu\bar\gamma} \wedge \partial G_{\beta\bar\tau} + \bar\partial\partial G_{\beta\bar\gamma}.$$

On the other hand,

$$\Omega^{\bar\alpha}_{\bar\gamma} = G^{\bar\tau\mu}G^{\bar\alpha\nu}\bar\partial G_{\mu\bar\gamma} \wedge \partial G_{\nu\bar\tau} - G^{\bar\alpha\tau}\bar\partial\partial G_{\tau\bar\gamma};$$

hence

$$G_{\alpha\bar\gamma}\Omega^{\alpha}_{\beta} = -G_{\beta\bar\alpha}\Omega^{\bar\alpha}_{\bar\gamma}.$$

In our case, this means that

$$\begin{aligned}\langle\Omega(K,\overline{H})M,L\rangle_v &= G_{\alpha\bar\gamma}\Omega^{\alpha}_{\beta}(K,\overline{H})M^{\beta}\overline{L^{\gamma}} = -G_{\beta\bar\alpha}\Omega^{\bar\alpha}_{\bar\gamma}(K,\overline{H})M^{\beta}\overline{L^{\gamma}}\\ &= \overline{G_{\alpha\bar\beta}\Omega^{\alpha}_{\gamma}(H,\overline{K})L^{\gamma}\overline{M^{\beta}}}\\ &= \overline{\langle\Omega(H,\overline{K})L,M\rangle_v},\end{aligned}$$

and (2.5.2) is proved.

Now, (2.5.3). First of all, $\bar\partial_H\theta = p_H^*(D''\theta)$, by (2.2.6). Corollary 2.3.7 says that $D''\theta = \eta^{\mathcal{H}} \wedge \Omega$; in local coordinates,

$$\begin{aligned}(\eta^{\mathcal{H}} \wedge \Omega)^{\alpha} = dz^{\sigma} \wedge \Omega^{\alpha}_{\sigma} =& R^{\alpha}_{\sigma;\mu\bar\nu}\,dz^{\sigma} \wedge dz^{\mu} \wedge d\bar z^{\nu} + R^{\alpha}_{\sigma\delta;\bar\nu}\,dz^{\sigma} \wedge \psi^{\delta} \wedge d\bar z^{\nu}\\ &+ R^{\alpha}_{\sigma\bar\gamma;\mu}\,dz^{\sigma} \wedge dz^{\mu} \wedge \overline{\psi^{\gamma}} + R^{\alpha}_{\sigma\delta\bar\gamma}\,dz^{\sigma} \wedge \psi^{\delta} \wedge \overline{\psi^{\gamma}};\end{aligned} \tag{2.5.4}$$

in particular, then, $\bar\partial_H\theta = 0$ iff $R^{\alpha}_{\sigma;\mu\bar\nu} = R^{\alpha}_{\mu;\sigma\bar\nu}$. Hence

$$\begin{aligned}\langle\Omega(L,\overline{K})H,M\rangle_v &= G_{\alpha\bar\tau}R^{\alpha}_{\sigma;\mu\bar\nu}L^{\mu}\overline{K^{\nu}}H^{\sigma}\overline{M^{\tau}}\\ &= G_{\alpha\bar\tau}R^{\alpha}_{\mu;\sigma\bar\nu}L^{\mu}\overline{K^{\nu}}H^{\sigma}\overline{M^{\tau}} = \langle\Omega(H,\overline{K})L,M\rangle_v.\end{aligned}$$

Finally,

$$\langle\Omega(H,\overline{M})L,K\rangle_v = \overline{\langle\Omega(M,\overline{H})K,L\rangle_v} = \overline{\langle\Omega(K,\overline{H})M,L\rangle_v} = \langle\Omega(H,\overline{K})L,M\rangle_v.$$

□

We remark that (2.5.2) is equivalent to

$$\langle\Omega(H,\overline{K})L,M\rangle_v = \langle L,\Omega(K,\overline{H})M\rangle_v \tag{2.5.5}$$

for all H, K, L, $M \in \mathcal{H}_v$ and $v \in \tilde M$.

In Hermitian geometry, quite a relevant role is played by the holomorphic sectional curvature; so it is natural to look for an analogue in our situation. A first attempt could be:

Definition 2.5.1: The *(horizontal) holomorphic flag curvature* $K_F(H)$ of a strongly pseudoconvex Finsler metric F along a horizontal vector $H \in \mathcal{H}_v$ is given by

$$K_F(H) = \frac{2}{\langle H,H\rangle_v^2}\langle\Omega(H,\overline{H})H,H\rangle_v.$$

Exactly as in the Hermitian case, if $\tilde{\partial}_H\theta = 0$ then the holomorphic flag curvature completely determines the horizontal part of the curvature tensor:

PROPOSITION 2.5.2: *Let $\mathcal{H}$ be a complex vector space, and let*

$$R,\ S\colon \mathcal{H}\times\overline{\mathcal{H}}\times\mathcal{H}\times\overline{\mathcal{H}}\to\mathbb{C}$$

be two quadrilinear maps satisfying (2.5.2) *and* (2.5.3), *that is*

$$R(K,\overline{H},M,\overline{L}) = \overline{R(H,\overline{K},L,\overline{M})} \tag{2.5.6}$$

and

$$R(L,\overline{K},H,\overline{M}) = R(H,\overline{K},L,\overline{M}) = R(H,\overline{M},L,\overline{K}) \tag{2.5.7}$$

for all H, K, L, $M\in\mathcal{H}$. Assume that

$$\forall H\in\mathcal{H} \qquad R(H,\overline{H},H,\overline{H}) = S(H,\overline{H},H,\overline{H}).$$

Then $R\equiv S$.

Proof: Up to replacing R and S by $R-S$ and 0, without loss of generality we can assume $R(H,\overline{H},H,\overline{H})=0$ for all $H\in\mathcal{H}$, and we should prove $R\equiv 0$.

Let G be the group of all permutations on the set of four letters $\{H,K,L,M\}$. The permutations $(H,K,L,M)\mapsto(K,H,M,L)$ and $(H,K,L,M)\mapsto(L,K,H,M)$ generate a subgroup G_1 of order 8; in particular, G/G_1 consists of 3 lateral classes. As representatives of these classes we can take: the identity; the transposition $(H,K,L,M)\mapsto(H,L,K,M)$; and $(H,K,L,M)\mapsto(L,H,K,M)$. Then (2.5.6) and (2.5.7) imply that the function

$$(H,K,L,M)\mapsto \operatorname{Re}\{R(H,\overline{K},L,\overline{M}) + R(H,\overline{L},K,\overline{M}) + R(L,\overline{H},K,\overline{M})\}$$

is symmetric. Since it vanishes on the diagonal, it is identically zero:

$$\operatorname{Re}\{R(H,\overline{K},L,\overline{M}) + R(H,\overline{L},K,\overline{M}) + R(L,\overline{H},K,\overline{M})\}\equiv 0.$$

Put $L=H$ and $M=K$; we get

$$\operatorname{Re}\{2R(H,\overline{H},K,\overline{K}) + R(H,\overline{K},H,\overline{K})\} = 0.$$

Replacing H by iH we obtain

$$\operatorname{Re}\{2R(H,\overline{H},K,\overline{K}) - R(H,\overline{K},H,\overline{K})\} = 0,$$

and hence

$$\forall H,\ K\in\mathcal{H} \qquad R(H,\overline{H},K,\overline{K}) = 0, \tag{2.5.8}$$

because $R(H,\overline{H},K,\overline{K})$ is real, by (2.5.6).

Now replacing K by $K+L$ in (2.5.8) yields

$$\begin{aligned} 0 = R(H,\overline{H},K+L,\overline{K}+\overline{L}) &= R(H,\overline{H},L,\overline{K}) + R(H,\overline{H},K,\overline{L}) \\ &= 2\operatorname{Re}\{R(H,\overline{H},L,\overline{K})\}, \end{aligned}$$

by (2.5.6). Replacing H by $H+M$ we get

$$0=\mathrm{Re}\{R(H+M,\overline{H}+\overline{M},L,\overline{K})\}=\mathrm{Re}\{R(H,\overline{M},L,\overline{K})+R(M,\overline{H},L,\overline{K})\}$$
$$=\mathrm{Re}\{R(H,\overline{K},L,\overline{M})+R(H,\overline{M},K,\overline{L})\};$$

hence

$$\mathrm{Re}\{R(H,\overline{K},L,\overline{M})\}=-\mathrm{Re}\{R(H,\overline{M},K,\overline{L})\}. \tag{2.5.9}$$

Replacing H by iH we get

$$-\mathrm{Im}\{R(H,\overline{K},L,\overline{M})\}=\mathrm{Im}\{R(H,\overline{M},K,\overline{L})\}.$$

Replacing here K by iK we obtain

$$\mathrm{Re}\{R(H,\overline{K},L,\overline{M})\}=\mathrm{Re}\{R(H,\overline{M},K,\overline{L})\};$$

so, comparing with (2.5.9) we have

$$\forall H,K,L,M\in\mathcal{H}\qquad \mathrm{Re}\{R(H,\overline{K},L,\overline{M})\}=0.$$

Replacing again K by iK we also get

$$\forall H,K,L,M\in\mathcal{H}\qquad \mathrm{Im}\{R(H,\overline{K},L,\overline{M})\}=0,$$

and we are done. □

Unfortunately, the holomorphic flag curvature is not the right object to study. In fact, as we have seen in the previous section, the geometrical meaningful objects (e.g., the variation formulas) involve contractions with χ. The geometry of the manifold still lives in the tangent bundle, and the radial vector fields χ and ι are just the means we use to read what we need in $T\tilde{M}$, where the algebra is easier. In a certain sense, we might even say that the holomorphic flag curvature contains too many informations — and, conversely, that the condition $\bar{\partial}_H\theta=0$ is too restricting. So in the next subsection we shall concentrate on a notion of holomorphic curvature on the tangent bundle (and not on the tangent-tangent bundle) associated to a strongly pseudoconvex Finsler metric, which is more suitable to our purposes.

2.5.2. The holomorphic curvature

DEFINITION 2.5.2: Let $F\colon T^{1,0}M\to\mathbb{R}^+$ be a strongly pseudoconvex Finsler metric on a complex manifold M, and take $v\in\tilde{M}$. Then the *holomorphic curvature* $K_F(v)$ of F along v is given by

$$K_F(v)=K_F(\chi(v))=\frac{2}{G(v)^2}\langle\Omega(\chi,\bar{\chi})\chi,\chi\rangle_v.$$

Clearly,

$$K_F(\mu_\zeta(v)) = K_F(v) \tag{2.5.10}$$

for all $\zeta \in \mathbb{C}^*$. Note that, by Proposition 2.5.1, the holomorphic curvature is necessarily real-valued.

In local coordinates we get

$$K_F = -\frac{2}{G^2} G_{\alpha} \delta_{\bar\nu} (\Gamma^{\alpha}_{;\mu}) v^{\mu} \overline{v^{\nu}}. \tag{2.5.11}$$

This definition is compatible with the usual notion of holomorphic sectional curvature in Hermitian geometry, as it is easily checked. But to better show its geometric content — as well as a variational aspect — we need to recall a couple of concepts.

DEFINITION 2.5.3: Let $\mu = g\, d\zeta \otimes d\bar\zeta$ be a Hermitian metric defined in a neighborhood of the origin in $\mathbb{C}$. Then the *Gaussian curvature* $K(\mu)(0)$ of μ at the origin is given by

$$K(\mu)(0) = -\frac{1}{2g(0)} (\Delta \log g)(0),$$

where Δ denotes the usual Laplacian

$$\Delta u = 4 \frac{\partial^2 u}{\partial \zeta \partial \bar\zeta}.$$

Now take a holomorphic map $\varphi\colon \Delta \to M$ (where Δ is the unit disk in $\mathbb{C}$) such that $\varphi'(0) \neq 0$. Then $\mu = \varphi^* G$ is a Hermitian metric in a neighborhood of the origin, and so $K(\varphi^* G)(0)$ makes sense. Wu (in [Wu]) has proved that if F comes from a Hermitian metric on M then the holomorphic curvature of F along $v \in \tilde M$ is the maximum value that $K(\varphi^* G)(0)$ can attain when φ varies among the holomorphic maps with $\varphi'(0) = \lambda v$, $\lambda \in \mathbb{C}^*$. This is true in our case too, as we shall see as a corollary of the following:

THEOREM 2.5.3: *Let $F\colon T^{1,0}M \to \mathbb{R}^+$ be a strongly pseudoconvex Finsler metric on a complex manifold M. Let $\varphi\colon \Delta \to M$ be a holomorphic map, and set $p = \varphi(0)$ and $v = \varphi'(0)$, with $v \neq o_p$. Then*

$$K(\varphi^* G)(0) = K_F(v) - \frac{2}{G(v)^2} \left\| \nabla_{(\varphi')^H} (\varphi')^H - \frac{\langle \nabla_{(\varphi')^H} (\varphi')^H, \chi \rangle_v}{\langle \chi, \chi \rangle_v} \chi \right\|_v^2. \tag{2.5.12}$$

Proof: Writing $\varphi^* G = g\, d\zeta \otimes d\bar\zeta$, we have

$$g(\zeta) = G(\varphi(\zeta); \varphi'(\zeta)), \qquad g(0) = G(v),$$

and

$$K(\varphi^* G)(0) = -\frac{1}{2g(0)} (\Delta \log g)(0) = -\frac{2}{G(v)} \frac{\partial^2 (\log g)}{\partial \bar\zeta \partial \zeta}(0).$$

The computation of the Laplacian yields

$$\begin{aligned}\frac{\partial^2(\log g)}{\partial\bar\zeta\partial\zeta} = &-\frac{1}{G(\varphi;\varphi')^2}\left|G_{;\mu}(\varphi;\varphi')(\varphi')^\mu + G_\alpha(\varphi;\varphi')(\varphi'')^\alpha\right|^2\\ &+\frac{1}{G(\varphi;\varphi')}\Big\{G_{;\mu\bar\nu}(\varphi;\varphi')(\varphi')^\mu\overline{(\varphi')^\nu} + G_{\alpha\bar\beta}(\varphi;\varphi')(\varphi'')^\alpha\overline{(\varphi'')^\beta}\\ &\qquad + 2\,\mathrm{Re}\big[G_{\bar\alpha;\mu}(\varphi;\varphi')(\varphi')^\mu\overline{(\varphi'')^\alpha}\big]\Big\}.\end{aligned}$$

Hence writing $w=\varphi''(0)$ we get

$$\begin{aligned}K(\varphi^*G)(0) = &-\frac{2}{G^2}\left[G_{;\mu\bar\nu} - \frac{G_{;\mu}G_{;\bar\nu}}{G}\right]v^\mu\overline{v^\nu}\\ &-\frac{2}{G^2}\left[G_{\alpha\bar\beta} - \frac{G_\alpha G_{\bar\beta}}{G}\right]w^\alpha\overline{w^\beta}\\ &-\frac{4}{G^2}\,\mathrm{Re}\left\{\left[G_{\bar\beta;\mu} - \frac{G_{\bar\beta}G_{;\mu}}{G}\right]v^\mu\overline{w^\alpha}\right\},\end{aligned}\tag{2.5.13}$$

where everything is evaluated at v.

Now,

$$\nabla_{(\varphi')^H}(\varphi')^H(v) = \left[w^\alpha + \Gamma^\alpha_{;\mu}v^\mu\right]\delta_\alpha \in \mathcal{H}_v;$$

hence

$$\begin{aligned}\left\|\nabla_{(\varphi')^H}(\varphi')^H - \frac{\langle\nabla_{(\varphi')^H}(\varphi')^H,\chi\rangle_v}{\langle\chi,\chi\rangle_v}\chi\right\|_v^2 &= \|\nabla_{(\varphi')^H}(\varphi')^H\|_v^2 - \frac{|\langle\nabla_{(\varphi')^H}(\varphi')^H,\chi\rangle_v|^2}{G}\\ = \left[\Gamma^{\bar\tau}_{;\bar\nu}G_{\bar\tau;\mu} - \frac{G_{;\mu}G_{;\bar\nu}}{G}\right]v^\mu\overline{v^\nu} &+ \left[G_{\alpha\bar\beta} - \frac{G_\alpha G_{\bar\beta}}{G}\right]w^\alpha\overline{w^\beta}\\ + 2\,\mathrm{Re}\left\{\left[G_{\bar\beta;\mu} - \frac{G_{;\mu}G_{\bar\beta}}{G}\right]v^\mu\overline{w^\beta}\right\}.\end{aligned}\tag{2.5.14}$$

Finally, by Lemma 2.3.4,

$$\begin{aligned}K_F(v) &= -\frac{2}{G^2}\,G_\alpha\delta_{\bar\nu}(\Gamma^\alpha_{;\mu})v^\mu\overline{v^\nu} = -\frac{2}{G^2}\,\delta_{\bar\nu}(G_\alpha\Gamma^\alpha_{;\mu})v^\mu\overline{v^\nu}\\ &= -\frac{2}{G^2}\left[G_{;\mu\bar\nu} - \Gamma^{\bar\tau}_{;\bar\nu}G_{\bar\tau;\mu}\right]v^\mu\overline{v^\nu}.\end{aligned}\tag{2.5.15}$$

The assertion then follows from (2.5.13), (2.5.14) and (2.5.15). □

COROLLARY 2.5.4: *Let $F\colon T^{1,0}M \to \mathbb{R}^+$ be a strongly pseudoconvex Finsler metric on a complex manifold M, and take $p\in M$ and $v\in \tilde M_p$. Then*

$$K_F(v) = \sup\{K(\varphi^*G)(0)\},\tag{2.5.16}$$

where the supremum is taken with respect to the family of all holomorphic maps $\varphi\colon\Delta\to M$ with $\varphi(0)=p$ and $\varphi'(0)=\lambda v$ for some $\lambda\in\mathbb{C}^*$. The supremum is achieved by the maps φ such that

$$\nabla_{(\varphi')^H}(\varphi')^H(\varphi'(0)) = a\chi(\varphi'(0)),\tag{2.5.17}$$

for some $a \in \mathbb{C}$.

Proof: It suffices to show that a holomorphic map $\varphi\colon \Delta \to M$ with $\varphi(0) = p$, $\varphi'(0) = \lambda v$ and satisfying (2.5.17) exists. Now, we can certainly find a small disk $\Delta_r = \{\zeta \in \mathbb{C} \mid |\zeta| < r\}$ and a holomorphic map $\psi\colon \Delta_r \to M$ with $\psi(0) = p$, $\psi'(0) = v$ and $\nabla_{(\psi')^H}(\psi')^H(v) = 0$. Then $\varphi\colon \Delta \to M$ given by $\varphi(\zeta) = \psi(r\zeta)$ satisfies $\varphi(0) = p$, $\varphi'(0) = rv$, $\nabla_{(\varphi')^H}(\varphi')^H(rv) = 0$ and

$$K(\varphi^*G)(0) = K(\psi^*G)(0) = K_F(v),$$

by (2.5.10) and (2.5.12). □

In the third chapter we shall study holomorphic maps $\varphi\colon \Delta \to M$ satisfying

$$\nabla_{(\varphi')^H}(\varphi')^H \equiv A\chi(\varphi')$$

for suitable functions A; we shall see that their existence implies that the holomorphic curvature is constant.

2.6. The Cartan connection vs. the Chern-Finsler connection

2.6.1 The radial vector fields

Let $F\colon T^{1,0}M \to \mathbb{R}^+$ be a complex Finsler metric on a complex manifold M. As we have already remarked at the beginning of subsection 2.4.1, using F we may define a function $F^o\colon T_{\mathbb{R}}M \to \mathbb{R}^+$ by setting

$$\forall u \in T_{\mathbb{R}}M \qquad F^o(u) = F(u_o). \tag{2.6.1}$$

Since $_o$ is a $\mathbb{R}$-isomorphism preserving the complex structure J, setting $\lambda = \xi + \eta J$ with ξ, $\eta \in \mathbb{R}$ we have

$$\forall u \in T_{\mathbb{R}}M \qquad F^o(\lambda u) = |\lambda| F^o(u),$$

where $|\lambda| = (\xi^2 + \eta^2)^{1/2}$. In particular, F^o satisfies all the properties of real Finsler metrics but possibly the last one: the indicatrices might be not strongly convex.

DEFINITION 2.6.1: We shall say that the complex Finsler metric F is *strongly convex* if F^o is a real Finsler metric.

In this section we shall always assume that $F\colon T^{1,0}M \to \mathbb{R}^+$ is a strongly convex Finsler metric. Thanks to (2.6.1), the F-length of any curve in M is the same as its F^o-length; in particular, F and F^o have the same geodesics and induce the same distance function on M. On the other hand, the Cartan connection induced by F^o and the Chern-Finsler connection induced by F might (and actually shall) differ:

this section is devoted to compare them. As we shall see, this is not a trivial task; so we shall pursue the comparison only up to obtaining the results we shall need in the next chapter. However, it will be enough to get an idea about what is going on.

We start recalling that, as we already remarked in subsection 2.1.3, the canonical isomorphism ${}^o\colon T^{1,0}\tilde{M} \to T_{\mathbb{R}}\tilde{M}$ induces an isomorphism ${}^o\colon \mathcal{V} \to \mathcal{V}_{\mathbb{R}}$ such that

$$(\dot{\partial}_\alpha)^o = \dot{\partial}^o_\alpha \qquad \text{and} \qquad (i\dot{\partial}_\alpha)^o = \dot{\partial}^o_{\alpha+n}. \tag{2.6.2}$$

In particular, if $V \in \mathcal{V}$ is expressed in local coordinates as $V = V^\alpha \dot{\partial}_\alpha$, then $V^o = U^a \dot{\partial}^o_a$, where

$$U^a = \begin{cases} \frac{1}{2}(V^\alpha + \overline{V^\alpha}) = \operatorname{Re} V^\alpha & \text{if } 1 \le a \le n, \\ \frac{1}{2i}(V^\alpha - \overline{V^\alpha}) = \operatorname{Im} V^\alpha & \text{if } n+1 \le a \le 2n, \end{cases} \tag{2.6.3}$$

where we recall that greek indeces run from 1 to $n = \dim_{\mathbb{C}} M$, latin indeces run from 1 to $2n = \dim_{\mathbb{R}} M$, and we used (and shall thoroughly use) the following convention: if in an equality there is a free latin index on one side and a free greek index on the other side, *the greek index is equal to the corresponding latin index taken mod n*. For example, in (2.6.3), the α on the right-hand side and the a on the left-hand side are related by

$$\alpha = \begin{cases} a & \text{if } 1 \le a \le n, \\ a-n & \text{if } n+1 \le a \le 2n. \end{cases}$$

Conversely, if $U \in \mathcal{V}_{\mathbb{R}}$ with $U = U^a \dot{\partial}^o_a$, then $U_o = V^\alpha \dot{\partial}_\alpha$ with

$$V^\alpha = U^\alpha + iU^{\alpha+n}.$$

We also recall that both the real vertical radial vector field $\iota^o\colon T_{\mathbb{R}}M \to \mathcal{V}_{\mathbb{R}}$ and the (complex) vertical radial vector field $\iota\colon T^{1,0}M \to \mathcal{V}$ are well-defined and satisfy

$$\forall v \in T^{1,0}M \qquad \iota^o(v^o) = \big(\iota(v)\big)^o.$$

Let $G\colon \tilde{M} \to \mathbb{R}^+$ be the square of the Finsler metric F (or F^o; by (2.6.1), it is the same). We have seen how using G we can define both a Hermitian metric $\langle\, ,\rangle$ on $\mathcal{V}$ and a Riemannian metric $\langle\,|\,\rangle$ on $\mathcal{V}_{\mathbb{R}}$ by setting

$$\forall V_1, V_2 \in \mathcal{V}_v \qquad \langle V_1, V_2\rangle_v = G_{\alpha\bar\beta}(v) V_1^\alpha \overline{V_2^\beta},$$

and

$$\forall U_1, U_2 \in (\mathcal{V}_{\mathbb{R}})_u \qquad \langle U_1 \mid U_2\rangle_u = \tfrac{1}{2} G_{ab}(u) U_1^a U_2^b.$$

Our first aim is to compare these two metrics. Of course, it is tantamount to comparing the Hessian and the Levi form of G. Using (2.6.2) we easily find that

$$G_a = \begin{cases} G_\alpha + G_{\bar\alpha} & \text{if } 1 \le a \le n, \\ i(G_\alpha - G_{\bar\alpha}) & \text{if } n+1 \le a \le 2n, \end{cases} \tag{2.6.4}$$

and

$$G_{ab} = \begin{cases} G_{\alpha\beta} + G_{\bar\alpha\beta} + G_{\alpha\bar\beta} + G_{\bar\alpha\bar\beta} & \text{if } 1 \le a, b \le n, \\ i(G_{\alpha\beta} + G_{\bar\alpha\beta} - G_{\alpha\bar\beta} - G_{\bar\alpha\bar\beta}) & \text{if } 1 \le a \le n \text{ and } n+1 \le b \le 2n, \\ i(G_{\alpha\beta} - G_{\bar\alpha\beta} + G_{\alpha\bar\beta} - G_{\bar\alpha\bar\beta}) & \text{if } n+1 \le a \le 2n \text{ and } 1 \le b \le n, \\ -(G_{\alpha\beta} - G_{\bar\alpha\beta} - G_{\alpha\bar\beta} + G_{\bar\alpha\bar\beta}) & \text{if } n+1 \le a, b \le 2n. \end{cases} \qquad (2.6.5)$$

Now take V_1, $V_2 \in \mathcal{V}_v$, and set $U_j = V_j^o$ for $j = 1, 2$. Then (2.6.3) and (2.6.5) yield

$$\begin{aligned} G_{ab}U_1^aU_2^b &= G_{\alpha\beta}V_1^\alpha V_2^\beta + G_{\bar\alpha\beta}\overline{V_1^\alpha}V_2^\beta + G_{\alpha\bar\beta}V_1^\alpha\overline{V_2^\beta} + G_{\bar\alpha\bar\beta}\overline{V_1^\alpha}\overline{V_2^\beta} \\ &= 2\,\mathrm{Re}\big[G_{\alpha\bar\beta}V_1^\alpha\overline{V_2^\beta} + G_{\alpha\beta}V_1^\alpha V_2^\beta\big]. \end{aligned}$$

So we have proved the

PROPOSITION 2.6.1: *Let $F: T^{1,0}M \to \mathbb{R}^+$ be a strongly convex Finsler metric on a complex manifold M. Then*

$$\forall V, W \in \mathcal{V} \qquad \langle V^o \mid W^o \rangle = \mathrm{Re}\big[\langle V, W\rangle + \langle\!\langle V, W\rangle\!\rangle\big].$$

Note that if F comes from a Hermitian metric then $G_{\alpha\beta} \equiv 0$ and so

$$\langle V^o \mid W^o \rangle = \mathrm{Re}\langle V, W\rangle.$$

In general, however, the symmetric product is not identically zero; as a consequence, the Cartan and the Chern-Finsler connection are less related than in the Hermitian case. Nevertheless, there is an important case where this relation still holds: since $\langle\!\langle \cdot, \iota\rangle\!\rangle \equiv 0$, we have

$$\forall V \in \mathcal{V} \qquad \langle V^o \mid \iota^o \rangle = \mathrm{Re}\,\langle V, \iota\rangle. \qquad (2.6.6)$$

To compare the Cartan and the Chern-Finsler connection we need to compare the corresponding horizontal bundles; we start by considering the horizontal radial vector fields. We recall that the complex horizontal radial vector field $\chi \in \mathcal{X}(\mathcal{H})$ is given by

$$\chi(v) = v^\alpha\delta_\alpha = v^\alpha(\partial_\alpha - \Gamma^\beta_{;\alpha}\dot\partial_\beta),$$

where $\Gamma^\beta_{;\alpha} = G^{\bar\mu\beta}G_{\bar\mu;\alpha}$. Using χ we can define a section χ^o of $T_{\mathbb{R}}\tilde M$ just by setting

$$\forall u \in T_{\mathbb{R}}M \qquad \chi^o(u) = \big(\chi(u_o)\big)^o.$$

On the other hand, the real Finsler structure F^o determines a real horizontal radial vector field, that we shall denote here by $\hat\chi \in \mathcal{X}(\mathcal{H}_{\mathbb{R}})$, given by

$$\forall u \in T_{\mathbb{R}}M \qquad \hat\chi(u) = u^a\hat\delta_a = u^a(\partial_a^o - \hat\Gamma^b_a\dot\partial_b^o),$$

where the $\hat\Gamma^b_a$ are the Christoffel symbols of the non-linear Cartan connection, and so are given by equation (1.4.14). We anticipate that, as we shall see in the next subsection, in general the elements $\hat\delta_a = \partial^o_a - \hat\Gamma^b_a \dot\partial^o_b$ of the standard local frame of $\mathcal{H}_{\mathbb{R}}$ are not of the form $(\delta_\alpha)^o$ or $(i\delta_\alpha)^o$.

So we must compare $\hat\chi$ and χ^o. Now, if $u = v^o \in \tilde M$ we have

$$\begin{aligned}\chi^o(u) &= \bigl(\chi(v)\bigr)^o = (v^\alpha\partial_\alpha)^o - (v^\alpha\Gamma^\beta_{;\alpha}\dot\partial_\beta)^o \\ &= u^a\partial^o_a - \tfrac{1}{2}(v^\alpha\Gamma^\beta_{;\alpha} + \overline{v^\alpha}\Gamma^{\bar\beta}_{;\bar\alpha})\dot\partial^o_\beta - \tfrac{1}{2i}(v^\alpha\Gamma^\beta_{;\alpha} - \overline{v^\alpha}\Gamma^{\bar\beta}_{;\bar\alpha})\dot\partial^o_{\beta+n}.\end{aligned}$$

So to compare $\hat\chi$ and χ^o we ought to compare

$$u^a\hat\Gamma^b_a \qquad \text{with} \qquad \begin{cases}\tfrac{1}{2}(v^\alpha\Gamma^\beta_{;\alpha} + \overline{v^\alpha}\Gamma^{\bar\beta}_{;\bar\alpha}) & \text{if } 1 \le b \le n,\\ \tfrac{1}{2i}(v^\alpha\Gamma^\beta_{;\alpha} - \overline{v^\alpha}\Gamma^{\bar\beta}_{;\bar\alpha}) & \text{if } n+1 \le b \le 2n.\end{cases}$$

Since (G_{bc}) is an invertible matrix, we can multiply everything by G_{bc} and then compare the outcomes. Using (1.4.14), (2.6.3), (2.6.4) and (2.6.5) we first get

$$G_{bc}u^a\hat\Gamma^b_a = G_{c;d}u^d - G_{;c} = \begin{cases}2\,\mathrm{Re}\bigl[G_{\gamma;\delta}v^\delta + G_{\gamma;\bar\delta}\overline{v^\delta} - G_{;\gamma}\bigr] & \text{if } 1 \le c \le n,\\ -2\,\mathrm{Im}\bigl[G_{\gamma;\delta}v^\delta + G_{\gamma;\bar\delta}\overline{v^\delta} - G_{;\gamma}\bigr] & \text{if } n+1 \le c \le 2n.\end{cases} \tag{2.6.7}$$

On the other hand, we have

$$\begin{aligned}&\tfrac{1}{2}G_{\beta c}(v^\alpha\Gamma^\beta_{;\alpha} + \overline{v^\alpha}\Gamma^{\bar\beta}_{;\bar\alpha}) + \tfrac{1}{2i}G_{\beta+n,c}(v^\alpha\Gamma^\alpha_{;\beta} - \overline{v^\alpha}\Gamma^{\bar\beta}_{;\bar\alpha})\\ &\qquad = \begin{cases}2\,\mathrm{Re}\bigl[G_{\beta\gamma}v^\delta\Gamma^\beta_{;\delta} + G_{\gamma;\bar\delta}\overline{v^\delta}\bigr] & \text{if } 1 \le c \le n,\\ -2\,\mathrm{Im}\bigl[G_{\beta\gamma}v^\delta\Gamma^\beta_{;\delta} + G_{\gamma;\bar\delta}\overline{v^\delta}\bigr] & \text{if } n+1 \le c \le 2n.\end{cases}\end{aligned} \tag{2.6.8}$$

It turns out that we have proved the following

PROPOSITION 2.6.2: *Let $F\colon T^{1,0}M \to \mathbb{R}^+$ be a strongly convex Finsler metric on a complex manifold M. Then*

$$\forall V \in \mathcal{V} \qquad \langle \chi^o - \hat\chi \mid V^o\rangle = \mathrm{Re}\bigl\langle \theta(\Theta(V),\chi),\chi\bigr\rangle. \tag{2.6.9}$$

In particular, $\chi^o \equiv \hat\chi$ iff F is weakly Kähler.

Proof: First of all, note that $\chi^o - \hat\chi$ is a section of $\mathcal{V}_{\mathbb{R}}$, and so the left-hand side of (2.6.9) makes sense. Then (2.6.9) itself follows subtracting (2.6.7) from (2.6.8) and recalling (2.3.22). Finally, we get the last assertion writing (2.6.9) both for V and for iV. □

So we have obtained a new characterization of weakly Kähler metrics. Note that this is coherent with the fact that F and F^o have the same geodesics: indeed, the geodesic equation depends only on $T^H = \chi(\dot\sigma)$. Finally, we explicitly remark that, as a consequence of Proposition 2.6.2, F is weakly Kähler iff

$$u^a\hat\Gamma^b_a\dot\partial^o_b = (v^\alpha\Gamma^\beta_{;\alpha}\dot\partial_\beta)^o. \tag{2.6.10}$$

2.6.2. The horizontal bundles

The next step consists in comparing the horizontal bundle $\mathcal{H} \subset T^{1,0}\tilde{M}$ associated to the Chern-Finsler connection, and the horizontal bundle $\mathcal{H}_{\mathbb{R}} \subset T_{\mathbb{R}}\tilde{M}$ associated to the Cartan connection. Let $\Theta: \mathcal{V} \to \mathcal{H}$ denote the complex horizontal map associated to $\mathcal{H}$, and let $\hat{\Theta}: \mathcal{V}_{\mathbb{R}} \to \mathcal{H}_{\mathbb{R}}$ denote the real horizontal map associated to $\mathcal{H}_{\mathbb{R}}$. Since ${}^o: \mathcal{V} \to \mathcal{V}_{\mathbb{R}}$ is a $\mathbb{R}$-isomorphism, we get a $\mathbb{R}$-isomorphism $\hat{}: \mathcal{H} \to \mathcal{H}_{\mathbb{R}}$ given by

$$\forall H \in \mathcal{H} \qquad \hat{H} = \hat{\Theta}\big((\Theta^{-1}(H))^o\big).$$

Note that $\widehat{\delta_\alpha} = \hat{\delta}_\alpha$ and $\widehat{i\delta_\alpha} = \hat{\delta}_{\alpha+n}$.

Now, the canonical isomorphism ^{o} sends $\mathcal{H}$ into $T_{\mathbb{R}}\tilde{M}$; if the image would be $\mathcal{H}_{\mathbb{R}}$, then for all practical purposes we could identify $\mathcal{H}$ and $\mathcal{H}_{\mathbb{R}}$. So the comparison between $\mathcal{H}$ and $\mathcal{H}_{\mathbb{R}}$ boils down to comparing H^o and $\hat{H}$ for any $H \in \mathcal{H}$ (note that both have the same coordinates with respect to the corresponding usual local frames). The first step is:

PROPOSITION 2.6.3: *Let $F: T^{1,0}M \to \mathbb{R}^+$ be a strongly convex Finsler metric on a complex manifold M. Then:*

(i) *$\hat{\Theta}$ commutes with the complex structure J iff $\widehat{iH} = J\hat{H}$ for all $H \in \mathcal{H}$ iff $\mathcal{H}_{\mathbb{R}}$ is J-invariant iff*

$$\hat{\Gamma}^{\beta}_{\alpha+n} = -\hat{\Gamma}^{\beta+n}_{\alpha} \qquad \text{and} \qquad \hat{\Gamma}^{\beta+n}_{\alpha+n} = \hat{\Gamma}^{\beta}_{\alpha} \tag{2.6.11}$$

for all $1 \leq \alpha, \beta \leq n$;

(ii) *$\hat{H} = H^o$ for all $H \in \mathcal{H}$ iff $\mathcal{H}_{\mathbb{R}}$ is J-invariant and*

$$\Gamma^{\beta}_{;\alpha} = \hat{\Gamma}^{\beta}_{\alpha} + i\hat{\Gamma}^{\beta+n}_{\alpha} \tag{2.6.12}$$

for all $1 \leq \alpha, \beta \leq n$.

Proof: (i) Since Θ^{-1} commutes with i, and ^{o} transforms multiplication by i into the complex structure J, it is clear that $\hat{\Theta}$ commutes with J iff $\widehat{iH} = J\hat{H}$ for all $H \in \mathcal{H}$. In turn, this happens iff

$$\partial^o_{\alpha+n} - \hat{\Gamma}^{b}_{\alpha+n}\dot{\partial}^o_b = \hat{\delta}_{\alpha+n} = \widehat{i\delta_\alpha} = J\hat{\delta}_\alpha = J(\partial^o_\alpha - \hat{\Gamma}^c_\alpha\dot{\partial}^o_c) = \partial^o_{\alpha+n} - \hat{\Gamma}^c_\alpha J\dot{\partial}^o_c$$

for all $1 \leq \alpha \leq n$, that is iff (2.6.11) holds — and this clearly implies that $\mathcal{H}_{\mathbb{R}}$ is J-invariant. Finally, if $\mathcal{H}_{\mathbb{R}}$ is J-invariant we should have

$$J(\partial^o_\alpha - \hat{\Gamma}^c_\alpha\dot{\partial}^o_c) = \partial^o_{\alpha+n} - \hat{\Gamma}^c_\alpha(J\dot{\partial}^o_c) \in \mathcal{H}_{\mathbb{R}},$$

and then (2.6.11) again holds, because $\hat{\delta}_{\alpha+n}$ is the unique element of $\mathcal{H}_{\mathbb{R}}$ whose image under $d\pi$ is $\partial/\partial x^{\alpha+n}$, and $d\pi$ commutes with J.

(ii) We have $\hat{H} = H^o$ for all $H \in \mathcal{H}$ iff $(\hat{\delta}_\alpha)_o = \delta_\alpha$ and $(\widehat{i\delta_\alpha})_o = i\delta_\alpha$ for all $1 \leq \alpha \leq n$. Now,

$$(\hat{\delta}_\alpha)_o = \partial_\alpha - (\hat{\Gamma}^{\beta}_{\alpha} + i\hat{\Gamma}^{\beta+n}_{\alpha})\dot{\partial}_\beta,$$
$$(\widehat{i\delta_\alpha})_o = i\big(\partial_\alpha - (\hat{\Gamma}^{\beta+n}_{\alpha+n} - i\hat{\Gamma}^{\beta}_{\alpha+n})\dot{\partial}_\beta\big),$$

and so $\hat{H} = H^o$ for all $H \in \mathcal{H}$ iff (2.6.11) and (2.6.12) hold. □

Set

$$\tilde{\Gamma}^b_a = \begin{cases} \operatorname{Re}\Gamma^\beta_{;\alpha} & \text{if } 1 \le a,b \le n, \\ \operatorname{Im}\Gamma^\beta_{;\alpha} & \text{if } 1 \le a \le n \text{ and } n+1 \le b \le 2n, \\ -\operatorname{Im}\Gamma^\beta_{;\alpha} & \text{if } n+1 \le a \le 2n \text{ and } 1 \le b \le n, \\ \operatorname{Re}\Gamma^\beta_{;\alpha} & \text{if } n+1 \le a,b \le 2n. \end{cases}$$

Then Proposition 2.6.3.(ii) says that $\hat{H} = H^o$ for all $H \in \mathcal{H}$ iff

$$\hat{\Gamma}^b_a = \tilde{\Gamma}^b_a$$

for all $1 \le a, b \le 2n$. Arguing as before, we must compare $G_{bc}\hat{\Gamma}^c_a$ with $G_{bc}\tilde{\Gamma}^c_a$ — and the computations begin to become long and tedious. First of all,

$$G_{bc}\hat{\Gamma}^b_a = \frac{1}{2}\{[G_{ab;d}u^d + G_{b;a} - G_{a;b}] - G_{abd}(\hat{\Gamma}^d_e u^e)\}.$$

Using (2.6.3), (2.6.4), (2.6.5) and their equivalents for third order derivatives, and assuming that F is weakly Kähler, that is using (2.6.10), we get:

$$G_{ab;d}u^d$$

$$= \begin{cases} 2\operatorname{Re}\big[(G_{\alpha\beta;\delta} + G_{\bar{\alpha}\beta;\delta})v^\delta + (G_{\alpha\beta;\bar{\delta}} + G_{\bar{\alpha}\beta;\bar{\delta}})\overline{v^\delta}\big] & \text{if } 1 \le a,b \le n, \\ -2\operatorname{Im}\big[(G_{\alpha\beta;\delta} + G_{\bar{\alpha}\beta;\delta})v^\delta + (G_{\alpha\beta;\bar{\delta}} + G_{\bar{\alpha}\beta;\bar{\delta}})\overline{v^\delta}\big] & \text{if } 1 \le a \le n \\ & \quad\text{and } n+1 \le b \le 2n, \\ -2\operatorname{Im}\big[(G_{\alpha\beta;\delta} - G_{\bar{\alpha}\beta;\delta})v^\delta + (G_{\alpha\beta;\bar{\delta}} - G_{\bar{\alpha}\beta;\bar{\delta}})\overline{v^\delta}\big] & \text{if } n+1 \le a \le 2n \\ & \quad\text{and } 1 \le b \le n, \\ -2\operatorname{Re}\big[(G_{\alpha\beta;\delta} - G_{\bar{\alpha}\beta;\delta})v^\delta + (G_{\alpha\beta;\bar{\delta}} - G_{\bar{\alpha}\beta;\bar{\delta}})\overline{v^\delta}\big] & \text{if } n+1 \le a,b \le 2n. \end{cases}$$

$$G_{b;a} = \begin{cases} 2\operatorname{Re}\big[G_{\beta;\alpha} + G_{\beta;\bar{\alpha}}\big] & \text{if } 1 \le a,b \le n, \\ -2\operatorname{Im}\big[G_{\beta;\alpha} + G_{\beta;\bar{\alpha}}\big] & \text{if } 1 \le a \le n \text{ and } n+1 \le b \le 2n, \\ -2\operatorname{Im}\big[G_{\beta;\alpha} - G_{\beta;\bar{\alpha}}\big] & \text{if } n+1 \le a \le 2n \text{ and } 1 \le b \le n, \\ -2\operatorname{Re}\big[G_{\beta;\alpha} - G_{\beta;\bar{\alpha}}\big] & \text{if } n+1 \le a,b \le 2n. \end{cases}$$

$$G_{a;b} = \begin{cases} 2\operatorname{Re}\big[G_{\alpha;\beta} + G_{\bar{\alpha};\beta}\big] & \text{if } 1 \le a,b \le n, \\ -2\operatorname{Im}\big[G_{\alpha;\beta} + G_{\bar{\alpha};\beta}\big] & \text{if } 1 \le a \le n \text{ and } n+1 \le b \le 2n, \\ -2\operatorname{Im}\big[G_{\alpha;\beta} - G_{\bar{\alpha};\beta}\big] & \text{if } n+1 \le a \le 2n \text{ and } 1 \le b \le n, \\ -2\operatorname{Re}\big[G_{\alpha;\beta} - G_{\bar{\alpha};\beta}\big] & \text{if } n+1 \le a,b \le 2n. \end{cases}$$

$$G_{abd}(\hat{\Gamma}^d_e u^e)$$

$$= \begin{cases} 2\operatorname{Re}\big[(G_{\alpha\beta\delta} + G_{\bar{\alpha}\beta\delta})\Gamma^\delta_{;\varepsilon}v^\varepsilon + (G_{\alpha\beta\bar{\delta}} + G_{\bar{\alpha}\beta\bar{\delta}})\Gamma^{\bar{\delta}}_{;\bar{\varepsilon}}\overline{v^\varepsilon}\big] & \text{if } 1 \le a,b \le n, \\ -2\operatorname{Im}\big[(G_{\alpha\beta\delta} + G_{\bar{\alpha}\beta\delta})\Gamma^\delta_{;\varepsilon}v^\varepsilon + (G_{\alpha\beta\bar{\delta}} + G_{\bar{\alpha}\beta\bar{\delta}})\Gamma^{\bar{\delta}}_{;\bar{\varepsilon}}\overline{v^\varepsilon}\big] & \text{if } 1 \le a \le n \\ & \quad\text{and } n+1 \le b \le 2n, \\ -2\operatorname{Im}\big[(G_{\alpha\beta\delta} - G_{\bar{\alpha}\beta\delta})\Gamma^\delta_{;\varepsilon}v^\varepsilon + (G_{\alpha\beta\bar{\delta}} - G_{\bar{\alpha}\beta\bar{\delta}})\Gamma^{\bar{\delta}}_{;\bar{\varepsilon}}\overline{v^\varepsilon}\big] & \text{if } n+1 \le a \le 2n \\ & \quad\text{and } 1 \le b \le n, \\ -2\operatorname{Re}\big[(G_{\alpha\beta\delta} - G_{\bar{\alpha}\beta\delta})\Gamma^\delta_{;\varepsilon}v^\varepsilon + (G_{\alpha\beta\bar{\delta}} - G_{\bar{\alpha}\beta\bar{\delta}})\Gamma^{\bar{\delta}}_{;\bar{\varepsilon}}\overline{v^\varepsilon}\big] & \text{if } n+1 \le a,b \le 2n. \end{cases}$$

Furthermore,

$$G_{bc}\tilde{\Gamma}^c_a = \begin{cases} 2\,\mathrm{Re}\big[G_{\beta\gamma}\Gamma^\gamma_{;\alpha} + G_{\beta;\bar\alpha}\big] & \text{if } 1 \le a, b \le n, \\ -2\,\mathrm{Im}\big[G_{\beta\gamma}\Gamma^\gamma_{;\alpha} + G_{\beta;\bar\alpha}\big] & \text{if } 1 \le a \le n \text{ and } n+1 \le b \le 2n, \\ -2\,\mathrm{Im}\big[G_{\beta\gamma}\Gamma^\gamma_{;\alpha} - G_{\beta;\bar\alpha}\big] & \text{if } n+1 \le a \le 2n \text{ and } 1 \le b \le n, \\ -2\,\mathrm{Re}\big[G_{\beta\gamma}\Gamma^\gamma_{;\alpha} - G_{\beta;\bar\alpha}\big] & \text{if } n+1 \le a, b \le 2n. \end{cases}$$

Now set

$$K_\nu = [G_{\mu;\nu} - G_{\nu;\mu} + G_{\nu\sigma}\Gamma^\sigma_{;\mu}]v^\mu.$$

Since F is weakly Kähler, $K_\nu \equiv 0$ by (2.3.22). So we get

$$\begin{aligned} 0 = (K_\beta)_\alpha &= G_{\alpha;\beta} - G_{\beta;\alpha} + G_{\beta\gamma}\Gamma^\gamma_{;\alpha} + [-G_{\alpha\beta;\delta} + G_{\alpha\beta\gamma}\Gamma^\gamma_{;\delta} + G_{\beta\gamma}\Gamma^\gamma_{\alpha;\delta}]v^\delta, \\ 0 = (K_\beta)_{\bar\alpha} &= G_{\bar\alpha;\beta} + [-G_{\bar\alpha\beta;\delta} + G_{\bar\alpha\beta\gamma}\Gamma^\gamma_{;\delta} + G_{\beta\gamma}\Gamma^\gamma_{\bar\alpha;\delta}]v^\delta. \end{aligned} \tag{2.6.13}$$

Moreover,

$$G_{\beta\bar\gamma}\Gamma^{\bar\gamma}_{\alpha;\bar\delta}\overline{v^\delta} = G_{\alpha\beta;\bar\delta}\overline{v^\delta} - G_{\alpha\beta\bar\gamma}\Gamma^{\bar\gamma}_{;\bar\delta}\overline{v^\delta},$$

$$G_{\beta\bar\gamma}\Gamma^{\bar\gamma}_{\bar\delta;\bar\alpha}\overline{v^\delta} = G_{\beta;\bar\alpha}, \qquad G_{\beta\bar\gamma}\Gamma^{\bar\gamma}_{\bar\alpha;\bar\delta}\overline{v^\delta} = G_{\bar\alpha\beta;\bar\delta}\overline{v^\delta} - G_{\bar\alpha\beta\bar\gamma}\Gamma^{\bar\gamma}_{;\bar\delta}\overline{v^\delta}.$$

Summing up we have obtained

$$\begin{aligned} &G_{bc}(\tilde{\Gamma}^c_a - \hat{\Gamma}^c_a) \\ &= \begin{cases} \mathrm{Re}\big[G_{\beta\gamma}(\Gamma^\gamma_{\delta;\alpha} - \Gamma^\gamma_{\alpha;\delta})v^\delta + G_{\beta\bar\gamma}(\Gamma^{\bar\gamma}_{\bar\delta;\bar\alpha} - \Gamma^{\bar\gamma}_{\bar\alpha;\bar\delta})\overline{v^\delta} - G_{\beta\gamma}\Gamma^\gamma_{\bar\alpha;\delta}v^\delta - G_{\beta\bar\gamma}\Gamma^{\bar\gamma}_{\alpha;\bar\delta}\overline{v^\delta}\big] \\ \hfill \text{if } 1 \le a, b \le n, \\ -\mathrm{Im}\big[G_{\beta\gamma}(\Gamma^\gamma_{\delta;\alpha} - \Gamma^\gamma_{\alpha;\delta})v^\delta + G_{\beta\bar\gamma}(\Gamma^{\bar\gamma}_{\bar\delta;\bar\alpha} - \Gamma^{\bar\gamma}_{\bar\alpha;\bar\delta})\overline{v^\delta} - G_{\beta\gamma}\Gamma^\gamma_{\bar\alpha;\delta}v^\delta - G_{\beta\bar\gamma}\Gamma^{\bar\gamma}_{\alpha;\bar\delta}\overline{v^\delta}\big] \\ \hfill \text{if } 1 \le a \le n \text{ and } n+1 \le b \le 2n, \\ -\mathrm{Im}\big[G_{\beta\gamma}(\Gamma^\gamma_{\delta;\alpha} - \Gamma^\gamma_{\alpha;\delta})v^\delta - G_{\beta\bar\gamma}(\Gamma^{\bar\gamma}_{\bar\delta;\bar\alpha} - \Gamma^{\bar\gamma}_{\bar\alpha;\bar\delta})\overline{v^\delta} + G_{\beta\gamma}\Gamma^\gamma_{\bar\alpha;\delta}v^\delta - G_{\beta\bar\gamma}\Gamma^{\bar\gamma}_{\alpha;\bar\delta}\overline{v^\delta}\big] \\ \hfill \text{if } n+1 \le a \le 2n \text{ and } 1 \le b \le n, \\ -\mathrm{Re}\big[G_{\beta\gamma}(\Gamma^\gamma_{\delta;\alpha} - \Gamma^\gamma_{\alpha;\delta})v^\delta - G_{\beta\bar\gamma}(\Gamma^{\bar\gamma}_{\bar\delta;\bar\alpha} - \Gamma^{\bar\gamma}_{\bar\alpha;\bar\delta})\overline{v^\delta} + G_{\beta\gamma}\Gamma^\gamma_{\bar\alpha;\delta}v^\delta - G_{\beta\bar\gamma}\Gamma^{\bar\gamma}_{\alpha;\bar\delta}\overline{v^\delta}\big] \\ \hfill \text{if } n+1 \le a, b \le 2n. \end{cases} \end{aligned} \tag{2.6.14}$$

So we have proved most of the following

Theorem 2.6.4: *Let $F: T^{1,0}M \to \mathbb{R}^+$ be a strongly convex weakly Kähler Finsler metric on a complex manifold M. Then:*

$$\begin{aligned} \langle \hat{H} - H^o \mid V^o \rangle = \frac{1}{2}\mathrm{Re}\Big[&\langle \theta(H,\chi), \Theta(V) \rangle + \langle\!\langle \theta(H,\chi), \Theta(V) \rangle\!\rangle \\ &+ \langle \tau(\chi, \overline{\Theta^{-1}(H)}), V \rangle + \langle\!\langle \tau(\chi, \overline{\Theta^{-1}(H)}), V \rangle\!\rangle\Big], \end{aligned} \tag{2.6.15}$$

for all $H \in \mathcal{H}$ and $V \in \mathcal{V}$. In particular, $\hat{H} = H^o$ for all $H \in \mathcal{H}$ iff

$$\theta(H,\chi) = 0 \qquad \text{and} \qquad \tau(\chi, \overline{W}) = 0 \tag{2.6.16}$$

for all $H \in \mathcal{H}$ and $W \in \mathcal{V}$.

Proof: First of all, $\hat{H} - H^o$ belongs to $\mathcal{V}_{\mathbb{R}}$, and so the left-hand side of (2.6.15) makes sense. Moreover, if $K = H^o$ and $U = V^o$ we have

$$\langle \hat{H} - H^o \mid V^o \rangle = \tfrac{1}{2} G_{bc}(\tilde{\Gamma}^c_a - \hat{\Gamma}^c_a) K^a U^b;$$

so (2.6.15) follows from (2.6.14) and (2.3.19). We are left with the final assertion. If (2.6.16) holds, then (2.6.15) immediately yields $\hat{H} = H^o$ for all $H \in \mathcal{H}$. Conversely, assume that $\hat{H} = H^o$ for all $H \in \mathcal{H}$. Then Proposition 2.6.1 and (2.6.15) imply that

$$\langle \Theta^{-1}\big(\theta(H,\chi)\big)^o + \tau\big(\chi, \overline{\Theta^{-1}(H)}\big)^o \mid V^o \rangle = 0$$

for all $H \in \mathcal{H}$ and $V \in \mathcal{V}$, and so

$$\Theta^{-1}\big(\theta(H,\chi)\big) + \tau\big(\chi, \overline{\Theta^{-1}(H)}\big) \equiv 0.$$

Replacing H by iH we get the assertion. □

In particular, if F is Kähler we find

$$\forall H \in \mathcal{H} \qquad \hat{H} = H^o + \tfrac{1}{2}\tau\big(\chi, \overline{\Theta^{-1}(H)}\big)^o,$$

that is

$$\forall V \in \mathcal{V} \qquad \hat{\Theta}(V^o) = \big(\Theta(V)\big)^o + \tfrac{1}{2}\tau(\chi, \overline{V})^o.$$

If F comes from a Hermitian metric, then $\tau(\chi, \overline{V}) \equiv 0$, and so we do not see this difference between the Levi-Civita and the Chern connections; otherwise, the mixed part of the (1,1)-torsion measures how far apart are the Cartan and the Chern-Finsler connections.

Finally, we should remark that there is one instance where the behaviors of $\hat{H}$ and H^o agree:

COROLLARY 2.6.5: *Let $F\colon T^{1,0}M \to \mathbb{R}^+$ be a strongly convex weakly Kähler Finsler metric on a complex manifold M. Then*

$$\forall H \in \mathcal{H} \qquad \langle \hat{H} \mid \chi^o \rangle = \langle H^o \mid \chi^o \rangle = \operatorname{Re}\langle H, \chi \rangle, \tag{2.6.17}$$

(where we recall that $\hat{\chi} = \chi^o$ since F is weakly Kähler).

Proof: Indeed, since F is weakly Kähler we have

$$\langle \theta(H,\chi), \chi \rangle = 0.$$

Furthermore, we know that $\langle\!\langle \cdot, \iota \rangle\!\rangle \equiv 0$. Finally,

$$\langle \tau(\chi, \overline{V}), \iota \rangle = -G_{\alpha\bar{\beta}} \Gamma^\alpha_{\bar{\gamma};\delta} \overline{V^\gamma} v^\delta \overline{v^\beta} = -(G_\alpha \Gamma^\alpha_{\bar{\gamma};\delta}) \overline{V^\gamma} v^\delta = 0,$$

and (2.6.17) follows from (2.6.15). □

2.6.3. The covariant derivatives

The next step in our comparison program concerns the covariant derivative. Let $\hat{\nabla}$ denote the covariant derivative associated to the Cartan connection, and ∇ denote the covariant derivative associated to the Chern-Finsler connection. We are not interested here in comparing $\hat{\nabla}$ and ∇ in their full generality; for our aims, it is enough to compare the covariant derivatives appearing in the second variation formula, namely $\hat{\nabla}_{\hat{\chi}}$ and $\nabla_{\chi+\bar{\chi}}$.

So we shall compare $\hat{\nabla}_{\hat{\chi}}V^o$ and $(\nabla_{\chi+\bar{\chi}}V)^o$ for any $V \in \mathcal{X}(\mathcal{V})$. As usual, this is equivalent to comparing

$$\langle V_1^o \mid \hat{\nabla}_{\hat{\chi}}V_2^o\rangle \qquad \text{and} \qquad \langle V_1^o \mid (\nabla_{\chi+\bar{\chi}}V_2)^o\rangle$$

for all V_1, $V_2 \in \mathcal{X}(\mathcal{V})$.

First of all, by (2.3.14) $G_{\alpha\bar{\gamma}}\Gamma^{\bar{\gamma}}_{\bar{\beta};\bar{\mu}}\overline{v^\mu} = \bar{\chi}(G_{\alpha\bar{\beta}})$, and so

$$\begin{aligned}\langle V_1^o|(\nabla_{\chi+\bar{\chi}}V_2)^o\rangle &= \mathrm{Re}\big[\langle V_1, \nabla_\chi V_2\rangle + \langle\!\langle V_1, \nabla_\chi V_2\rangle\!\rangle + \langle V_1, \nabla_{\bar{\chi}} V_2\rangle + \langle\!\langle V_1, \nabla_{\bar{\chi}} V_2\rangle\!\rangle\big]\\
&= \frac{1}{2}\Big\{G_{\alpha\bar{\beta}}V_1^\alpha\overline{\chi(V_2^\beta)} + G_{\bar{\alpha}\beta}\overline{V_1^\alpha}\chi(V_2^\beta) + \bar{\chi}(G_{\alpha\bar{\beta}})V_1^\alpha\overline{V_2^\beta} + \chi(G_{\bar{\alpha}\beta})\overline{V_1^\alpha}V_2^\beta\\
&\qquad + G_{\alpha\beta}V_1^\alpha\chi(V_2^\beta) + G_{\bar{\alpha}\bar{\beta}}\overline{V_1^\alpha}\,\overline{\chi(V_2^\beta)}\\
&\qquad + G_{\alpha\gamma}\Gamma^\gamma_{\beta;\mu}v^\mu V_1^\alpha V_2^\beta + G_{\bar{\alpha}\bar{\gamma}}\Gamma^{\bar{\gamma}}_{\bar{\beta};\bar{\mu}}\overline{v^\mu}\,\overline{V_1^\alpha}\,\overline{V_2^\beta}\\
&\qquad + G_{\alpha\bar{\beta}}V_1^\alpha\overline{\bar{\chi}(V_2^\beta)} + G_{\bar{\alpha}}\overline{V_1^\alpha}\bar{\chi}(V_2^\beta)\\
&\qquad + G_{\alpha\beta}V_1^\alpha\bar{\chi}(V_2^\beta) + G_{\bar{\alpha}\bar{\beta}}\overline{V_1^\alpha}\,\overline{\bar{\chi}(V_2^\beta)}\Big\}.\end{aligned}$$

On the other hand, writing as usual $U_j = V_j^o$ for $j = 1$, 2 and recalling (1.4.12), (1.4.4) and (1.4.6), we have

$$\begin{aligned}\langle V_1^o \mid \hat{\nabla}_{\hat{\chi}}V_2^o\rangle &= \tfrac{1}{2}G_{ac}\big[\hat{\chi}(U_2^c)\hat{\Gamma}^c_{b;j}u^jU_2^b\big]U_1^a\\
&= \frac{1}{2}G_{ac}U_1^a\hat{\chi}(U_2^c) + \frac{1}{4}\big[\hat{\delta}_j(G_{ab}) + \hat{\delta}_b(G_{aj}) - \hat{\delta}_a(G_{bj})\big]u^jU_1^aU_2^b\\
&= \frac{1}{2}G_{ac}U_1^a\hat{\chi}(U_2^c) + \frac{1}{4}\big[\hat{\chi}(G_{ab}) + G_{a;b} - G_{b;a}\big]U_1^aU_2^b\\
&= \frac{1}{2}\Big\{G_{\alpha\beta}V_1^\alpha\hat{\chi}(V_2^\beta) + G_{\alpha\bar{\beta}}V_1^\alpha\overline{\hat{\chi}(V_2^\beta)} + G_{\bar{\alpha}\beta}\overline{V_1^\alpha}\hat{\chi}(V_2^\beta) + G_{\bar{\alpha}\bar{\beta}}\overline{V_1^\alpha}\,\overline{\hat{\chi}(V_2^\beta)}\Big\}\\
&\quad + \frac{1}{4}\big[\hat{\chi}(G_{\alpha\beta}) + G_{\alpha;\beta} - G_{\beta;\alpha}\big]V_1^\alpha V_2^\beta + \frac{1}{4}\big[\hat{\chi}(G_{\alpha\bar{\beta}}) + G_{\alpha;\bar{\beta}} - G_{\bar{\beta};\alpha}\big]V_1^\alpha\overline{V_2^\beta}\\
&\quad + \frac{1}{4}\big[\hat{\chi}(G_{\bar{\alpha}\beta}) + G_{\bar{\alpha};\beta} - G_{\beta;\bar{\alpha}}\big]\overline{V_1^\alpha}V_2^\beta + \frac{1}{4}\big[\hat{\chi}(G_{\bar{\alpha}\bar{\beta}}) + G_{\bar{\alpha};\bar{\beta}} - G_{\bar{\beta};\bar{\alpha}}\big]\overline{V_1^\alpha}\,\overline{V_2^\beta}.\end{aligned}$$

We now assume that F is weakly Kähler, that is that $\hat{\chi} \equiv \chi^o = \chi + \bar{\chi}$. Then

$$\begin{aligned}&\langle V_1^o \mid (\nabla_{\chi+\bar{\chi}}V_2)^o - \hat{\nabla}_{\hat{\chi}}V_2^o\rangle\\
&\qquad = \frac{1}{2}\mathrm{Re}\big\{\big[2G_{\alpha\gamma}\Gamma^\gamma_{\beta;\mu}v^\mu + G_{\beta;\alpha} - G_{\alpha;\beta} - (\chi+\bar{\chi})(G_{\alpha\beta})\big]V_1^\alpha V_2^\beta\\
&\qquad\qquad + \big[(\bar{\chi}-\chi)(G_{\alpha\bar{\beta}}) + G_{\bar{\beta};\alpha} - G_{\alpha;\bar{\beta}}\big]V_1^\alpha\overline{V_2^\beta}\big\}.\end{aligned}$$

Now since F is weakly Kähler, recalling (2.6.13) and (2.3.22) we get

$$
\begin{gathered}
0 = G_{\beta;\alpha} - G_{\alpha;\beta} + G_{\alpha\gamma}(\Gamma^{\gamma}_{\mu;\beta} + \Gamma^{\gamma}_{\beta;\mu})v^{\mu} - \chi(G_{\alpha\beta}), \\
0 = G_{\bar\beta;\alpha} + G_{\alpha\gamma}\Gamma^{\gamma}_{\bar\beta;\mu}v^{\mu} - \chi(G_{\alpha\bar\beta}), \qquad (2.6.18) \\
\bar\chi(G_{\alpha\beta}) = G_{\alpha\bar\gamma}\Gamma^{\bar\gamma}_{\beta;\bar\mu}\overline{v^{\mu}}, \qquad \bar\chi(G_{\alpha\bar\beta}) = G_{\alpha;\bar\beta} + G_{\alpha\bar\gamma}(\Gamma^{\bar\gamma}_{\bar\beta;\bar\mu} - \Gamma^{\bar\gamma}_{\bar\mu;\bar\beta})\overline{v^{\mu}}.
\end{gathered}
$$

Therefore

$$
\begin{aligned}
\langle V_1^o \mid (\nabla_{\chi+\bar\chi}V_2)^o - \hat\nabla_{\hat\chi}V_2^o\rangle = \frac{1}{2}\operatorname{Re}\big\{ &\big[G_{\alpha\gamma}(\Gamma^{\gamma}_{\beta;\mu} - \Gamma^{\gamma}_{\mu;\beta})v^{\mu} - G_{\alpha\bar\gamma}\Gamma^{\bar\gamma}_{\beta;\bar\mu}\overline{v^{\mu}}\big]V_1^{\alpha}V_2^{\beta} \\
&+ \big[G_{\alpha\bar\gamma}(\Gamma^{\bar\gamma}_{\bar\beta;\bar\mu} - \Gamma^{\bar\gamma}_{\bar\mu;\bar\beta})\overline{v^{\mu}} - G_{\alpha\gamma}\Gamma^{\gamma}_{\bar\beta;\mu}v^{\mu}\big]V_1^{\alpha}\overline{V_2^{\beta}}\big\},
\end{aligned}
\qquad (2.6.19)
$$

and we have proved the

THEOREM 2.6.6: *Let $F\colon T^{1,0}M \to \mathbb{R}^+$ be a strongly convex weakly Kähler Finsler metric on a complex manifold M. Then*

$$
\hat\nabla_{\hat\chi}V^o = (\nabla_{\chi+\bar\chi}V)^o - \frac{1}{2}\big[\Theta^{-1}(\theta(\Theta(V),\chi)) + \tau(\chi,\overline{V})\big]^o,
$$

for all $V \in \mathcal{X}(\mathcal{V})$. In particular, F is Kähler iff

$$
\hat\nabla_{\hat\chi}V^o = (\nabla_{\chi^o}V)^o - \tfrac{1}{2}\tau(\chi,\overline{V})^o
$$

for all $V \in \mathcal{X}(\mathcal{V})$.

Proof: Indeed (2.6.19) exactly says

$$
\langle V_1^o \mid (\nabla_{\chi+\bar\chi}V_2)^o - \hat\nabla_{\hat\chi}V_2^o\rangle = \frac{1}{2}\langle V_1^o \mid \big[\Theta^{-1}(\theta(\Theta(V_2),\chi)) + \tau(\chi,\overline{V})\big]^o\rangle,
$$

and we are done. □

So once again it is the mixed part of the (1,1)-torsion which distinguishes between the Cartan and the Chern-Finsler connections, even for Kähler metrics. So it could be useful to register a couple of facts about it:

PROPOSITION 2.6.7: *Let $F\colon T^{1,0}M \to \mathbb{R}^+$ be a strongly pseudoconvex Finsler metric on a complex manifold. Then*

(i) *for any V, $W \in \mathcal{V}$ we have*

$$
\langle V, \tau(\chi,\overline{W})\rangle = \langle W, \tau(\chi,\overline{V})\rangle;
$$

in particular,

$$
\langle \tau(\chi,\overline{V}), \iota\rangle \equiv 0.
$$

(ii) *if F is weakly Kähler, for any V, $W \in \mathcal{V}$ we have*

$$
\langle\!\langle \tau(\chi,\overline{V}), W\rangle\!\rangle = \langle \theta(\Theta(W),\chi), \Theta(V)\rangle;
$$

in particular, if F is Kähler we have

$$\langle\!\langle \tau(\chi,\overline{V}), W\rangle\!\rangle \equiv 0.$$

Proof: (i) Indeed

$$\langle V, \tau(\chi,\overline{W})\rangle_v = -G_{\alpha\bar\gamma}\Gamma^{\bar\gamma}_{\bar\beta;\bar\mu}\overline{v^\mu}V^\alpha W^\beta = -\bar\chi(G_{\alpha\beta})V^\alpha W^\beta = \langle W, \tau(\chi,\overline{V})\rangle_v.$$

The second assertion follows from $\bar\chi(G_{\alpha\beta})v^\alpha = \bar\chi(G_{\alpha\beta}v^\alpha) = 0$.

(ii) Indeed

$$G_{\alpha\gamma}\Gamma^{\gamma}_{\bar\beta;\mu}v^\mu = \chi(G_{\alpha\bar\beta}) - G_{\bar\beta;\alpha} = G_{\gamma\bar\beta}(\Gamma^{\gamma}_{\alpha;\mu} - \Gamma^{\gamma}_{\mu;\alpha})v^\mu, \tag{2.6.20}$$

by (2.6.18) because F is weakly Kähler, and we are done. □

In particular, we have

$$\langle \hat\nabla_{\hat\chi}V^o \mid \hat\chi\rangle = \langle (\nabla_{\chi^o}V)^o \mid \chi^o\rangle = \operatorname{Re}\langle \nabla_{\chi^o}V, \chi\rangle.$$

Proceeding with our task of comparing the two second variation formulas — the real one Theorem 1.5.4 and the complex one Theorem 2.4.4 — we now want to compare

$$\langle \hat\nabla_{\hat H}V^o \mid \iota^o\rangle \qquad \text{and} \qquad \langle (\nabla_{H^o}V)^o \mid \iota^o\rangle$$

for any $H \in \mathcal{H}$ and $V \in \mathcal{X}(\mathcal{V})$. As usual, let us start with some computations in local coordinates.

$$\begin{aligned}\langle (\nabla_{H^o}V)^o \mid \iota^o\rangle &= \operatorname{Re}\big[\langle \nabla_H V, \iota\rangle + \langle \nabla_{\overline{H}}V, \iota\rangle\big]\\ &= \operatorname{Re}\big[G_\alpha H(V^\alpha) + G_\alpha \Gamma^\alpha_{\beta;\gamma}V^\beta H^\gamma + G_\alpha \overline{H}(V^\alpha)\big]\\ &= \operatorname{Re}\big[G_\alpha H^o(V^\alpha) + (G_{\beta;\gamma} - G_{\beta\sigma}\Gamma^\sigma_{;\gamma})V^\beta H^\gamma\big].\end{aligned}$$

On the other hand, setting $U = V^o$ and $\hat H = K^a\hat\delta_a$, we have

$$\begin{aligned}\langle \hat\nabla_{\hat H}V^o \mid \iota^o\rangle &= \tfrac12 G_a[\hat H(U^a) + \hat\Gamma^a_{b;c}K^cU^b]\\ &= \tfrac12 G_a\hat H(U^a) + \tfrac14[\hat\delta_c(G_{ab}) + \hat\delta_b(G_{ac}) - \hat\delta_a(G_{bc})]u^aK^cU^b\\ &= \tfrac12 G_a\hat H(U^a) + \tfrac14[G_{b;c} + G_{c;b} - \hat\chi(G_{bc})]K^cU^b.\end{aligned}$$

Now,

$$G_a\hat H(U^a) = G_\alpha\hat H(V^\alpha) + G_{\bar\alpha}\hat H(\overline{V^\alpha}) = 2\operatorname{Re}[G_\alpha\hat H(V^\alpha)],$$

$$\begin{aligned}[G_{b;c} + G_{c;b} - \hat\chi(G_{bc})]K^cU^b = 2\operatorname{Re}\Big\{&[G_{\beta;\gamma} + G_{\gamma;\beta} - \hat\chi(G_{\beta\gamma})]V^\beta H^\gamma\\ &+ [G_{\beta;\bar\gamma} + G_{\bar\gamma;\beta} - \hat\chi(G_{\beta\bar\gamma})]V^\beta\overline{H^\gamma}\Big\}.\end{aligned}$$

So subtracting we get

$$\begin{aligned}\langle\hat{\nabla}_{\hat{H}}V^o-(\nabla_{H^o}V)^o\mid\iota^o\rangle=&\operatorname{Re}\big[G_\alpha(\hat{H}-H^o)(V^\alpha)\big]\\&-\frac{1}{2}\operatorname{Re}\Big\{[G_{\beta;\gamma}-G_{\gamma;\beta}-2G_{\beta\sigma}\Gamma^\sigma_{;\gamma}+\hat{\chi}(G_{\beta\gamma})]V^\beta H^\gamma\\&\qquad+[\hat{\chi}(G_{\beta\bar{\gamma}})-G_{\beta;\bar{\gamma}}-G_{\bar{\gamma};\beta}]V^\beta\overline{H^\gamma}\Big\}.\end{aligned}$$

If we assume that F is weakly Kähler then $\hat{\chi}=\chi+\bar{\chi}$ and, by (2.6.18), we get

$$\begin{aligned}\langle\hat{\nabla}_{\hat{H}}V^o-(\nabla_{H^o}V)^o\mid\iota^o\rangle=&\operatorname{Re}[G_\alpha(\hat{H}-H^o)(V^\alpha)]\\&-\frac{1}{2}\operatorname{Re}\Big\{[G_{\beta\sigma}(\Gamma^\sigma_{\gamma;\mu}-\Gamma^\sigma_{\mu;\gamma})v^\mu+G_{\beta\bar{\tau}}\Gamma^{\bar{\tau}}_{\gamma;\bar{\nu}}\overline{v^\nu}]V^\beta H^\gamma\\&\qquad+[G_{\beta\sigma}\Gamma^\sigma_{\bar{\gamma};\mu}v^\mu+G_{\beta\bar{\tau}}(\Gamma^{\bar{\tau}}_{\bar{\gamma};\bar{\nu}}-\Gamma^{\bar{\tau}}_{\bar{\nu};\bar{\gamma}})\overline{v^\nu}]V^\beta\overline{H^\gamma}\Big\}.\end{aligned}\tag{2.6.21}$$

As a consequence, we have proved the following:

THEOREM 2.6.8: *Let $F\colon T^{1,0}M\to\mathbb{R}^+$ be a strongly convex weakly Kähler Finsler metric on a complex manifold M, and take $H\in\mathcal{H}$ and $V\in\mathcal{X}(\mathcal{V})$. Then:*
(i) *if V does not depend on the vector variables, i.e., if $V=\xi^V$ for some vector field $\xi\in\mathcal{X}(T^{1,0}M)$, then*

$$\langle\hat{\nabla}_{\hat{H}}V^o-(\nabla_{H^o}V)^o\mid\iota^o\rangle=\langle\hat{H}-H^o\mid V^o\rangle;$$

(ii) *if $V\in\mathcal{X}(\mathcal{V})$ is any vertical vector field then*

$$\langle\hat{\nabla}_{\hat{H}}V^o-(\nabla_{H^o}V)^o\mid\iota^o\rangle=(\hat{H}-H^o)(\langle V^o\mid\iota^o\rangle).$$

Proof: Recalling Theorem 2.6.4, and setting $W^o=\hat{H}-H^o\in\mathcal{V}_{\mathbb{R}}$, we see that (2.6.21) can be written as

$$\langle\hat{\nabla}_{\hat{H}}V^o-(\nabla_{H^o}V)^o\mid\iota^o\rangle=\operatorname{Re}[G_\alpha W^o(V^\alpha)]+\langle W^o\mid V^o\rangle,$$

and (i) is proved. In general, we have

$$G_\alpha W^o(V^\alpha)=W^o(G_\alpha V^\alpha)-W^o(G_\alpha)V^\alpha=W^o(\langle V,\iota\rangle)-(\langle V,W\rangle+\langle\!\langle V,W\rangle\!\rangle),$$

and (ii) follows from Proposition 2.6.1. □

So in general the Cartan and the Chern-Finsler connections actually differ, and the difference is not a tensor: it depends on first-order derivatives — unless F is Kähler, hypothesis that we are willing to accept, and $\tau(\chi,\overline{V})\equiv0$, hypothesis that we are not willing to accept, because as we shall see better in the next chapter we are able to study the geometry of Kähler Finsler manifold without assuming it. At this point, then, one might even wonder whether there exists another canonical connection on real Finsler manifold which agrees with the Chern-Finsler connection on Kähler Finsler manifolds. However, we stop our direct comparison of Cartan and Chern-Finsler connections here, devoting the last subsection to the curvatures.

2.6.4. The horizontal curvatures

We are left to compare the last piece in the second variation formulas: the horizontal flag curvature. We should compare $\langle\hat{\Omega}(\hat{\chi},\hat{H})\hat{H}|\hat{\chi}\rangle$ (where $\hat{\Omega}$ is of course the curvature operator of the Cartan connection) with

$$\langle\bigl(\Omega(H,\bar{\chi})\chi\bigr)^o \mid H^o\rangle - \langle\bigl(\Omega(\chi,\overline{H})\chi\bigr)^o \mid H^o\rangle,$$

where this expression is suggested by Theorem 2.4.4 together with Lemma 2.3.8 and (2.5.2).

Let us start computing the expressions involved. First of all, $\hat{\nabla}_{\hat{H}}\hat{\chi}\equiv 0$ for any $\hat{H}\in\mathcal{H}_{\mathbb{R}}$, because $\hat{\nabla}$ is a good connection; hence, recalling Propositions 1.4.5 and 1.3.2, we have

$$\langle\hat{\Omega}(\hat{\chi},\hat{H})\hat{H}\mid\hat{\chi}\rangle = -\langle\hat{\Omega}(\hat{\chi},\hat{H})\hat{\chi}\mid\hat{H}\rangle = \langle\hat{\nabla}_{[\hat{\chi},\hat{H}]}\hat{\chi}\mid\hat{H}\rangle.$$

Now $\hat{\nabla}_{[\hat{\chi},\hat{H}]}\iota^o$ is the vertical part of $[\hat{\chi},\hat{H}]$; therefore (1.1.20) yields

$$\langle\hat{\Omega}(\hat{\chi},\hat{H})\hat{H}\mid\hat{\chi}\rangle = \tfrac{1}{2}G_{ab}[\hat{\delta}_c(\hat{\Gamma}^a_d)-\hat{\delta}_e(\hat{\Gamma}^a_c)]K^c u^e K^b,$$

where $\hat{H}=K^b\hat{\delta}_b\in\mathcal{H}_{\mathbb{R}}$.

From now on we assume F Kähler, so that $\hat{H}=H^o+\frac{1}{2}\tau\bigl(\chi,\overline{\Theta^{-1}(H)}\bigr)^o$. In particular,

$$\hat{\Gamma}^a_e=\begin{cases}\operatorname{Re}(\Gamma^\alpha_{;\varepsilon}+\frac{1}{2}\Gamma^\alpha_{\bar{\varepsilon};\mu}v^\mu) & \text{if } 1\le a,e\le n;\\ \operatorname{Im}(\Gamma^\alpha_{;\varepsilon}+\frac{1}{2}\Gamma^\alpha_{\bar{\varepsilon};\mu}v^\mu) & \text{if } n+1\le a\le 2n \text{ and } 1\le e\le n;\\ -\operatorname{Im}(\Gamma^\alpha_{;\varepsilon}-\frac{1}{2}\Gamma^\alpha_{\bar{\varepsilon};\mu}v^\mu) & \text{if } 1\le a\le n \text{ and } n+1\le e\le 2n;\\ \operatorname{Re}(\Gamma^\alpha_{;\varepsilon}-\frac{1}{2}\Gamma^\alpha_{\bar{\varepsilon};\mu}v^\mu) & \text{if } n+1\le a,e\le 2n.\end{cases}$$

A long but trivial computation then yields

$$\begin{aligned}
&\tfrac{1}{2}G_{ab}\hat{\delta}_c(\hat{\Gamma}^a_e)u^eK^bK^c\\
&\quad=\operatorname{Re}\Bigl\{\bigl[G_{\alpha\beta}\delta_\gamma(\Gamma^\alpha_{;\varepsilon})-\tfrac{1}{2}G_{\alpha\beta}\Gamma^{\bar{\tau}}_{\gamma;\bar{\nu}}\overline{v^\nu}\Gamma^\alpha_{\bar{\tau};\varepsilon}+\tfrac{1}{2}G_{\bar{\alpha}\beta}\delta_\gamma(\Gamma^{\bar{\alpha}}_{\varepsilon;\bar{\nu}}\overline{v^\nu})\\
&\qquad\quad-\tfrac{1}{4}G_{\bar{\alpha}\beta}\Gamma^{\bar{\tau}}_{\gamma;\bar{\nu}}\overline{v^\nu}(\Gamma^{\bar{\alpha}}_{\varepsilon\bar{\tau};\bar{\nu}}\overline{v^\nu}+\Gamma^{\bar{\alpha}}_{\varepsilon;\bar{\tau}})\bigr]v^\varepsilon H^\beta H^\gamma\\
&\qquad+\bigl[G_{\alpha\beta}\delta_{\bar{\gamma}}(\Gamma^\alpha_{;\varepsilon})-\tfrac{1}{2}G_{\alpha\beta}\Gamma^\sigma_{\bar{\gamma};\mu}v^\mu\Gamma^\alpha_{\sigma;\varepsilon}+\tfrac{1}{2}G_{\bar{\alpha}\beta}\delta_{\bar{\gamma}}(\Gamma^{\bar{\alpha}}_{\varepsilon;\bar{\nu}}\overline{v^\nu})\\
&\qquad\quad-\tfrac{1}{4}G_{\bar{\alpha}\beta}\Gamma^\sigma_{\bar{\gamma};\mu}v^\mu\Gamma^{\bar{\alpha}}_{\varepsilon\sigma;\bar{\nu}}\overline{v^\nu}\bigr]v^\varepsilon H^\beta\overline{H^\gamma}\\
&\qquad+\bigl[G_{\bar{\alpha}\beta}\delta_\gamma(\Gamma^{\bar{\alpha}}_{;\bar{\varepsilon}})-\tfrac{1}{2}G_{\bar{\alpha}\beta}\Gamma^{\bar{\tau}}_{\gamma;\bar{\nu}}\overline{v^\nu}\Gamma^{\bar{\alpha}}_{\bar{\tau};\bar{\varepsilon}}+\tfrac{1}{2}G_{\alpha\beta}\delta_\gamma(\Gamma^\alpha_{\bar{\varepsilon};\mu}v^\mu)\\
&\qquad\quad-\tfrac{1}{4}G_{\alpha\beta}\Gamma^{\bar{\tau}}_{\gamma;\bar{\nu}}\overline{v^\nu}\Gamma^\alpha_{\bar{\varepsilon}\bar{\tau};\mu}v^\mu\bigr]\overline{v^\varepsilon}H^\beta H^\gamma\\
&\qquad+\bigl[G_{\bar{\alpha}\beta}\delta_{\bar{\gamma}}(\Gamma^{\bar{\alpha}}_{;\bar{\varepsilon}})-\tfrac{1}{2}G_{\bar{\alpha}\beta}\Gamma^\sigma_{\bar{\gamma};\mu}v^\mu\Gamma^{\bar{\alpha}}_{\sigma;\bar{\varepsilon}}+\tfrac{1}{2}G_{\alpha\beta}\delta_{\bar{\gamma}}(\Gamma^\alpha_{\bar{\varepsilon};\mu}v^\mu)\\
&\qquad\quad-\tfrac{1}{4}G_{\alpha\beta}\Gamma^\sigma_{\bar{\gamma};\mu}v^\mu(\Gamma^\alpha_{\bar{\varepsilon}\sigma;\mu}v^\mu+\Gamma^\alpha_{\bar{\varepsilon};\sigma})\bigr]\overline{v^\varepsilon}H^\beta\overline{H^\gamma}\Bigr\},
\end{aligned}\tag{2.6.22}$$

where $\Gamma^{\bar\alpha}_{\epsilon\bar\tau;\bar\nu}=\dot\partial_\epsilon\dot\partial_{\bar\tau}(\Gamma^{\bar\alpha}_{;\bar\nu})$ and so on. Analogously

$$\begin{aligned}
&\tfrac12 G_{ab}\hat\delta_e(\hat\Gamma^a_c)u^eK^bK^c\\
&=\mathrm{Re}\Big\{\big[G_{\alpha\beta}\delta_\epsilon(\Gamma^\alpha_{;\gamma})-\tfrac12 G_{\alpha\beta}\Gamma^{\bar\tau}_{\epsilon;\bar\nu}\overline{v^\nu}\Gamma^\alpha_{\bar\tau;\gamma}+\tfrac12 G_{\bar\alpha\beta}\delta_\epsilon(\Gamma^{\bar\alpha}_{\gamma;\bar\nu}\overline{v^\nu})\\
&\quad-\tfrac14 G_{\bar\alpha\beta}\Gamma^{\bar\tau}_{\epsilon;\bar\nu}\overline{v^\nu}(\Gamma^{\bar\alpha}_{\gamma\bar\tau;\bar\nu}\overline{v^\nu}+\Gamma^{\bar\alpha}_{\gamma;\bar\tau})\big]v^\epsilon H^\beta H^\gamma\\
&\quad+\big[G_{\bar\alpha\beta}\delta_\epsilon(\Gamma^{\bar\alpha}_{;\bar\gamma})-\tfrac12 G_{\bar\alpha\beta}\Gamma^{\bar\tau}_{\epsilon;\bar\nu}\overline{v^\nu}\Gamma^{\bar\alpha}_{\bar\tau;\bar\gamma}+\tfrac12 G_{\alpha\beta}\delta_\epsilon(\Gamma^\alpha_{\bar\gamma;\mu}v^\mu)\\
&\quad-\tfrac14 G_{\alpha\beta}\Gamma^{\bar\tau}_{\epsilon;\bar\nu}\overline{v^\nu}\Gamma^\alpha_{\bar\gamma\bar\tau;\mu}v^\mu\big]v^\epsilon H^\beta\overline{H^\gamma}\\
&\quad+\big[G_{\alpha\beta}\delta_{\bar\epsilon}(\Gamma^\alpha_{;\gamma})-\tfrac12 G_{\alpha\beta}\Gamma^\sigma_{\bar\epsilon;\mu}v^\mu\Gamma^\alpha_{\sigma;\gamma}+\tfrac12 G_{\bar\alpha\beta}\delta_{\bar\epsilon}(\Gamma^{\bar\alpha}_{\gamma;\bar\nu}\overline{v^\nu})\\
&\quad-\tfrac14 G_{\bar\alpha\beta}\Gamma^\sigma_{\bar\epsilon;\mu}v^\mu\Gamma^{\bar\alpha}_{\gamma\sigma;\bar\nu}\overline{v^\nu}\big]\overline{v^\epsilon}H^\beta H^\gamma\\
&\quad+\big[G_{\bar\alpha\beta}\delta_{\bar\epsilon}(\Gamma^{\bar\alpha}_{;\bar\gamma})-\tfrac12 G_{\bar\alpha\beta}\Gamma^\sigma_{\bar\epsilon;\mu}v^\mu\Gamma^{\bar\alpha}_{\sigma;\bar\gamma}+\tfrac12 G_{\alpha\beta}\delta_{\bar\epsilon}(\Gamma^\alpha_{\bar\gamma;\mu}v^\mu)\\
&\quad-\tfrac14 G_{\alpha\beta}\Gamma^\sigma_{\bar\epsilon;\mu}v^\mu(\Gamma^\alpha_{\bar\gamma\sigma;\mu}v^\mu+\Gamma^\alpha_{\bar\gamma;\sigma})\big]\overline{v^\epsilon}H^\beta\overline{H^\gamma}\Big\}.
\end{aligned}\tag{2.6.23}$$

On the other hand,

$$\begin{aligned}
\langle(\Omega(H,\bar\chi)\chi)^o\mid H^o\rangle&=\langle(\Omega(\chi,\overline{H})\chi)^o\mid H^o\rangle\\
&=\mathrm{Re}\Big\{[G_{\alpha\beta}\delta_{\bar\gamma}(\Gamma^\alpha_{;\epsilon})-G_{\bar\alpha\beta}\delta_\epsilon(\Gamma^{\bar\alpha}_{;\bar\gamma})]v^\epsilon H^\beta\overline{H^\gamma}\\
&\qquad+[G_{\bar\alpha\beta}\delta_\gamma(\Gamma^{\bar\alpha}_{;\bar\epsilon})-G_{\alpha\beta}\delta_{\bar\epsilon}(\Gamma^\alpha_{;\gamma})]\overline{v^\epsilon}H^\beta H^\gamma\Big\}.
\end{aligned}\tag{2.6.24}$$

We must subtract (2.6.23) and (2.6.24) from (2.6.22). Using (2.3.16), $v^\epsilon\Gamma^{\bar\alpha}_{\epsilon;\bar\tau}=0$, (2.6.18), $v^\epsilon\Gamma^{\bar\alpha}_{\epsilon\bar\tau;\bar\nu}=0$ and $v^\epsilon\Gamma^{\bar\alpha}_{\epsilon\sigma;\bar\nu}=-\Gamma^{\bar\alpha}_{\sigma;\bar\nu}$, we obtain

$$\begin{aligned}
\mathrm{Re}\Big\{&\big[\tfrac12 G_{\bar\alpha\beta}\delta_\gamma(\Gamma^{\bar\alpha}_{\epsilon;\bar\nu}\overline{v^\nu})-\tfrac12\delta_\epsilon(\Gamma^\alpha_{\gamma;\bar\nu}\overline{v^\nu})\big]v^\epsilon H^\beta H^\gamma\\
&+\big[\tfrac14 G_{\bar\alpha\beta}\Gamma^\sigma_{\bar\gamma;\mu}v^\mu\Gamma^{\bar\alpha}_{\sigma;\bar\nu}\overline{v^\nu}-\tfrac12 G_{\alpha\beta}\Gamma^\sigma_{\bar\gamma;\mu}v^\mu\Gamma^\alpha_{;\sigma}-\tfrac12 G_{\alpha\beta}\chi(\Gamma^\alpha_{\bar\gamma;\mu}v^\mu)\big]H^\beta\overline{H^\gamma}\\
&+\big[-\tfrac12 G_{\bar\alpha\beta}\Gamma^{\bar\tau}_{\gamma;\bar\nu}\overline{v^\nu}\Gamma^{\bar\alpha}_{;\bar\tau}-\tfrac12 G_{\bar\alpha}\bar\chi(\Gamma^{\bar\alpha}_{\gamma;\bar\nu}\overline{v^\nu})\big]H^\beta H^\gamma\\
&+\big[\tfrac12 G_{\alpha\beta}\delta_{\bar\gamma}(\Gamma^\alpha_{\bar\epsilon;\mu}v^\mu)-\tfrac12 G_{\bar\alpha}\Gamma^\sigma_{\bar\gamma;mu}v^\mu\Gamma^{\bar\alpha}_{\sigma;\bar\epsilon}-\tfrac12 G_{\alpha\beta}\delta_{\bar\epsilon}(\Gamma^\alpha_{\bar\gamma;\mu}v^\mu)\big]\overline{v^\epsilon}H^\beta\overline{H^\gamma}\Big\}.
\end{aligned}\tag{2.6.25}$$

The Bianchi identities say that $D''\tau=0$. Now,

$$\begin{aligned}
D''\tau=&[\Gamma^\alpha_{\bar\gamma;\mu}\Gamma^{\bar\gamma}_{\bar\beta;\bar\nu}-\delta_{\bar\nu}(\Gamma^\alpha_{\bar\beta;\mu})]d\bar z^\nu\wedge dz^\mu\wedge\overline{\psi^\beta}\otimes\dot\partial_\alpha\\
&-\Gamma^\alpha_{\bar\gamma\bar\beta;\mu}\overline{\psi^\gamma}\wedge dz^\mu\wedge\overline{\psi^\beta}\otimes\dot\partial_\alpha+\Gamma^\alpha_{\bar\beta;\mu}\delta_{\bar\nu}(\Gamma^{\bar\beta}_{;\bar\gamma})dz^\mu\wedge d\bar z^\nu\wedge d\bar z^\gamma\otimes\dot\partial_\alpha;
\end{aligned}$$

In particular,

$$\delta_\gamma(\Gamma^{\bar\alpha}_{\epsilon;\bar\nu})=\Gamma^{\bar\alpha}_{\sigma;\bar\nu}\Gamma^\sigma_{\epsilon;\gamma}.\tag{2.6.26}$$

Since F is Kähler, (2.6.26) yields

$$\delta_\gamma(\Gamma^{\bar\alpha}_{\epsilon;\bar\nu}\overline{v^\nu})v^\epsilon=\Gamma^{\bar\alpha}_{\sigma;\bar\nu}\overline{v^\nu}\Gamma^\sigma_{\epsilon;\gamma}v^\epsilon=\Gamma^{\bar\alpha}_{\sigma;\bar\nu}\overline{v^\nu}\Gamma^\sigma_{\gamma;\epsilon}v^\epsilon=\delta_\epsilon(\Gamma^{\bar\alpha}_{\gamma;\bar\nu}\overline{v^\nu})v^\epsilon,$$

and the first term in (2.6.25) vanishes.

Furthermore, since by (2.6.20) $G_{\alpha\beta}\Gamma^{\alpha}_{\bar\tau;\mu}v^{\mu}=0$, (2.6.26) and (2.6.18) imply

$$\begin{aligned}0=G_{\alpha\beta}\Gamma^{\alpha}_{\bar\tau;\mu}\Gamma^{\bar\tau}_{\bar\varepsilon;\bar\gamma}v^{\mu}\overline{v^{\varepsilon}}&=G_{\alpha\beta}\delta_{\bar\gamma}(\Gamma^{\alpha}_{\bar\varepsilon;\mu}v^{\mu})\overline{v^{\varepsilon}}=G_{\alpha\beta}\delta_{\bar\varepsilon}(\Gamma^{\alpha}_{\bar\gamma;\mu}v^{\mu})\overline{v^{\varepsilon}},\\ G_{\bar\alpha\beta}\Gamma^{\bar\alpha}_{\sigma;\bar\varepsilon}\overline{v^{\varepsilon}}\Gamma^{\sigma}_{\bar\gamma;\mu}v^{\mu}=\bar\chi(G_{\beta\sigma})\Gamma^{\sigma}_{\bar\gamma;\mu}v^{\mu}&=-G_{\beta\sigma}\delta_{\bar\varepsilon}(\Gamma^{\sigma}_{\bar\gamma;\mu}v^{\mu})\overline{v^{\varepsilon}}=0,\end{aligned}$$

and so the last term and the first addend in the second term of (2.6.25) vanish.

Now set $\varphi(\overline{W})=\tau^{\mathcal{H}}(\chi,\overline{W})$, so that

$$\varphi=-\Gamma^{\alpha}_{\bar\beta;\mu}v^{\mu}\overline{\psi^{\beta}}\otimes\delta_{\alpha}.$$

Then

$$\begin{aligned}\nabla_{\chi}\varphi&=-\chi(\Gamma^{\alpha}_{\bar\gamma;\mu})\overline{\psi^{\gamma}}\otimes\delta_{\alpha}-\Gamma^{\alpha}_{\bar\gamma;\mu}\overline{\psi^{\gamma}}\otimes\nabla_{\chi}\delta_{\alpha}\\ &=-\big[\chi(\Gamma^{\alpha}_{\bar\gamma;\mu}v^{\mu})+\Gamma^{\sigma}_{\bar\gamma;\mu}v^{\mu}\Gamma^{\alpha}_{;\sigma}\big]\overline{\psi^{\gamma}}\otimes\delta_{\alpha}.\end{aligned}$$

The covariant derivative commutes with contractions; so, since $\nabla_{\chi}\chi=0$, we have

$$(\nabla_{\chi}\tau^{\mathcal{H}})(\chi,\overline{W})=(\nabla_{\chi}\varphi)(\overline{W})=-\big[\chi(\Gamma^{\alpha}_{\bar\gamma;\mu}v^{\mu})+\Gamma^{\sigma}_{\bar\gamma;\mu}v^{\mu}\Gamma^{\alpha}_{;\sigma}\big]\overline{W^{\gamma}}\delta_{\alpha}.$$

In particular, the second term in (2.6.25) is given by

$$\frac{1}{2}\langle\!\langle H,(\nabla_{\chi}\tau^{\mathcal{H}})(\chi,\overline{\Theta^{-1}(H)})\rangle\!\rangle,$$

while the third term in (2.6.25) is given by

$$\frac{1}{2}\langle H,(\nabla_{\chi}\tau^{\mathcal{H}})(\chi,\overline{\Theta^{-1}(H)})\rangle.$$

In conclusion, we have proved the following

Theorem 2.6.9: *Let $F\colon T^{1,0}M\to\mathbb{R}^{+}$ be a strongly convex Kähler Finsler metric on a complex manifold M. Then for any $H\in\mathcal{H}$ one has*

$$\begin{aligned}&\langle\hat\Omega(\hat\chi,\hat H)\hat H\mid\hat\chi\rangle\\ &=\langle(\Omega(H,\bar\chi)\chi)^{o}\mid H^{o}\rangle-\langle(\Omega(\chi,\overline{H})\chi)^{o}\mid H^{o}\rangle+\frac{1}{2}\langle((\nabla_{\chi}\tau^{\mathcal{H}})(\chi,\overline{\Theta^{-1}(H)}))^{o}\mid H^{o}\rangle.\end{aligned}$$

So we have the following situation: the integrands in the second variation formulas Theorem 1.5.4 and Theorem 2.4.4 are different (unless $\tau(\chi,\bar\cdot)\equiv0$, of course) but the integrals themselves agree (at least for fixed variations). So if we are able (and in the next chapter we shall) to control the integrand in the complex form of the second variation formula then we are able to control the real Morse index form, even though we are still not able to control the real horizontal flag curvature. This is the reason that suggested the unusual formulation of the Cartan-Hadamard Theorem 1.7.17; and maybe it also suggests that there could be a better connection than the Cartan connnection on real Finsler manifolds: it seems that the Cartan connection contains pieces useless from a geometric point of view (i.e., disappearing in the computation of the first and second variation formulas).

CHAPTER 3

Manifolds with constant holomorphic curvature

3.0. Introduction

In order to study intrinsic metrics and distances on complex manifolds Vesentini [V] introduced the notion of complex geodesic for the Kobayashi and Carathéodory metrics and distances. Roughly speaking a complex geodesic is a holomorphic embedding of the unit disk equipped with the hyperbolic metric which is an isometry with respect to the intrinsic metric or distance (or both) defined on the manifold under consideration.

The problem of existence of complex geodesics is satisfactorily solved only for convex domains by the work of Lempert [L], who moreover proved uniqueness for strongly convex domains (see also [A, chapter 2.6]). In this case the Kobayashi and Caratéodory metrics coincide and are smooth strongly pseudoconvex complex Finsler metrics.

In order to find a different approach to this problem, which may eventually lead to an understanding of it on a larger class of complex manifolds, it is very natural to ask whether it is possible from a differential geometric point of view to find optimal conditions on an abstract complex Finsler metric implying the existence and uniqueness of complex geodesics. In this general framework we shall investigate the problem of finding isometric holomorphic embeddings into a complex Finsler manifold of the unit disk Δ with Poincaré metric, of $\mathbb{C}$ with the euclidean metric and of $\mathbb{P}_1$ with the Fubini-Study metric. Not surprisingly the machinery devised in the previous chapters will be used extensively and the necessary and sufficient conditions for the existence of the isometric embeddings will involve Kählerianity, symmetries of the curvature and constancy of the holomorphic curvature.

As a byproduct of our work we shall indicate how these result may contribute to the understanding of the function theory and the classification of complex Finsler metrics of constant holomorphic curvature. In particular when the curvature is negative constant we shall show that the metric must be the Kobayashi metric of the manifold and that the geometric picture is quite alike to the one of strongly convex domains. When the curvature vanishes we shall show how this may be used to characterize $\mathbb{C}^n$. In both cases it is very interesting to discover that it appears very naturally a relation with the homogeneous complex Monge-Ampère equation. All these applications and remarks suggest the possibility of further investigations.

The content of this chapter is the following. In the first section we introduce the notion of geodesic complex curves (roughly speaking, totally geodesic holomorphic disks) and we describe conditions ensuring their existence and uniqueness: it turns out that one needs constant holomorphic curvature, the weak Kähler condition and a symmetry property of the curvature operator. In the second section we focus on Kähler Finsler manifolds with nonpositive constant holomorphic curvature, and we prove that they are foliated by geodesic complex curves, reproducing in this setting the situation described by Lempert for strongly convex domains. The proof of this theorem requires more or less all the material discussed in the previous chapters.

3.1. Geodesic complex curves

3.1.1. Definitions and properties

We begin by giving a precise notion of geodesic complex curves of a complex Finsler manifold. Let us fix some notations. First of all, set

$$\Delta_r = \{z \in \mathbb{C} \mid |z| < r\};$$

in particular, $\Delta_1 = \Delta$ is the unit disk in $\mathbb{C}$.

DEFINITION 3.1.1: For $c \in \mathbb{R}$ let μ_c denote the Hermitian metric

$$\mu_c = \frac{1}{(1 + (c/2)|\zeta|^2)^2}\, d\zeta \otimes d\bar{\zeta},$$

which is defined over the disk Δ_r with $r = \sqrt{2/|c|}$ if $c < 0$, on the complex plane $\mathbb{C}$ if $c = 0$ and on the projective line $\mathbb{P}_1 = \mathbb{C} \cup \{\infty\}$ if $c > 0$. In particular, for $c = -2$ we have the *Poincaré metric* μ_{-2} on Δ

$$\mu_{-2} = \frac{1}{(1 - |\zeta|^2)^2}\, d\zeta \otimes d\bar{\zeta}; \tag{3.1.1}$$

for $c = 0$ we have the *euclidean metric* μ_0 on $\mathbb{C}$

$$\mu_0 = d\zeta \otimes d\bar{\zeta}, \tag{3.1.2}$$

and for $c = 2$ we have the *Fubini-Study metric* μ_2 on $\mathbb{P}_1$

$$\mu_2 = \frac{1}{(1 + |\zeta|^2)^2}\, d\zeta \otimes d\bar{\zeta}. \tag{3.1.3}$$

We recall a couple of facts concerning these metrics. First of all, they have constant Gaussian curvature $K(\mu_c) \equiv 2c$. Furthermore, if we introduce a complex-valued function A_c defined where μ_c is defined and given by

$$A_c(\zeta) = \frac{c\bar{\zeta}}{1 + (c/2)|\zeta|^2}, \tag{3.1.4}$$

then the geodesic equation for μ_c is

$$\ddot{\sigma} = A_c(\sigma)(\dot{\sigma})^2. \tag{3.1.5}$$

Now let M be a complex manifold of dimension n and let $F: T^{1,0}M \to \mathbb{R}^+$ be a strongly pseudoconvex Finsler metric on M.

DEFINITION 3.1.2: A holomorphic map $\varphi: \Delta_r \to M$ is called a *segment of infinitesimal c-geodesic complex curve* if φ is an isometry from Δ_r endowed with μ_c to M endowed with the strongly pseudoconvex Finsler metric F. We say that φ is a *segment of c-geodesic complex curve* iff φ maps geodesics parametrized by arc length for the metric μ_c restricted to Δ_r into geodesics parametrized by arc length for F on M. If $c < 0$ and $r = \sqrt{2/|c|}$ (respectively, $c = 0$ and $r = +\infty$, or $c > 0$ and φ is defined on $\mathbb{P}_1$) we call φ an *(infinitesimal) c-geodesic complex curve*. If $c = -2$ (respectively, $c = 0$ or $c = 2$) we shall sometimes say *hyperbolic* (respectively, *parabolic* or *elliptic*) *geodesic* instead of c-geodesic. Finally we shall say that φ is a (segment) of (infinitesimal) c-geodesic complex curve *through* $(p; v) \in T^{1,0}M$ iff $\varphi(0) = p$ and $\varphi'(0) = v$.

Our definition of geodesic complex curve is of course related to Vesentini's notion of complex geodesic for intrinsic metrics and in fact it is a bit more flexible. For instance there are no complex geodesics in the sense of Vesentini for an annulus $A \subset \mathbb{C}$ endowed with the Kobayashi metric (i.e., holomorphic maps $\varphi: \Delta \to A$ such that for all $z, w \in \Delta$ the Kobayashi distance between $\varphi(z)$ and $\varphi(w)$ equals the Poincaré distance between z and w) as such a φ would necessarily be biholomorphic. On the other hand there are plenty of hyperbolic geodesic complex curves in the sense explained above.

Furthermore the definitions given above have the advantage that they implicitly contain a recipe to find differential equations for geodesic complex curves. As a first step towards the solution of the problem of existence of such curves we recover the system of PDE's which they must satisfy.

Let M be a complex manifold with a strongly pseudoconvex Finsler metric defined on it. Then, if, as usual the prime stands for $\partial/\partial\zeta$ and A_c are the functions defined in (3.1.4), we have (as first proved, in a different way, in [AP1]):

PROPOSITION 3.1.1: *Let $F: T^{1,0}M \to \mathbb{R}^+$ be a strongly pseudoconvex Finsler metric on a complex manifold M, and $\varphi: \Delta_r \to M$ a holomorphic map. Then φ is a segment of c-geodesic complex curve, for $c \in \mathbb{R}$ (and $r < \sqrt{2/|c|}$ when $c < 0$), iff*

$$\begin{cases} D_c^\alpha(\varphi) = (\varphi'')^\alpha + A_c(\varphi')^\alpha + \Gamma^\alpha_{;\mu}(\varphi')(\varphi')^\mu = 0 \\ N_{c,\alpha}(\varphi) = \left[G_{\mu;\alpha}(\varphi;\varphi') - G_{\alpha;\mu}(\varphi') + G_{\alpha\beta}(\varphi')\Gamma^\beta_{;\mu}(\varphi')\right](\varphi')^\mu = 0, \end{cases} \tag{3.1.6}$$

for all $\alpha = 1, \ldots, n$, that is iff

$$\begin{cases} \nabla_{(\varphi')^H}(\varphi')^H = -A_c\chi(\varphi'), \\ \langle\theta(K, (\varphi')^H), (\varphi')^H\rangle_{\varphi'} \equiv 0, \quad \text{for all } K \in \mathcal{H}_{\varphi'}. \end{cases} \tag{3.1.7}$$

Proof: For any $\zeta \in \Delta_r$ and $\xi \in \mathbb{C} \equiv T^{1,0}_\zeta\Delta_r$ with $\mu_c(\zeta;\xi) = 1$, let $\sigma_{\zeta,\xi}$ be the unique geodesic for μ_c, parametrized by arc length, starting from ζ tangent to ξ. Then φ is a segment of c-geodesic complex curve iff for all such $\sigma_{\zeta,\xi}$ the curve $\gamma_{\zeta,\xi} = \varphi \circ \sigma_{\zeta,\xi}$ satisfies the geodesic equation (2.4.11) for F, that is iff in local coordinates

$$\begin{aligned} &\ddot\gamma^\alpha_{\zeta,\xi} + \Gamma^\alpha_{;\mu}(\dot\gamma_{\zeta,\xi})\dot\gamma^\mu_{\zeta,\xi} \\ &\qquad = G^{\bar\nu\alpha}(\dot\gamma_{\zeta,\xi})\left[G_{\bar\mu;\bar\nu}(\dot\gamma_{\zeta,\xi}) - G_{\bar\nu;\bar\mu}(\dot\gamma_{\zeta,\xi}) + G_{\bar\nu\bar\beta}(\dot\gamma_{\zeta,\xi})\overline{\Gamma^\beta_{;\mu}(\dot\gamma_{\zeta,\xi})}\right]\overline{\dot\gamma^\mu_{\zeta,\xi}}, \end{aligned} \tag{3.1.8}$$

for $\alpha = 1, \ldots, n$.

Now, being φ holomorphic, one has

$$\begin{aligned} \dot\gamma_{\zeta,\xi} &= (\varphi' \circ \sigma_{\zeta,\xi})\dot\sigma_{\zeta,\xi}, \\ \ddot\gamma_{\zeta,\xi} &= [(\varphi'' + A_c\varphi') \circ \sigma_{\zeta,\xi}]\dot\sigma^2_{\zeta,\xi}, \end{aligned} \tag{3.1.9}$$

where we used (3.1.5). In particular,

$$\dot\gamma_{\zeta,\xi}(0) = \varphi'(\zeta)\xi$$

and

$$\ddot\gamma_{\zeta,\xi}(0) = [\varphi''(\zeta) + A_c(\zeta)\varphi'(\zeta)]\xi^2.$$

Assume now that φ is a segment of c-geodesic complex curve . Then evaluating (3.1.8) in $t = 0$ we get

$$\begin{aligned} &\xi^2(\varphi'' + A_c\varphi')^\alpha + \Gamma^\alpha_{;\mu}(\xi\varphi')\xi(\varphi')^\mu \\ &\qquad = G^{\bar\nu\alpha}(\xi\varphi')\left[G_{\bar\mu;\bar\nu}(\xi\varphi') - G_{\bar\nu;\bar\mu}(\xi\varphi') + G_{\bar\nu\bar\beta}(\xi\varphi')\overline{\Gamma^\beta_{;\mu}(\xi\varphi')}\right]\overline{\xi(\varphi')^\mu}, \end{aligned}$$

for $\alpha = 1, \ldots, n$. Using the homogeneity of G (and of its derivatives) we get

$$\begin{aligned} &\xi^2\left[(\varphi'' + A_c\varphi')^\alpha + \Gamma^\alpha_{;\mu}(\varphi')(\varphi')^\mu\right] \\ &\qquad = |\xi|^2 G^{\bar\nu\alpha}(\varphi')\left[G_{\bar\mu;\bar\nu}(\varphi') - G_{\bar\nu;\bar\mu}(\varphi') + G_{\bar\nu\bar\beta}(\varphi')\overline{\Gamma^\beta_{;\mu}(\varphi')}\right]\overline{(\varphi')^\mu}, \end{aligned}$$

for all $\alpha = 1, \ldots, n$, and, being $\arg\xi$ arbitrary and $G^{\bar\nu\alpha}$ invertible, it follows that φ satisfies the required system.

Conversely, assume that φ satisfies the system (3.1.6) of PDE's. Fix $\zeta \in \Delta_r$ and $\xi \in \mathbb{C}$ with $\mu_c(\zeta;\xi) = 1$. Evaluating the system in $\sigma_{\zeta,\xi}(t)$, and then adding for all α the first equation multiplied by $\left(\dot\sigma_{\zeta,\xi}(t)\right)^2$ to the conjugated of the second one multiplied by $G^{\bar\nu\alpha}$ and $|\dot\sigma_{\zeta,\xi}(t)|^2$, using the homogeneity of G and recalling (3.1.9) we get exactly (3.1.8).

Finally, it is clear that (3.1.6) is the expression in local coordinates of (3.1.7); cf. (2.3.22). □

In particular, then, if φ is a segment of c-geodesic complex curve, the metric F is weakly Kähler in the directions tangent to φ or, as we shall say, that *the metric F is weakly Kähler along φ*. In other words we have

COROLLARY 3.1.2: *Let $F\colon T^{1,0}M \to \mathbb{R}^+$ be a strongly pseudoconvex Finsler metric on a complex manifold M, and fix $c \in \mathbb{R}$. Then a holomorphic solution $\varphi\colon \Delta_r \to M$ (with $r < \sqrt{2/|c|}$ if $c < 0$) of*

$$\nabla_{(\varphi')^H}(\varphi')^H = -A_c\chi(\varphi')$$

is a segment of c-geodesic complex curve iff F is weakly Kähler along φ. In particular if F is weakly Kähler, then a holomorphic map $\varphi\colon \Delta_r \to M$ is a segment of c-geodesic complex curve iff $D_c^\alpha(\varphi) = 0$ holds for $\alpha = 1, \ldots, n$.

At this point it is clear that it is crucial to study the system $D_c^\alpha(\varphi) = 0$ looking for conditions for its resolubility by holomorphic functions. Recalling Corollary 2.5.4 it is clear that we should look for conditions concerning the holomorphic curvature of F.

DEFINITION 3.1.3: We shall say that a holomorphic map $\varphi\colon \Delta_r \to M$ *realizes the holomorphic curvature at* 0 if

$$K(\varphi^*G)(0) = K_F(\varphi'(0)).$$

More generally, if $c \in \mathbb{R}$ (and $r < \sqrt{2/|c|}$ when $c < 0$) we shall say that φ *realizes the holomorphic curvature at $\zeta_0 \in \Delta_r$ with respect to μ_c* if $\varphi \circ \gamma_{\zeta_0}^c$ realizes it at 0, where

$$\gamma_{\zeta_0}^c(\zeta) = \frac{\zeta + \zeta_0}{1 - (c/2)\overline{\zeta_0}\zeta}$$

so that $\gamma_{\zeta_0}^c$ is an isometry with respect to μ_c with $\gamma_{\zeta_0}^c(0) = \zeta_0$ and $(\gamma_{\zeta_0}^c)'(0)$ is a positive real number.

Then we have:

PROPOSITION 3.1.3: *Let $F\colon T^{1,0}M \to \mathbb{R}^+$ be a strongly pseudoconvex Finsler metric on a complex manifold M, and take $c \in \mathbb{R}$. Let $\varphi\colon \Delta_r \to M$ be a holomorphic solution of $D_c^\alpha(\varphi) = 0$ with $\varphi'(0) \neq 0$ (and $r < \sqrt{2/|c|}$ if $c < 0$). Then*

(i) *φ realizes the holomorphic curvature at 0;*
(ii) *φ realizes the holomorphic curvature with respect to μ_c at every point of Δ_r;*

(iii) *if* $F(\varphi'(0)) = 1$, *then* φ *is a segment of infinitesimal c-geodesic complex curve for* F.

Proof: (i) As we have seen in Corollary 2.5.4, a holomorphic map $\varphi\colon \Delta_r \to M$ realizes the holomorphic curvature at 0 iff

$$(\varphi'')^\alpha(0) + \Gamma^\alpha_{;\mu}(\varphi'(0))(\varphi')^\mu(0) = a(\varphi')^\alpha(0),$$

for some $a \in \mathbb{C}$ and all $\alpha = 1, \ldots, n$. But this is exactly$D^\alpha_c(\varphi) = 0$ evaluated at $\zeta = 0$ (and with $a = 0$); hence (i) follows at once.

(ii) Take $\zeta_0 \in \Delta_r$, and set $\psi = \varphi \circ \gamma^c_{\zeta_0}$. Then

$$\begin{aligned} \psi(0) &= \varphi(\zeta_0); \\ \psi'(0) &= (1 + (c/2)|\zeta_0|^2)\varphi'(\zeta_0); \\ \psi''(0) &= (1 + (c/2)|\zeta_0|^2)^2 (\varphi''(\zeta_0) + A_c(\zeta_0)\varphi'(\zeta_0)). \end{aligned}$$

Therefore φ realizes the holomorphic curvature at ζ_0 with respect to μ_c iff

$$\left(\varphi''(\zeta_0) + A_c(\zeta_0)\varphi'(\zeta_0)\right)^\alpha + \Gamma^\alpha_{;\mu}(\varphi'(\zeta_0))\left(\varphi'(\zeta_0)\right)^\mu = \tilde{a}\left(\varphi'(\zeta_0)\right)^\alpha,$$

and hence we have (ii).

(iii) If $D^\alpha_c(\varphi) = 0$ holds then, recalling the well known homogeneity properties of G, we get

$$\begin{aligned} G_\alpha(\varphi')(\varphi'')^\alpha + A_c G(\varphi') &= G_\alpha(\varphi')\left[(\varphi'')^\alpha + A_c(\varphi')^\alpha\right] \\ &= -G_\alpha(\varphi')\Gamma^\alpha_{;\mu}(\varphi')(\varphi')^\mu = -G_{;\mu}(\varphi')(\varphi')^\mu. \end{aligned}$$

Therefore

$$\frac{\partial}{\partial\zeta}[G(\varphi')] = G_{;\mu}(\varphi')(\varphi')^\mu + G_\alpha(\varphi')(\varphi'')^\alpha = -A_c G(\varphi').$$

Now, along the curve $t \mapsto e^{i\theta}t$ we have

$$\frac{\partial}{\partial\zeta} = \frac{1}{2}e^{-i\theta}\frac{d}{dt};$$

Thus the function $t \mapsto G(\varphi'(e^{i\theta}t))$ is a solution of the Cauchy problem

$$\begin{cases} f'(t) = \dfrac{-2ct}{1 + (c/2)t^2} f(t), \\ f(0) = 1. \end{cases}$$

But $f(t) = \left(1 + (c/2)t^2\right)^{-2}$ is a solution of the same problem; therefore

$$G(\varphi(\zeta); d\varphi_\zeta(d/dt)) = G(\varphi(\zeta); \varphi'(\zeta)) \equiv \frac{1}{\left(1 + (c/2)|\zeta|^2\right)^2} = \mu_c(\zeta; d/dt),$$

and we are done. □

3.1.2. The existence theorem

In complete analogy with the case of usual real geodesics one likes to solve the Cauchy problem for the operators D_c. Here of course there are difficulties, as we have at hands a system of PDE's and we look for holomorphic solutions. In particular one expects compatibility conditions. In fact we shall see shortly that these are also sufficient. Let

$$S^{1,0}M = \{v \in T^{1,0}M \mid F(v) = 1\}$$

denote the unit sphere bundle. Then we can summarize our result as follows:

THEOREM 3.1.4: *Let $F\colon T^{1,0}M \to \mathbb{R}^+$ be a strongly pseudoconvex Finsler metric on a complex manifold M, and take $c \in \mathbb{R}$. Then the Cauchy problem*

$$\begin{cases} D_c(\varphi) = 0, \\ \varphi(0) = p,\ \varphi'(0) = v_0, \end{cases} \tag{3.1.10}$$

has a holomorphic solution for all $(p; v_0) \in S^{1,0}M$ iff the following condition holds:

$$\tau(\chi, \bar{\chi}) = c\, G\iota. \tag{3.1.11}$$

Furthermore, when a holomorphic solution of (3.1.10) exists, it is unique.

Proof: To show the necessity of (3.1.11) is an easy matter. Assume $\varphi\colon \Delta_r \to M$ is such a solution, and apply $\partial/\partial\bar{\zeta}$ to $D_c(\varphi) = 0$. We obtain

$$\begin{aligned} 0 &= \frac{c}{\big(1 + (c/2)|\zeta|^2\big)^2}(\varphi')^\alpha (\Gamma^\alpha_{;\mu})_{;\bar{\nu}}(\varphi')(\varphi')^\mu \overline{(\varphi')^\nu} + \Gamma^\alpha_{\bar{\beta};\mu}(\varphi')(\varphi')^\mu \overline{(\varphi'')^\beta} \\ &= \frac{c}{\big(1 + (c/2)|\zeta|^2\big)^2}(\varphi')^\alpha + \delta_{\bar{\nu}}(\Gamma^\alpha_{;\mu})(\varphi')(\varphi')^\mu \overline{(\varphi')^\nu} \\ &\quad + \Gamma^\alpha_{\bar{\beta};\mu}(\varphi')(\varphi')^\mu \left[\overline{(\varphi'')^\beta} + \Gamma^{\bar{\beta}}_{;\bar{\nu}}(\varphi')\overline{(\varphi')^\nu}\right] \\ &= \frac{c}{\big(1 + (c/2)|\zeta|^2\big)^2}(\varphi')^\alpha + \delta_{\bar{\nu}}(\Gamma^\alpha_{;\mu})(\varphi')(\varphi')^\mu \overline{(\varphi')^\nu} - \overline{A_c}\Gamma^\alpha_{\bar{\beta};\mu}(\varphi')(\varphi')^\mu \overline{(\varphi')^\beta} \\ &= \frac{c}{\big(1 + (c/2)|\zeta|^2\big)^2}(\varphi')^\alpha + \delta_{\bar{\nu}}(\Gamma^\alpha_{;\mu})(\varphi')(\varphi')^\mu \overline{(\varphi')^\nu}, \end{aligned} \tag{3.1.12}$$

where we used $D_c(\varphi) = 0$ and

$$\Gamma^\alpha_{\bar{\beta};\mu}\overline{v^\beta} = G^{\alpha\bar{\nu}}_{\bar{\beta}} G_{\bar{\nu};\mu}\overline{v^\beta} + G^{\alpha\bar{\nu}} G_{\bar{\beta}\bar{\nu};\mu}\overline{v^\beta} = 0, \tag{3.1.13}$$

by (2.3.10) and (2.3.3).

Now, since $\varphi'(0) \in S^{1,0}M$, Proposition 3.1.3.(iii) yields

$$G(\varphi'(\zeta)) = \frac{1}{\big(1 + (c/2)|\zeta|^2\big)^2};$$

therefore — setting $v = \varphi'(\zeta)$ — (3.1.12) becomes

$$c\, G(v) v^\alpha = -\delta_{\bar{\nu}}(\Gamma^\alpha_{;\mu}) v^\mu \overline{v^\nu},$$

that is, by (2.3.19), (3.1.11) holds.

So (3.1.11) is a necessary condition for $D_c(\varphi)=0$ to have a holomorphic solution. Now our task will be to show that it is sufficient too; so let us assume that (3.1.11) holds.

For any $e^{i\theta}\in S^1$, let us consider the Cauchy problem

$$\begin{cases} \ddot g^\alpha(t) = -A_c(t)\dot g^\alpha(t) - \Gamma^\alpha_{;\mu}\big(\dot g(t)\big)\dot g^\mu(t), & \text{for } \alpha=1,\dots,n,\\ g(0)=p, \quad \dot g(0)=e^{i\theta}v_0. \end{cases} \tag{3.1.14}$$

The ODE theory gives an $\varepsilon>0$ and uniquely determined maps $g_{e^{i\theta}}\colon(-\varepsilon,\varepsilon)\to M$ solving (3.1.14). Define $\varphi\colon\Delta_\varepsilon\to M$ by

$$\varphi(\zeta)=g_{\zeta/|\zeta|}(|\zeta|). \tag{3.1.15}$$

Writing $\zeta=te^{i\theta}$ we have

$$\frac{\partial}{\partial\zeta}=-\frac{ie^{-i\theta}}{2t}\left[\frac{\partial}{\partial\theta}+it\frac{\partial}{\partial t}\right] \quad\text{and}\quad \frac{\partial}{\partial\bar\zeta}=\frac{ie^{i\theta}}{2t}\left[\frac{\partial}{\partial\theta}-it\frac{\partial}{\partial t}\right]; \tag{3.1.16}$$

so if φ were holomorphic, it would follow that

$$\varphi'(\zeta)=e^{-i\theta}\dot g_{e^{i\theta}}(t), \qquad \varphi''(\zeta)=e^{-2i\theta}\ddot g_{e^{i\theta}}(t),$$

and thus φ would be a holomorphic solution of (3.1.10). Therefore, we must prove that, assuming (3.1.11), the map φ defined by (3.1.15) is holomorphic. This will complete the proof of the theorem since the uniqueness of holomorphic solutions is immediate. In fact, since a holomorphic map on a disk is uniquely determined by its restriction to the real axis, the uniqueness statement for (3.1.14) implies that φ is the unique possible holomorphic solution of (3.1.10).

Now let $f_c\colon\mathbb{R}\to\mathbb{R}$ be the unique solution of the Cauchy problem

$$\begin{cases} \ddot f = A_c(f)\dot f^2,\\ f(0)=0, \quad \dot f(0)=1; \end{cases}$$

explicitly, the function f_c is given by

$$f_c(t)=\sqrt{-\frac{2}{c}}\,\frac{e^{(-c/2)^{1/2}t}-e^{(-c/2)^{1/2}t}}{e^{(-c/2)^{1/2}t}+e^{(-c/2)^{1/2}t}}=\begin{cases}\sqrt{\frac{2}{|c|}}\tanh\left(\sqrt{\frac{|c|}{2}}\,t\right) & \text{if } c<0,\\ t & \text{if } c=0,\\ \sqrt{\frac{2}{c}}\tan\left(\sqrt{\frac{c}{2}}\,t\right) & \text{if } c>0,\end{cases}$$

where, here and in the sequel, the case $c=0$ (where some expressions could make no sense) should be considered as a limit for $c\to0$, and we have chosen the obvious determination for square roots of negative numbers. Then, if we define

$$\sigma_\theta(t)=g_{e^{i\theta}}(f_c(t)), \tag{3.1.17}$$

we have

$$\dot\sigma_\theta=\dot f_c(\dot g_{e^{i\theta}}\circ f_c) \quad\text{and}\quad \ddot\sigma_\theta=\dot f_c^2[(\ddot g_{e^{i\theta}}+A_c\dot g_{e^{i\theta}})\circ f_c];$$

so σ_θ satisfies

$$\begin{cases} \ddot\sigma_\theta^\alpha = -\Gamma^\alpha_{;\mu}(\dot\sigma_\theta)\,\dot\sigma_\theta^\mu, & \text{for } \alpha = 1,\dots,n,\\ \sigma_\theta(0) = p, \quad \dot\sigma_\theta(0) = e^{i\theta}v_0. \end{cases}$$

If $h = G(\dot\sigma_\theta)$, then

$$\begin{aligned} \dot h &= G_{;\mu}(\dot\sigma_\theta)\,\dot\sigma_\theta^\mu + G_\alpha(\dot\sigma_\theta)\,\ddot\sigma_\theta^\alpha\\ &= G_{;\mu}(\dot\sigma_\theta)\,\dot\sigma_\theta^\mu + G_{\alpha\bar\beta}(\dot\sigma_\theta)\,\ddot\sigma_\theta^\alpha\,\overline{\dot\sigma_\theta^\beta}\\ &= G_{;\mu}(\dot\sigma_\theta)\,\dot\sigma_\theta^\mu - G_{\bar\beta;\mu}(\dot\sigma_\theta)\,\dot\sigma_\theta^\mu\overline{\dot\sigma_\theta^\beta} = 0. \end{aligned}$$

So $h(t) \equiv h(0) = 1$, and the curve $\dot\sigma_\theta = d^{1,0}\sigma_\theta(d/dt)$ actually lives in $S^{1,0}M$, that is the curve

$$\tilde\sigma_\theta = (\dot\sigma_\theta)^o = d\sigma_\theta(d/dt)$$

lives in $S_{\mathbb{R}}M = (S^{1,0}M)^o \subset T_{\mathbb{R}}M$.

Now let us consider the restriction to $S^{1,0}M$ of the horizontal radial vector field χ:

$$X = \chi|_{S^{1,0}M} \in \mathcal{X}(T^{1,0}(S^{1,0}M)),$$

which is locally given by

$$X = v^\mu\delta_\mu = v^\mu\partial_\mu - \Gamma^\alpha_{;\mu}v^\mu\dot\partial_\alpha. \tag{3.1.18}$$

Then, by construction, $\tilde\sigma_\theta$ is the integral curve in $S_{\mathbb{R}}M$ starting at $(p; e^{i\theta}v_0)$ of the vector field X^o, the real part of X. If we denote by e^{tX^o} the local one-parameter group of diffeomorphisms induced by X^o on $S_{\mathbb{R}}M$, then we may write

$$\tilde\sigma_\theta(t) = e^{tX^o}(e^{i\theta}u_0) \qquad \text{and} \qquad \sigma_\theta(t) = \pi(e^{tX^o}e^{i\theta}u_0), \tag{3.1.19}$$

where $\pi\colon S_{\mathbb{R}}M \to M$ is the canonical projection, $u_0 = v_0^o \in S_{\mathbb{R}}M$ and, for simplicity, we write $e^{i\theta}u_0$ for $(\cos\theta + J\sin\theta)u_0$.

Using the vertical radial vector field ι, we define another vector field on $S^{1,0}M$ by

$$Z = i\,\iota|_{S^{1,0}M} \in \mathcal{X}(T^{1,0}(S^{1,0}M)),$$

so that in local coordinates

$$Z = iv^\alpha\dot\partial_\alpha. \tag{3.1.20}$$

We observe that $d\pi(Z) = 0$. Since, by direct inspection, one immediately sees that the map $(e^{i\theta}, u) \mapsto e^{i\theta}u$ is the one-parameter group of diffeomorphisms of $S_{\mathbb{R}}M$ induced by the vector field Z^o, the real part of Z, it follows that (3.1.19) becomes

$$\tilde\sigma_\theta(t) = e^{tX^o}e^{\theta Z^o}u_0 \qquad \text{and} \qquad \sigma_\theta(t) = \pi(e^{tX^o}e^{\theta Z^o}u_0). \tag{3.1.21}$$

Because of (3.1.15), (3.1.16), (3.1.17) and (3.1.21), the holomorphicity of φ is equivalent to

$$\frac{\partial}{\partial\theta}\pi\left(e^{f_c^{-1}(t)X^o}e^{\theta Z^o}u_0\right)\bigg|_{\theta=0} = tJ\,\frac{\partial}{\partial t}\pi\left(e^{f_c^{-1}(t)X^o}e^{\theta Z^o}u_0\right)\bigg|_{\theta=0}, \tag{3.1.22}$$

where we may choose $\theta = 0$ because u_0 is a generic vector in $S_{\mathbb{R}}M$.

We start by computing

$$[X^o, JX^o] = i[X+\overline{X}, X-\overline{X}] = -2i[X,\overline{X}],$$

and

$$[X^o, Z^o] = [X+\overline{X}, Z+\overline{Z}] = [X,Z]^o + [X,\overline{Z}]^o.$$

Working in local coordinates (recalling that $\Gamma^{\alpha}_{\beta;\mu}v^{\beta} = \Gamma^{\alpha}_{;\mu}$ and the Lemma 2.3.3), we have

$$\begin{aligned}[X,Z] &= i[v^{\mu}\delta_{\mu}, v^{\alpha}\dot{\partial}_{\alpha}] = i\{v^{\alpha}v^{\mu}[\delta_{\mu},\dot{\partial}_{\alpha}] + v^{\mu}\delta_{\mu}(v^{\alpha})\dot{\partial}_{\alpha} - v^{\alpha}\delta_{\alpha}\} \\ &= i\{v^{\alpha}v^{\mu}\Gamma^{\sigma}_{\alpha;\mu}\dot{\partial}_{\sigma} - v^{\mu}\Gamma^{\alpha}_{;\mu}\dot{\partial}_{\alpha} - v^{\alpha}\delta_{\alpha}\} = -iX.\end{aligned} \tag{3.1.23}$$

Also, using (3.1.13) and again Lemma 2.3.3, we find that

$$[X,\overline{Z}] = -i[v^{\mu}\delta_{\mu}, \overline{v^{\alpha}}\dot{\partial}_{\bar{\alpha}}] = -iv^{\mu}\overline{v^{\alpha}}[\delta_{\mu},\dot{\partial}_{\bar{\alpha}}] = -iv^{\mu}\overline{v^{\alpha}}\Gamma^{\sigma}_{\bar{\alpha};\mu}\dot{\partial}_{\sigma} = 0.$$

Finally, using again Lemma 2.3.3 together with (2.3.19), we get

$$\begin{aligned}[X,\overline{X}] &= [v^{\mu}\delta_{\mu}, \overline{v^{\nu}}\delta_{\bar{\nu}}] \\ &= v^{\mu}\overline{v^{\nu}}[\delta_{\mu},\delta_{\bar{\nu}}] = v^{\mu}\overline{v^{\nu}}\{\delta_{\bar{\nu}}(\Gamma^{\sigma}_{;\mu})\dot{\partial}_{\sigma} - \delta_{\mu}(\Gamma^{\bar{\tau}}_{;\bar{\nu}})\dot{\partial}_{\bar{\tau}}\} = -\tau(\chi,\bar{\chi}) + \overline{\tau(\chi,\bar{\chi})}.\end{aligned}$$

Therefore using the hypothesis (3.1.11) for vectors in $S^{1,0}M$, we get

$$[X,\overline{X}] = -c\iota + c\bar{\iota} = icZ^o.$$

So we can conclude

$$[X^o, JX^o] = 2cZ^o \qquad \text{and} \qquad [X^o, Z^o] = -JX^o. \tag{3.1.24}$$

Now for $r > 0$ define $u_r = e^{rX^o}u_0$. Put

$$u(t) = d(e^{tX^o})_{e^{-tX^o}u_r}(Z^o) \in T_{u_r}(S_{\mathbb{R}}M).$$

Then

$$\begin{aligned}\frac{du}{dt}(t) &= \frac{d}{dt}\left(d(e^{tX^o})_{e^{-tX^o}u_r}(Z^o)\right) = -d(e^{tX^o})_{e^{-tX^o}u_r}(\mathcal{L}_{X^o}Z^o) \\ &= d(e^{tX^o})_{e^{-tX^o}u_r}(JX^o),\end{aligned}$$

where $\mathcal{L}_{X^o}$ is the Lie derivative, and

$$\begin{aligned}\frac{d^2u}{dt^2}(t) &= \frac{d}{dt}\left(d(e^{tX^o})_{e^{-tX^o}u_r}(JX^o)\right) = -d(e^{tX^o})_{e^{-tX^o}u_r}(\mathcal{L}_{X^o}(JX^o)) \\ &= -2c\, d(e^{tX^o})_{e^{-tX^o}u_r}(Z^o) = -2cu(t).\end{aligned}$$

In other words, $u(t)$ is a solution of the Cauchy problem

$$\begin{cases}\ddot{u} = -2cu, \\ u(0) = Z^o(u_r), \quad \dot{u}(0) = JX^o(u_r).\end{cases} \tag{3.1.25}$$

Therefore

$$u(t) = \frac{e^{\sqrt{-2c}\,t} + e^{-\sqrt{-2c}\,t}}{2} Z^o(u_r) + \frac{e^{\sqrt{-2c}\,t} - e^{-\sqrt{-2c}\,t}}{2\sqrt{-2c}} JX^o(u_r),$$

and, in particular,

$$\begin{aligned} d(\pi \circ e^{rX^o})_{u_0}(Z^o) = d\pi_{u_r}(u(r)) &= \frac{e^{\sqrt{-2c}\,r} - e^{-\sqrt{-2c}\,r}}{2\sqrt{-2c}} d\pi_{u_r}(JX^o) \\ &= \frac{e^{\sqrt{-2c}\,r} - e^{-\sqrt{-2c}\,r}}{2\sqrt{-2c}} J d\pi_{u_r}(X^o). \end{aligned}$$

Now,

$$e^{\sqrt{-2c} f_c^{-1}(t)} = \frac{1 + \sqrt{-c/2}\,t}{1 - \sqrt{-c/2}\,t};$$

thus we get

$$\frac{\partial}{\partial \theta} \pi \left(e^{f_c^{-1}(t) X^o} e^{\theta Z^o} u_0 \right) \bigg|_{\theta=0} = d(\pi \circ e^{f_c^{-1}(t) X^o})_{u_0}(Z^o) = \frac{t}{1 + (c/2)t^2} J d\pi_{u_{f_c^{-1}(t)}}(X^o),$$

whereas

$$\begin{aligned} \frac{\partial}{\partial t} \pi \left(e^{f_c^{-1}(t) X^o} e^{\theta Z^o} u_0 \right) \bigg|_{\theta=0} &= \frac{\partial}{\partial t} \pi \left(e^{f_c^{-1}(t) X^o} u_0 \right) = \frac{df_c^{-1}}{dt} d\pi_{u_{f_c^{-1}(t)}}(X^o) \\ &= \frac{1}{1 + (c/2)t^2} d\pi_{u_{f_c^{-1}(t)}}(X^o), \end{aligned}$$

and the assertion finally follows from (3.1.22). □

Theorem 3.1.4 is satisfactory from a PDE point of view since condition (3.1.11) is a natural compatibility condition of Frobenius type (in fact, in the case $c = 0$ Theorem 3.1.4 could be directly derived as a consequence of Frobenius' theorem). Furthermore, condition (3.1.11) on the (1,1)-torsion τ of the metric F can be rephrased in a more geometric way, involving the constancy of the holomorphic curvature and suitable symmetries of the curvature operator. But for this we need a couple of preliminary computational lemmas.

LEMMA 3.1.5: *Let $F: T^{1,0}M \to \mathbb{R}^+$ be a strongly pseudoconvex Finsler metric on a complex manifold M. Then*

$$\langle (\nabla_{\overline{W}} \Omega)(H, \overline{K})\chi, \chi \rangle = \langle \tau\big(H, \overline{\theta(K, W)}\big), \iota \rangle$$

for all $W \in \mathcal{V}$ and H, $K \in \mathcal{H}$. In particular,

$$\langle (\nabla_{\overline{W}} \Omega)(H, \bar{\chi})\chi, \chi \rangle = 0$$

for all $W \in \mathcal{V}$ and $H \in \mathcal{H}$.

Proof: Since we are interested only in the horizontal part, we may replace Ω by

$$\Omega^{\mathcal{H}} = \Omega^{\alpha}_{\beta} \otimes dz^{\beta} \otimes \delta_{\alpha}.$$

Since $\nabla_{\overline{W}} dz^{\beta} = 0$ and $\nabla_{\overline{W}} \delta_{\alpha} = 0$, we have

$$\nabla_{\overline{W}} \Omega^{\mathcal{H}} = (\nabla_{\overline{W}} \Omega^{\alpha}_{\beta}) \otimes dz^{\beta} \otimes \delta_{\alpha}.$$

Again, we only need the horizontal part, that is

$$p_H^*(\nabla_{\overline{W}} \Omega^{\alpha}_{\beta}) = \overline{W}(R^{\alpha}_{\beta;\mu\bar{\nu}})\, dz^{\mu} \wedge d\bar{z}^{\nu} - R^{\alpha}_{\beta;\mu\bar{\rho}} \overline{\omega^{\rho}_{\nu}(W)}\, dz^{\mu} \wedge d\bar{z}^{\nu}.$$

Recalling (2.3.21), taking H, K, $L \in \mathcal{H}$ we get

$$\begin{aligned}
\langle (\nabla_{\overline{W}} \Omega^{\mathcal{H}}(H, \overline{K})L, \chi\rangle &= G_{\alpha}\big[\overline{W}(R^{\alpha}_{\beta;\mu\bar{\nu}}) - R^{\alpha}_{\beta;\mu\bar{\rho}} \overline{\omega^{\rho}_{\nu}(W)}\big] H^{\mu}\overline{K^{\nu}} L^{\beta} \\
&= -G_{\alpha}\big[\dot{\partial}_{\bar{\gamma}}\delta_{\bar{\nu}}(\Gamma^{\alpha}_{\beta;\mu}) - \delta_{\bar{\rho}}(\Gamma^{\alpha}_{\beta;\mu})\Gamma^{\bar{\rho}}_{\bar{\nu}\bar{\gamma}} + \dot{\partial}_{\bar{\gamma}}(\Gamma^{\alpha}_{\beta\sigma}\delta_{\bar{\nu}}(\Gamma^{\sigma}_{;\mu})) \\
&\qquad - \Gamma^{\alpha}_{\beta\sigma}\delta_{\bar{\rho}}(\Gamma^{\sigma}_{;\mu})\Gamma^{\bar{\rho}}_{\bar{\nu}\bar{\gamma}}\big] H^{\mu}\overline{K^{\nu}} L^{\beta}\overline{W^{\gamma}} \\
&= -G_{\alpha}\big[\delta_{\bar{\nu}}\dot{\partial}_{\bar{\gamma}}(\Gamma^{\alpha}_{\beta;\mu}) - \Gamma^{\bar{\tau}}_{\bar{\gamma};\bar{\nu}}\dot{\partial}_{\bar{\tau}}(\Gamma^{\alpha}_{\beta;\mu}) + \dot{\partial}_{\bar{\gamma}}(\Gamma^{\alpha}_{\beta\sigma})\delta_{\bar{\nu}}(\Gamma^{\sigma}_{;\mu}) + \delta_{\bar{\nu}}(\Gamma^{\sigma}_{\bar{\gamma};\mu})\Gamma^{\alpha}_{\beta\sigma} \\
&\qquad - \Gamma^{\alpha}_{\beta\sigma}\Gamma^{\bar{\tau}}_{\bar{\gamma};\bar{\nu}}\Gamma^{\sigma}_{\bar{\tau};\mu} - \delta_{\bar{\rho}}(\Gamma^{\alpha}_{\beta;\mu})\Gamma^{\bar{\rho}}_{\bar{\nu}\bar{\gamma}} - \Gamma^{\alpha}_{\beta\sigma}\delta_{\bar{\rho}}(\Gamma^{\sigma}_{;\mu})\Gamma^{\bar{\rho}}_{\bar{\nu}\bar{\gamma}}\big] H^{\mu}\overline{K^{\nu}} L^{\beta}\overline{W^{\gamma}}
\end{aligned}$$

(where we used Lemma 2.3.3),

$$\begin{aligned}
&= -\big[\delta_{\bar{\nu}}(G_{\alpha}\dot{\partial}_{\beta}(\Gamma^{\alpha}_{\bar{\gamma};\mu})) - \Gamma^{\bar{\tau}}_{\bar{\gamma};\bar{\nu}} G_{\alpha}\dot{\partial}_{\beta}(\Gamma^{\alpha}_{\bar{\tau};\mu}) + G_{\beta\sigma}\delta_{\bar{\nu}}(\Gamma^{\sigma}_{\bar{\gamma};\mu}) - G_{\beta\sigma}\Gamma^{\bar{\tau}}_{\bar{\gamma};\bar{\nu}}\Gamma^{\sigma}_{\bar{\tau};\mu} \\
&\qquad - \delta_{\bar{\rho}}(G_{\alpha}\Gamma^{\alpha}_{\beta;\mu})\Gamma^{\bar{\rho}}_{\bar{\nu}\bar{\gamma}} - G_{\beta\sigma}\delta_{\bar{\rho}}(\Gamma^{\sigma}_{;\mu})\Gamma^{\bar{\rho}}_{\bar{\nu}\bar{\gamma}}\big] H^{\mu}\overline{K^{\nu}} L^{\beta}\overline{W^{\gamma}}
\end{aligned}$$

(where we used $G_{\alpha}\Gamma^{\alpha}_{\beta\sigma} = G_{\beta\sigma}$, (2.3.15), $G_{\alpha}\dot{\partial}_{\bar{\gamma}}(\Gamma^{\alpha}_{\beta\sigma}) = G_{\beta\sigma\bar{\gamma}} - G_{\alpha\bar{\gamma}}\Gamma^{\alpha}_{\beta\sigma} = 0$ and Lemma 2.3.4),

$$\begin{aligned}
&= -\big[-\delta_{\bar{\nu}}(G_{\alpha\beta}\Gamma^{\alpha}_{\bar{\gamma};\mu}) + \Gamma^{\bar{\tau}}_{\bar{\gamma};\bar{\nu}} G_{\alpha\beta}\Gamma^{\alpha}_{\bar{\tau};\mu} + G_{\beta\sigma}\delta_{\bar{\nu}}(\Gamma^{\sigma}_{\bar{\gamma};\mu}) - G_{\beta\sigma}\Gamma^{\bar{\tau}}_{\bar{\gamma};\bar{\nu}}\Gamma^{\sigma}_{\bar{\tau};\mu} \\
&\qquad - \delta_{\bar{\rho}}(G_{\alpha}\Gamma^{\alpha}_{\beta;\mu})\Gamma^{\bar{\rho}}_{\bar{\nu}\bar{\gamma}} - G_{\beta\sigma}\delta_{\bar{\rho}}(\Gamma^{\sigma}_{;\mu})\Gamma^{\bar{\rho}}_{\bar{\nu}\bar{\gamma}}\big] H^{\mu}\overline{K^{\nu}} L^{\beta}\overline{W^{\gamma}}
\end{aligned}$$

(where we used $G_{\alpha}\Gamma^{\alpha}_{\bar{\tau};\mu} = 0$),

$$\begin{aligned}
&= -\big[-\delta_{\bar{\nu}}(G_{\alpha\beta}\Gamma^{\alpha}_{\bar{\gamma};\mu}) + G_{\beta\sigma}\delta_{\bar{\nu}}(\Gamma^{\sigma}_{\bar{\gamma};\mu}) - \delta_{\bar{\rho}}(G_{\alpha}\Gamma^{\alpha}_{\beta;\mu})\Gamma^{\bar{\rho}}_{\bar{\nu}\bar{\gamma}} \\
&\qquad - G_{\beta\sigma}\delta_{\bar{\rho}}(\Gamma^{\sigma}_{;\mu})\Gamma^{\bar{\rho}}_{\bar{\nu}\bar{\gamma}}\big] H^{\mu}\overline{K^{\nu}} L^{\beta}\overline{W^{\gamma}}.
\end{aligned}$$

Hence

$$\begin{aligned}
&\langle (\nabla_{\overline{W}}\Omega)(H, \overline{K})\chi, \chi\rangle \\
&\quad = -\big[-\delta_{\bar{\nu}}(G_{\alpha\beta}\Gamma^{\alpha}_{\bar{\gamma};\mu}) + G_{\beta\sigma}\delta_{\bar{\nu}}(\Gamma^{\sigma}_{\bar{\gamma};\mu}) - \delta_{\bar{\rho}}(G_{\alpha}\Gamma^{\alpha}_{\beta;\mu})\Gamma^{\bar{\rho}}_{\bar{\nu}\bar{\gamma}} \\
&\qquad\quad - G_{\beta\sigma}\delta_{\bar{\rho}}(\Gamma^{\sigma}_{;\mu})\Gamma^{\bar{\rho}}_{\bar{\nu}\bar{\gamma}}\big] H^{\mu}\overline{K^{\nu}} v^{\beta}\overline{W^{\gamma}}, \\
&\quad = \delta_{\bar{\rho}}(G_{\alpha}\Gamma^{\alpha}_{;\mu})\Gamma^{\bar{\rho}}_{\bar{\nu}\bar{\gamma}} H^{\mu}\overline{K^{\nu}}\,\overline{W^{\gamma}},
\end{aligned}$$

(where we used (2.3.3) and $v^{\beta}\Gamma^{\alpha}_{\beta;\mu} = \Gamma^{\alpha}_{;\mu}$),

$$= G_{\alpha}\delta_{\bar{\rho}}(\Gamma^{\alpha}_{;\mu})\Gamma^{\bar{\rho}}_{\bar{\nu}\bar{\gamma}} H^{\mu}\overline{K^{\nu}}\,\overline{W^{\gamma}} = \langle \tau(H, \overline{\theta(K, W)}), \iota\rangle,$$

because $\theta(K, W) = -\Gamma^{\rho}_{\nu\gamma} K^{\nu} W^{\gamma}\delta_{\rho}$, by (2.3.19).

Finally,

$$\theta(\chi, W) = -\Gamma^{\rho}_{\nu\gamma} v^{\nu} W^{\gamma} = 0,$$

and the last assertion follows. □

In the sequel we shall need some symmetry properties of the curvature operator, which are particular instances of (2.5.3) and hold under weaker hypotheses.

LEMMA 3.1.6: *Let $F: T^{1,0}M \to \mathbb{R}^+$ be a strongly pseudoconvex Finsler metric on a complex manifold M. Then*
(i) $\langle \Omega(H, \bar{\chi})\chi, \chi \rangle = \langle \Omega(\chi, \bar{\chi})H, \chi \rangle$ *for all $H \in \mathcal{H}$ iff*

$$\langle \bar{\partial}_H \theta(H, \chi, \bar{\chi}), \chi \rangle = 0 \tag{3.1.26}$$

for all $H \in \mathcal{H}$;
(ii) $\langle \Omega(H, \overline{K})\chi, \chi \rangle = \langle \Omega(\chi, \overline{K})H, \chi \rangle$ *for all H, $K \in \mathcal{H}$ iff*

$$\langle \bar{\partial}_H \theta(H, \chi, \overline{K}), \chi \rangle = 0 \tag{3.1.27}$$

for all H, $K \in \mathcal{H}$.

Proof: First of all, $\bar{\partial}_H \theta = p_H^*(D''\theta)$, by (2.2.6). Corollary 2.3.7 says that

$$D''\theta = \eta^{\mathcal{H}} \wedge \Omega;$$

the assertion then follows immediately from (2.5.4). □

The next result begins to establish a relationship among constant holomorphic curvature, the symmetry condition (3.1.26) and the compatibility condition (3.1.11):

PROPOSITION 3.1.7: *Let $F: T^{1,0}M \to \mathbb{R}^+$ be a strongly pseudoconvex Finsler metric on a complex manifold M. Then*

$$\tau(\chi, \bar{\chi}) = c\,G\iota. \tag{3.1.28}$$

iff F has constant holomorphic curvature $K_F \equiv 2c$ and

$$\langle \bar{\partial}_H \theta(H, \chi, \bar{\chi}), \chi \rangle = 0 \tag{3.1.29}$$

for all $H \in \mathcal{H}$. Furthermore, they both imply

$$\langle \Omega(\chi, \overline{K})\chi, \chi \rangle = c\,G\langle \chi, K \rangle \tag{3.1.30}$$

for all $K \in \mathcal{H}$.

Proof: If (3.1.28) holds, we have

$$\langle \Omega(\chi, \bar{\chi})\chi, \chi \rangle = \langle \tau(\chi, \bar{\chi}), \iota \rangle = cG^2,$$

and so $K_F \equiv 2c$. So we may assume that F has constant holomorphic curvature $2c$, and prove that (3.1.28) and (3.1.29) are equivalent under this assumption.

For $W \in \mathcal{V}$, let $K = \Theta(W) \in \mathcal{H}$; note that $\nabla_{\overline{W}}\chi = 0$ and $\nabla_W \chi = \Theta(W) = K$. Then

$$\begin{aligned} \overline{W}(cG^2) &= 2cG\langle \chi, K\rangle; \\ \overline{W}\langle \Omega(\chi,\bar\chi)\chi,\chi\rangle &= \langle(\nabla_{\overline{W}}\Omega)(\chi,\bar\chi)\chi,\chi\rangle + \langle \Omega(\chi,\overline{K})\chi,\chi\rangle + \langle \Omega(\chi,\bar\chi)\chi, K\rangle \\ &= \langle \Omega(\chi,\overline{K})\chi,\chi\rangle + \langle \tau^H(\chi,\bar\chi), K\rangle, \end{aligned} \tag{3.1.31}$$

where we used Lemmas 3.1.6 and 2.3.8. Since F has constant holomorphic curvature $2c$, we have

$$\langle \Omega(\chi,\bar\chi)\chi,\chi\rangle = cG^2,$$

and hence (3.1.31) yields

$$\langle \Omega(\chi,\overline{K})\chi,\chi\rangle = 2cG\langle\chi,K\rangle - \langle \tau(\chi,\bar\chi),K\rangle. \tag{3.1.32}$$

Subtracting $\langle \tau(\chi,\bar\chi),K\rangle = \langle \Omega(\chi,\bar\chi)\chi,K\rangle$ to both sides, we find that (3.1.28) holds if and only if

$$\langle \Omega(\chi,\overline{K})\chi,\chi\rangle = \langle \Omega(\chi,\bar\chi)\chi,K\rangle$$

for all $K \in \mathcal{H}$, that is, recalling (2.5.2), iff

$$\langle \Omega(K,\bar\chi)\chi,\chi\rangle = \langle \Omega(\chi,\bar\chi)K,\chi\rangle,$$

and thus, by Lemma 3.1.6, iff (3.1.29) holds.

Finally, if (3.1.28) holds, (3.1.32) yields (3.1.30). □

And now we are able to give the more geometric version of the existence theorem for holomorphic solutions of $D_c(\varphi) = 0$:

THEOREM 3.1.8: *Let $F\colon T^{1,0}M \to \mathbb{R}^+$ be a strongly pseudoconvex Finsler metric on a complex manifold M. Then the Cauchy problem* (3.1.10)

$$\begin{cases} D_c(\varphi) = 0, \\ \varphi(0) = p, \ \varphi'(0) = v_0, \end{cases}$$

has holomorphic solution for all $(p; v_0) \in S^{1,0}M$ iff F has constant holomorphic curvature $K_F \equiv 2c$ and either

$$\forall H \in \mathcal{H} \qquad \langle \bar\partial_H \theta(H,\chi,\bar\chi),\chi\rangle = 0 \tag{3.1.33}$$

or, equivalently

$$\forall H \in \mathcal{H} \qquad \langle \Omega(H,\bar\chi)\chi,\chi\rangle = \langle \Omega(\chi,\bar\chi)H,\chi\rangle. \tag{3.1.34}$$

Furthermore, when a holomorphic solution of (3.1.10) *exists, it is unique.*

Proof: It follows from Proposition 3.1.7 and Lemma 3.1.6. □

It must be remarked that in [AP2] condition (3.1.33) was expressed in an equivalent way in terms of a tensor defined in local coordinates.

Before going on let us state what we got so far in terms of the main characters of this section, the c-geodesic complex curves. Putting together Corollary 3.1.2, Proposition 3.1.3 and Theorem 3.1.8, we have the following:

THEOREM 3.1.9: *Let $F: T^{1,0}M \to \mathbb{R}^+$ be a strongly pseudoconvex Finsler metric on a complex manifold M, and take $c \in \mathbb{R}$. Then*

(i) There exists a segment of infinitesimal c-geodesic complex curve through $(p; v)$ for any $p \in M$ and $v \in S^{1,0}_p M$ if either (3.1.11) holds, or $K_F \equiv 2c$ and (3.1.33) holds, or $K_F \equiv 2c$ and (3.1.34) holds;

(ii) There exists a unique segment of c-geodesic complex curve through $(p; v)$ for all $p \in M$ and $v \in S^{1,0}_p M$ iff F is weakly Kähler and either (3.1.11) holds, or $K_F \equiv 2c$ and (3.1.33) holds, or $K_F \equiv 2c$ and (3.1.34) holds.

We are left with a last question: the existence of "complete" geodesic complex curves, i.e., defined on the largest possible disk (or on the plane, or on $\mathbb{P}_1$), and not just on a segment. As it may be expected the key ingredient is of course to assume in addition the completeness of the metric. Precisely we have the following:

THEOREM 3.1.10: *Let $F: T^{1,0}M \to \mathbb{R}^+$ be a complete strongly pseudoconvex Finsler metric on a complex manifold M, and take $c \in \mathbb{R}$. Then*

(i) There exists an infinitesimal c-geodesic complex curve through $(p; v)$ for any $(p; v) \in S^{1,0}M$ if either (3.1.11) holds, or $K_F \equiv 2c$ and (3.1.33) holds, or $K_F \equiv 2c$ and (3.1.34) holds;

(ii) There exists a unique c-geodesic complex curve through $(p; v)$ for any pair $(p; v) \in S^{1,0}M$ iff F is weakly Kähler and either (3.1.11) holds, or $K_F \equiv 2c$ and (3.1.33) holds, or $K_F \equiv 2c$ and (3.1.34) holds.

Proof: By Proposition 3.1.3, it is enough to prove (ii) and in fact to show that the holomorphic solution of (3.1.10) φ is defined on Δ_r with $r = \sqrt{-2/c}$ if $c < 0$, on $\mathbb{C}$ if $c = 0$ and on $\mathbb{P}_1$ if $c > 0$. Consider the distribution $\mathcal{D} = \mathbb{C}X^o \oplus \mathbb{C}Z^o \subset T(S_{\mathbb{R}}M)$, where X^o and Z^o are the vector fields defined in the proof of Theorem 3.1.4. The distribution $\mathcal{D}$ is involutive, since we have (3.1.24) and the remaining brackets are easily computed from the definitions:

$$[X^o, JZ^o] = X^o = [JX^o, Z^o], \quad [JX^o, JZ^o] = JX^o, \quad [Z^o, JZ^o] = 0.$$

If $\tilde{L}$ is the maximal integral manifold of $\mathcal{D}$ passing through $(p; v)$, then, the proof of Theorem 3.1.4 shows that $L = \pi(\tilde{L}) \subset M$ is a Riemann surface locally parametrized by the holomorphic solutions of $D_c(\varphi) = 0$. Since F restricted to L is a complete Hermitian metric of constant Gaussian curvature $2c$, because of Proposition 3.1.3.(iii), using the Riemann uniformization theorem we have that there exists a unique simply connected Riemann surface $\mathcal{R}$ endowed with the metric μ_c and a unique isometric holomorphic covering map $\psi: \mathcal{R} \to L$ such that $\psi(0) = p$ and $\psi'(0) = v$, where $\mathcal{R} = \Delta_r$ with $r = \sqrt{-2/c}$ if $c < 0$, $\mathcal{R} = \mathbb{C}$ if $c = 0$ and $\mathcal{R} = \mathbb{P}_1$

if $c > 0$. But, for some $r > 0$, the holomorphic solution $\varphi\colon \Delta_r \to L$ of $D_c(\varphi) = 0$ with $\varphi(0) = p$ and $\varphi'(0) = v$ is another holomorphic isometry from Δ_r endowed with the restriction of the metric μ_c to L endowed with the restriction of F; it follows that $\varphi = \psi|_{\Delta_r}$, and necessarily then ψ is the unique solution of the Cauchy problem (3.1.10) — and it has the required properties. □

For further reference, we explicitly state here some properties of geodesic complex curves which were not underlined before to simplify statements but nevertheless will turn out useful later. The remarks are obvious consequences of the construction of geodesic complex curves (cf. the proof of Theorem 3.1.4) and of the smooth dependence on initial data of solutions of ODE Cauchy problem.

PROPOSITION 3.1.11: *In the hypothesis of Theorem 3.1.10, if we denote by φ_v the unique c-geodesic complex curve through the point p tangent to $v \in S^{1,0}_p M$, then*

$$\varphi_v(e^{i\theta}\zeta) = \varphi_{e^{i\theta}v}(\zeta). \tag{3.1.35}$$

Furthermore the mapping $(\zeta, v) \mapsto \varphi_v(\zeta)$ is of class C^∞.

3.1.3. A characterization of the Kobayashi metric

As an application of our results we present a characterization of Kobayashi metric among the (smooth) strongly pseudoconvex Finsler metrics (cf. [AP2]). Related, though a bit weaker, characterizations were obtained by Faran [F] and Pang [P1].

We need to introduce some additional tools to get our aim; in particular we shall need a suitable version of Ahlfors' Lemma.

DEFINITION 3.1.4: A *pseudohermitian metric μ_g of scale g on Δ* is the upper semicontinuous pseudometric on the tangent bundle of Δ defined by

$$\mu_g = g\, d\zeta \otimes d\bar{\zeta}, \tag{3.1.36}$$

where $g\colon \Delta \to \mathbf{R}^+$ is a non-negative upper semicontinuous function such that the set $S_g = g^{-1}(0)$ is a discrete subset of Δ.

If μ_g is a standard Hermitian metric on Δ, i.e., g is a C^2 positive function, we have already defined (in section 2.5.2) the Gaussian curvature of μ_g by setting

$$K(\mu_g) = -\frac{1}{2g}\Delta \log g, \tag{3.1.37}$$

where Δ denotes the usual Laplacian

$$\Delta u = 4\frac{\partial^2 u}{\partial\zeta\partial\bar{\zeta}}. \tag{3.1.38}$$

To define a Gaussian curvature for pseudohermitian metrics we need a weak Laplacian.

DEFINITION 3.1.5: The (*lower*) *generalized Laplacian* of an upper semicontinuous function u is defined by

$$\Delta u(\zeta) = 4 \liminf_{r\to 0} \frac{1}{r^2}\left\{\frac{1}{2\pi}\int_0^{2\pi} [u(\zeta + re^{i\theta}) - u(\zeta)]\, d\theta\right\}. \tag{3.1.39}$$

This definition is quite nice as it shares all the important features of the definition for regular functions. First of all, when u is a function of class C^2 in a neighborhood of the point ζ_0, (3.1.39) actually reduces to (3.1.38). In fact, for r small enough we can write

$$\begin{aligned} u(\zeta_0 + re^{i\theta}) = u(\zeta_0) &+ \frac{\partial u}{\partial \zeta}(\zeta_0)re^{i\theta} + \frac{\partial u}{\partial \bar\zeta}(\zeta_0)re^{-i\theta} \\ &+ \frac{1}{2}\frac{\partial^2 u}{\partial \zeta^2}(\zeta_0)r^2e^{2i\theta} + \frac{\partial^2 u}{\partial \zeta \partial \bar\zeta}(\zeta_0)r^2 + \frac{1}{2}\frac{\partial^2 u}{\partial \bar\zeta^2}(\zeta_0)r^2e^{-2i\theta} + o(r^2); \end{aligned}$$

hence

$$\frac{1}{2\pi}\int_0^{2\pi} [u(\zeta_0 + re^{i\theta}) - u(\zeta)]\, d\theta = r^2 \frac{\partial^2 u}{\partial \zeta \partial \bar\zeta}(\zeta_0) + o(r^2),$$

and the claim follows.

Second, if u is an upper semicontinuous function, it is easy to see that $\Delta u \ge 0$ is equivalent to the submean property; so $\Delta u \ge 0$ iff u is subharmonic.

Finally, if ζ_0 is a point of local maximum for u, then clearly $\Delta u(\zeta_0) \le 0$.

DEFINITION 3.1.6: Let μ_g be a pseudohermitian metric on Δ. Then the *Gaussian curvature* $K(\mu_g)$ of μ_g is the function defined on $\Delta \setminus S_g$ by (3.1.37) — using the generalized Laplacian (3.1.39); clearly, if μ_g is a standard Hermitian metric, $K(\mu_g)$ reduced to the usual Gaussian curvature.

We are able to prove the classical Ahlfors lemma in this more general situation:

PROPOSITION 3.1.12: *Let $\mu = g\, d\zeta \otimes d\bar\zeta$ be a pseudohermitian metric on Δ such that $K(\mu) \le -4$ on $\Delta \setminus S_g$ for some $a > 0$. Then $g \le g_{-2}$, where*

$$g_{-2}(\zeta) = \frac{1}{(1 - |\zeta|^2)^2}$$

is the scale of the Poincaré metric.

Proof: The proof is just an appropriate adaptation of the classical one due to Ahlfors. For the sake of completeness we report it here.

For $0 < r < 1$, define $g_{-2}^r\colon \Delta_r \to \mathbb{R}^+$ by

$$g_{-2}^r(\zeta) = \frac{1}{(1 - |\zeta|^2/r^2)^2} = g_{-2}(\zeta/r),$$

and set $f_r = g/g_{-2}^r\colon \Delta_r \to \mathbb{R}^+$. Being upper semicontinuous, g is bounded on $\overline{\Delta_r}$; since we clearly have $g_{-2}^r(\zeta) \to +\infty$ as $|\zeta| \to r$, there is a point $\zeta_0 \in \Delta_r$ of maximum for f_r. Clearly, $\zeta_0 \notin S_g$; hence

$$0 \ge \Delta \log f_r(\zeta_0) \ge \Delta \log g(\zeta_0) - \Delta \log g_{-2}^r(\zeta_0) = -2g(\zeta_0)K(\mu_g)(\zeta_0) - 8g_{-2}^r(\zeta_0). \tag{3.1.40}$$

By assumption, $K(\mu)(\zeta_0) \le -4$; therefore (3.1.40) yields $g(\zeta_0) \le g_{-2}^r(\zeta_0)$ and thus, by the choice of ζ_0,

$$\forall \zeta \in \Delta_r \qquad g(\zeta) \le g_{-2}(\zeta/r).$$

Letting $r \to 1^-$ we obtain the assertion. □

To get our characterization, we shall now need some considerations which are meaningful without any change in the more general contest of "weak" Finsler metrics — and thus we shall present them in this setting.

DEFINITION 3.1.7: An *upper semicontinuous complex Finsler metric* F on a complex manifold M is an upper semicontinuous map $F\colon T^{1,0}M \to \mathbb{R}^+$ satisfying

(i) $F(p; v) > 0$ for all $p \in M$ and $v \in T_p^{1,0}M$ with $v \ne 0$;
(ii) $F(p; \lambda v) = |\lambda| F(p; v)$ for all $p \in M$, $v \in T_p^{1,0}M$ and $\lambda \in \mathbb{C}$.

Typical examples of upper semicontinuous complex Finsler metrics are the Carathéodory and Kobayashi metrics defined at the beginning of section 2.3.

Using a variational approach, we can easily define a notion of holomorphic curvature for an upper semicontinuous complex Finsler metric F following ideas of Royden [Ro], Wong [W2], Suzuki [Su]. As usual, we shall denote by $G\colon T^{1,0}M \to \mathbb{R}^+$ the function $G = F^2$.

DEFINITION 3.1.8: Take $p \in M$ and $v \in T_p^{1,0}M$, $v \ne 0$. The *holomorphic curvature* $K_F(v)$ of F at v is given by

$$K_F(v) = \sup\{K(\varphi^* G)(0)\},$$

where the supremum is taken with respect to the family of all holomorphic maps $\varphi\colon \Delta \to M$ with $\varphi(0) = p$ and $\varphi'(0) = \lambda v$ for some $\lambda \in \mathbb{C}^*$, and $K(\varphi^* G)$ is the Gaussian curvature discussed so far of the pseudohermitian metric $\varphi^* G$ on Δ. Clearly, this holomorphic curvature depends only on the complex line in $T_p^{1,0}M$ spanned by v, and not on v itself. Furthermore, by Corollary 2.5.4 if F is smooth we recover our previous definition of holomorphic curvature.

This notion has been used for studying intrinsic metrics and, for instance, it is known that the Carathéodory has holomorphic curvature bounded above by -4, whereas the holomorphic curvature of the Kobayashi metric is bounded below by -4; see, e.g., [W2] or [Su].

It is worthwhile to remark that, though apparently on the unit disk we have given two notions of curvature, they (fortunately) coincide. If F is an upper semicontinuous complex Finsler metric on Δ — and so $G = F^2$ is a pseudohermitian metric $G = g\, d\zeta \otimes d\bar\zeta$ on Δ —, a priori we have defined two curvatures for F: $K_F(\zeta; 1)$ and $K(G)(\zeta)$. As a consequence of the following Lemma we have at once that $K_F(\zeta; 1) = K(G)(\zeta)$.

LEMMA 3.1.13: *Let $\mu_g = g\, d\zeta \otimes d\bar\zeta$ be a pseudohermitian metric on Δ, and $\varphi \colon \Delta \to \Delta$ a holomorphic self-map of Δ. Then on $\Delta \setminus [\varphi^{-1}(S_g) \cup S_{\varphi'}]$ we have*

$$K(\varphi^* \mu_g) = K(\mu_g) \circ \varphi.$$

Proof: Let $\{g_n\}$ be a sequence of C^2 functions such that $g_n \ge g_{n+1}$ and with $g_n(x) \to g(x)$, and let $\mu_n = g_n\, d\zeta \otimes d\bar\zeta$ be the corresponding metrics. Then on $\Delta \setminus [\varphi^{-1}(S_{g_n}) \cup S_{\varphi'}] \supset \Delta \setminus [\varphi^{-1}(S_g) \cup S_{\varphi'}]$ we have

$$\begin{aligned} K(\varphi^* \mu_n) &= -\frac{1}{2|\varphi'|^2 (g_n \circ \varphi)} \Delta \log(|\varphi'|^2 g_n \circ \varphi) \\ &= -\frac{2}{|\varphi'|^2 (g_n \circ \varphi)} \left[\frac{\partial^2 \log(g_n \circ \varphi)}{\partial \zeta \partial \bar\zeta} + \frac{\partial^2 \log |\varphi'|^2}{\partial \zeta \partial \bar\zeta} \right] \\ &= -\frac{2}{|\varphi'|^2 (g_n \circ \varphi)} \left[|\varphi'|^2 \left(\frac{\partial^2 \log g_n}{\partial \zeta \partial \bar\zeta} \right) \circ \varphi \right] \\ &= -\frac{1}{2(g_n \circ \varphi)} (\Delta \log g_n) \circ \varphi = K(\mu_n) \circ \varphi. \end{aligned}$$

Letting $n \to +\infty$ and applying the dominated convergence theorem we get the assertion. □

The holomorphic curvature defined in this way is clearly invariant under holomorphic isometries. More generally, if $f \colon M \to N$ is holomorphic and F is an upper semicontinuous complex Finsler metric on N, we have

$$K_F\big(f(z); df_z(v)\big) \ge K_{f^*F}(z; v),$$

that is $f^* K_F \ge K_{f^*F}$.

Furthermore, as it is well known, this notion of holomorphic curvature have a built-in Ahlfors' Lemma, and so it is suitable for producing criteria of hyperbolicity (see for example Grauert-Riekziegel [GR], Kobayashi [K], Abate-Patrizio [AP2]).

Bounds on the holomorphic curvature allow to compare a complex Finsler metric with the Kobayashi metric and to give conditions for a given Finsler metric to be the Kobayashi metric. We now provide the right set-up for the problem in the general case. We need a last auxiliary notion to formulate our observation.

DEFINITION 3.1.9: Let $F: T^{1,0}M \to \mathbb{R}^+$ be an upper semicontinuous Finsler metric on a complex manifold M, and take $(p; v) \in T^{1,0}M$. We say that F is *realizable* at $(p; v)$ if there is a holomorphic map $\varphi: \Delta \to M$ such that $\varphi(0) = p$ and $\lambda\varphi'(0) = v$ with $|\lambda| = F(p; v)$. In other words, φ is an isometry at the origin between the Poincaré metric of Δ and F.

Because of its very definition, the Kobayashi metric is realizable in any taut manifold; on the other hand, as a consequence of the next result, the Carathéodory metric is realizable iff it coincides with the Kobayashi metric.

PROPOSITION 3.1.14: *Let $F: T^{1,0}M \to \mathbb{R}^+$ be an upper semicontinuous Finsler metric on a complex manifold M, and choose $(p_0; v_0) \in T^{1,0}M$. Then:*

(i) *If F is realizable at $(p_0; v_0)$, then $F(p_0; v_0) \geq \kappa_M(p_0; v_0)$;*
(ii) *If $K_F \leq -4$, then $F \leq \kappa_M$;*
(iii) *If F is realizable at $(p_0; v_0)$ and $K_F \leq -4$, then $F(p_0; v_0) = \kappa_M(p_0; v_0)$.*

Proof: (i) Let $\varphi: \Delta \to M$ be a holomorphic map with $\varphi(0) = p_0$ and $v_0 = \lambda\varphi'(0)$ such that $F(p_0; v_0) = |\lambda|$. Then

$$F(p_0; v_0) = |\lambda| \geq \kappa_M(p_0; v_0).$$

(ii) Take $(p; v) \in T^{1,0}M$ and let $\varphi: \Delta \to M$ be a holomorphic map with $\varphi(0) = p$ and $v = \lambda\varphi'(0)$. Then φ^*G is a pseudohermitian metric on Δ with Gaussian curvature bounded above by -4 (by Lemma 3.1.14); it follows from Proposition 3.1.12 that $\varphi^*G \leq \mu_{-2}$. Thus

$$F(p; v) = F\big(\varphi(0); \lambda\varphi'(0)\big) = \varphi^*F(0; \lambda) \leq |\lambda|.$$

Since this holds for all such φ, we get $F \leq \kappa_M$.
(iii) This is just a consequence of (i) and (ii). □

Let us turn to the case of smooth Finsler metrics. In this case we have given the differential geometric conditions which guarantee the same good behavior that Lempert [L] discovered for the Kobayashi metric of strongly convex domains. It should not be surprising that we can now obtain the following:

THEOREM 3.1.15: *Let $F: T^{1,0}M \to \mathbb{R}^+$ be a strongly pseudoconvex smooth complete Finsler metric on a complex manifold M. Assume that one of these equivalent conditions hold:*

$$\tau(\chi, \bar{\chi}) = -2G\iota,$$

or

$$K_F \equiv -4 \qquad \text{and} \qquad \langle \bar{\partial}_H \theta(H, \chi, \bar{\chi}), \chi \rangle = 0 \quad \text{for all } H \in \mathcal{H},$$

or

$$K_F \equiv -4 \qquad \text{and} \qquad \langle \Omega(H, \bar{\chi})\chi, \chi \rangle = \langle \Omega(\chi, \bar{\chi})H, \chi \rangle \quad \text{for all } H \in \mathcal{H}.$$

Then F is the Kobayashi metric of M.

Proof: Under the hypotheses, using also Lemma 3.1.6 and Proposition 3.1.7, the holomorphic curvature of F is -4. Furthermore, by Theorem 3.1.10 F is realizable. Then Proposition 3.1.14 complete the proof. □

3.2. Manifolds with constant nonpositive holomorphic curvature

3.2.1. A complex Cartan-Hadamard theorem

Now we shall take a closer look to Kähler Finsler manifolds of nonpositive constant holomorphic curvature, with an eye toward the very natural problem of their classification. Examples show that in the Finsler setting this problem is much more complicated than in the Hermitian one; for instance, Lempert's work [L] and Theorem 3.1.10 imply that all strongly convex domains of $\mathbb{C}^n$ endowed with the Kobayashi metric are weakly Kähler Finsler manifolds with constant holomorphic curvature -4. As a step toward the classification, in this section we shall prove that a simply connected Kähler Finsler manifold of nonpositive constant holomorphic curvature is E-diffeomorphic to an euclidean space. Furthermore, in the case of constant negative holomorphic curvature our results show that the Finsler geometry of the manifold is pretty much the same as the one of strongly convex domains endowed with the Kobayashi metric.

The crucial point in our argument will be to estimate the curvature terms

$$\mathrm{Re}\Big[\langle\Omega(\chi,\overline{H})H,\chi\rangle_v-\langle\Omega(H,\bar{\chi})H,\chi\rangle_v+\langle\!\langle\tau^{\mathcal{H}}(H,\bar{\chi}),H\rangle\!\rangle_v-\langle\!\langle\tau^{\mathcal{H}}(\chi,\overline{H}),H\rangle\!\rangle_v\Big] \tag{3.2.1}$$

appearing in the second variation formula for complex Finsler metrics, using as much as possible the constancy of the holomorphic curvature.

Let $F\colon T^{1,0}M\to\mathbb{R}^+$ be a strongly pseudoconvex Finsler metric on a complex manifold M. Then we know that F has constant holomorphic curvature $K_F\equiv 2c\in\mathbb{R}$ iff

$$\langle\Omega(\chi,\bar{\chi})\chi,\chi\rangle\equiv cG^2. \tag{3.2.2}$$

In this kind of problems tradition advises that the key idea it is to differentiate (3.2.2) in such a smart way to get an expression of (3.2.1) whose sign may be estimated without much effort. This scheme, which works without great pain in usual Kähler geometry, is much more complicate to carry out in Kähler Finsler geometry, as the reader will notice in a while.

In subsection 1.2 of this chapter we have already proved a few technical result needed here (Lemma 3.1.5 and Lemma 3.1.6). Furthermore, in Proposition 3.1.7 it was shown that, if $K_F\equiv 2c$ and either

$$\forall H\in\mathcal{H}\qquad \langle\bar{\partial}_H\theta(H,\chi,\bar{\chi}),\chi\rangle=0 \tag{3.2.3}$$

or, equivalently,

$$\tau(\chi,\bar{\chi})=c\,G\iota, \tag{3.2.4}$$

then it follows that

$$\langle \Omega(\chi, \overline{K})\chi, \chi \rangle = cG\langle \chi, K \rangle \tag{3.2.5}$$

for all $K \in \mathcal{H}$. This is a first step on the way of giving a nice expression to the terms of (3.2.1). To proceed, we need a further computational result.

LEMMA 3.2.1: *Let $F\colon T^{1,0}M \to \mathbb{R}^+$ be a strongly pseudoconvex Finsler metric on a complex manifold M. Then*

$$\langle (\nabla_V \Omega)(H, \overline{K})\chi, \chi \rangle = \langle \tau(\theta(H, V), \overline{K}), \iota \rangle$$

for all $V \in \mathcal{V}$ and H, $K \in \mathcal{H}$. In particular,

$$\langle (\nabla_V \Omega)(\chi, \overline{K})\chi, \chi \rangle = 0$$

for all $V \in \mathcal{V}$ and $K \in \mathcal{H}$.

Proof: As in Lemma 3.1.5 it suffices to consider $\Omega^{\mathcal{H}} = \Omega^\alpha_\beta \otimes dz^\beta \otimes \delta_\alpha$; so

$$\nabla_V \Omega^{\mathcal{H}} = (\nabla_V \Omega^\alpha_\beta) \otimes dz^\beta \otimes \delta_\alpha - \Omega^\alpha_\gamma \otimes \omega^\gamma_\beta(V)\, dz^\beta \otimes \delta_\alpha + \Omega^\gamma_\beta \otimes dz^\beta \otimes \omega^\alpha_\gamma(V)\delta_\alpha.$$

We are interested only in the horizontal part. Taking H, $K \in \mathcal{H}$ we get

$$\begin{aligned}
G_\alpha(\nabla_V \Omega^\alpha_\beta)(H, \overline{K}) &= G_\alpha\big[V(R^\alpha_{\beta;\mu\bar\nu}) - R^\alpha_{\beta;\rho\bar\nu}\omega^\rho_\mu(V)\big] H^\mu \overline{K^\nu} \\
&= -G_\alpha\big[\dot\partial_\lambda \delta_{\bar\nu}(\Gamma^\alpha_{\beta;\mu}) - \delta_{\bar\nu}(\Gamma^\alpha_{\beta;\rho})\Gamma^\rho_{\mu\lambda} + \dot\partial_\lambda\big(\Gamma^\alpha_{\beta\sigma}\delta_{\bar\nu}(\Gamma^\sigma_{;\mu})\big) \\
&\qquad - \Gamma^\alpha_{\beta\sigma}\delta_{\bar\nu}(\Gamma^\sigma_{;\rho})\Gamma^\rho_{\mu\lambda}\big] V^\lambda H^\mu \overline{K^\nu} \\
&= -G_\alpha\big[\delta_{\bar\nu}\dot\partial_\lambda(\Gamma^\alpha_{\beta;\mu}) - \Gamma^{\bar\tau}_{\lambda;\bar\nu}\dot\partial_{\bar\tau}(\Gamma^\alpha_{\beta;\mu}) - \delta_{\bar\nu}(\Gamma^\alpha_{\beta;\rho})\Gamma^\rho_{\mu\lambda} + \dot\partial_\lambda(\Gamma^\alpha_{\beta\sigma})\delta_{\bar\nu}(\Gamma^\sigma_{;\mu}) \\
&\qquad + \Gamma^\alpha_{\beta\sigma}\delta_{\bar\nu}(\Gamma^\sigma_{\lambda;\mu}) - \Gamma^\alpha_{\beta\sigma}\Gamma^\sigma_{\bar\tau;\mu}\Gamma^{\bar\tau}_{\lambda;\bar\nu} - \Gamma^\alpha_{\beta\sigma}\delta_{\bar\nu}(\Gamma^\sigma_{;\rho})\Gamma^\rho_{\mu\lambda}\big] V^\lambda H^\mu \overline{K^\nu}
\end{aligned}$$

(where we used Lemma 2.3.3),

$$\begin{aligned}
&= -\big[\delta_{\bar\nu}\big(G_\alpha \dot\partial_\beta(\Gamma^\alpha_{\lambda;\mu})\big) - \Gamma^{\bar\tau}_{\lambda;\bar\nu} G_\alpha \dot\partial_\beta(\Gamma^\alpha_{\bar\tau;\mu}) - \delta_{\bar\nu}(G_\alpha \Gamma^\alpha_{\beta;\rho})\Gamma^\rho_{\mu\lambda} \\
&\qquad + G_\alpha \dot\partial_\lambda(\Gamma^\alpha_{\beta\sigma})\delta_{\bar\nu}(\Gamma^\sigma_{;\mu}) + G_{\beta\sigma}\delta_{\bar\nu}(\Gamma^\sigma_{\lambda;\mu}) - G_{\beta\sigma}\Gamma^\sigma_{\bar\tau;\mu}\Gamma^{\bar\tau}_{\lambda;\bar\nu} \\
&\qquad - G_{\beta\sigma}\delta_{\bar\nu}(\Gamma^\sigma_{;\rho})\Gamma^\rho_{\mu\lambda}\big] V^\lambda H^\mu \overline{K^\nu}
\end{aligned}$$

(where we used Lemma 2.3.4 and $G_\alpha \Gamma^\alpha_{\beta\sigma} = G_{\beta\sigma}$),

$$\begin{aligned}
&= -\big[-\delta_{\bar\nu}(G_{\alpha\beta}\Gamma^\alpha_{\lambda;\mu}) + \delta_{\bar\nu}\big(\dot\partial_\beta \delta_\mu(G_\lambda)\big) + \Gamma^{\bar\tau}_{\lambda;\nu} G_{\alpha\beta}\Gamma^\alpha_{\bar\tau;\mu} - \delta_{\bar\nu}(G_\alpha \Gamma^\alpha_{\beta;\rho})\Gamma^\rho_{\mu\lambda} \\
&\qquad + G_\alpha \dot\partial_\lambda(\Gamma^\alpha_{\beta\sigma})\delta_{\bar\nu}(\Gamma^\sigma_{;\mu}) + G_{\beta\sigma}\delta_{\bar\nu}(\Gamma^\sigma_{\lambda;\mu}) - G_{\beta\sigma}\Gamma^\sigma_{\bar\tau;\mu}\Gamma^{\bar\tau}_{\lambda;\bar\nu} \\
&\qquad - G_{\beta\sigma}\delta_{\bar\nu}(\Gamma^\sigma_{;\rho})\Gamma^\rho_{\mu\lambda}\big] V^\lambda H^\mu \overline{K^\nu}
\end{aligned}$$

(where we used $G_\alpha \Gamma^\alpha_{\bar\tau;\mu} = 0$ and $G_\alpha \Gamma^\alpha_{\lambda;\mu} = \delta_\mu(G_\lambda)$),

$$\begin{aligned}
&= -\big[-\delta_{\bar\nu}(G_{\alpha\beta}\Gamma^\alpha_{\lambda;\mu}) + \delta_{\bar\nu}\big(\dot\partial_\beta \delta_\mu(G_\lambda)\big) - \delta_{\bar\nu}(G_\alpha \Gamma^\alpha_{\beta;\rho})\Gamma^\rho_{\mu\lambda} + G_\alpha \dot\partial_\lambda(\Gamma^\alpha_{\beta\sigma})\delta_{\bar\nu}(\Gamma^\sigma_{;\mu}) \\
&\qquad + G_{\beta\sigma}\delta_{\bar\nu}(\Gamma^\sigma_{\lambda;\mu}) - G_{\beta\sigma}\delta_{\bar\nu}(\Gamma^\sigma_{;\rho})\Gamma^\rho_{\mu\lambda}\big] V^\lambda H^\mu \overline{K^\nu}.
\end{aligned}$$

Furthermore,

$$\begin{aligned}
G_\alpha \Omega^\alpha_\gamma(H, \overline{K})\omega^\gamma_\beta(V) &= -G_\alpha\big[\delta_{\bar\nu}(\Gamma^\alpha_{\gamma;\mu}) + \Gamma^\alpha_{\gamma\sigma}\delta_{\bar\nu}(\Gamma^\sigma_{;\mu})\big]\Gamma^\gamma_{\beta\lambda} V^\lambda H^\mu \overline{K^\nu} \\
&= -\big[\delta_{\bar\nu}\big(\delta_\mu(G_\gamma)\big) + G_{\gamma\sigma}\delta_{\bar\nu}(\Gamma^\sigma_{;\mu})\big]\Gamma^\gamma_{\beta\lambda} V^\lambda H^\mu \overline{K^\nu};
\end{aligned}$$

$$G_\alpha \omega^\alpha_\gamma(V)\Omega^\gamma_\beta(H,\overline{K}) = -G_\alpha\Gamma^\alpha_{\gamma\lambda}\big[\delta_{\bar\nu}(\Gamma^\gamma_{\beta;\mu}) + \Gamma^\gamma_{\beta\sigma}\delta_{\bar\nu}(\Gamma^\sigma_{;\mu})\big]V^\lambda H^\mu \overline{K^\nu}$$
$$= -G_{\gamma\lambda}\big[\delta_{\bar\nu}(\Gamma^\gamma_{\beta;\mu}) + \Gamma^\gamma_{\beta\sigma}\delta_{\bar\nu}(\Gamma^\sigma_{;\mu})\big]V^\lambda H^\mu \overline{K^\nu}.$$

Summing up we find

$$\begin{aligned}
&\langle(\nabla_V\Omega)(H,\overline{K})\chi,\chi\rangle\\
&= -\big[-\delta_{\bar\nu}(G_{\alpha\beta}\Gamma^\alpha_{\lambda;\mu}) + \delta_{\bar\nu}\big(\dot\partial_\beta\delta_\mu(G_\lambda)\big) - \delta_{\bar\nu}(G_\alpha\Gamma^\alpha_{\beta;\rho})\Gamma^\rho_{\mu\lambda} + G_\alpha\dot\partial_\lambda(\Gamma^\alpha_{\beta\sigma})\delta_{\bar\nu}(\Gamma^\sigma_{;\mu})\\
&\qquad + G_{\beta\sigma}\delta_{\bar\nu}(\Gamma^\sigma_{\lambda;\mu}) - G_{\beta\sigma}\delta_{\bar\nu}(\Gamma^\sigma_{;\rho})\Gamma^\rho_{\mu\lambda} + \delta_{\bar\nu}\big(\delta_\mu(G_\gamma)\big)\Gamma^\gamma_{\beta\lambda} + G_{\gamma\sigma}\delta_{\bar\nu}(\Gamma^\sigma_{;\mu})\Gamma^\gamma_{\beta\lambda}\\
&\qquad + G_{\gamma\lambda}\delta_{\bar\nu}(\Gamma^\gamma_{\beta;\mu}) + G_{\gamma\nu}\Gamma^\gamma_{\beta\sigma}\delta_{\bar\nu}(\Gamma^\sigma_{;\mu})\big]v^\beta V^\lambda H^\mu\overline{K^\nu}\\
&= -\big[\delta_{\bar\nu}\big(v^\beta\dot\partial_\beta\delta_\mu(G_\lambda)\big) - \delta_{\bar\nu}(G_\alpha v^\beta\Gamma^\alpha_{\beta;\rho})\Gamma^\rho_{\mu\lambda} + G_\alpha v^\beta\dot\partial_\lambda(\Gamma^\alpha_{\beta\sigma})\delta_{\bar\nu}(\Gamma^\sigma_{;\mu})\\
&\qquad + G_{\gamma\lambda}\delta_{\bar\nu}(\Gamma^\gamma_{;\mu})\big]V^\lambda H^\mu\overline{K^\nu}\\
&= -\big[-\delta_{\bar\nu}(G_{;\rho})\Gamma^\rho_{\mu\lambda} - G_{\lambda\sigma}\delta_{\bar\nu}(\Gamma^\sigma_{;\mu}) + G_{\lambda\gamma}\delta_{\bar\nu}(\Gamma^\gamma_{;\mu})\big]V^\lambda H^\mu\overline{K^\nu}
\end{aligned}$$

(where we used $v^\beta\dot\partial_\beta\delta_\mu(G_\lambda) = v^\beta(G_{\lambda\beta;\mu} - \Gamma^\sigma_{\beta;\mu}G_{\lambda\sigma} - \Gamma^\sigma_{;\mu}G_{\lambda\sigma\beta}) = 0$
and $G_\alpha v^\beta\dot\partial_\lambda(\Gamma^\alpha_{\beta\sigma}) = -G_{\lambda\sigma}$),

$$\begin{aligned}
&= \delta_{\bar\nu}(G_{;\rho})\Gamma^\rho_{\mu\lambda}V^\lambda H^\mu\overline{K^\nu} = \delta_{\bar\nu}(G_\sigma\Gamma^\sigma_{;\rho})\Gamma^\rho_{\mu\lambda}V^\lambda H^\mu\overline{K^\nu}\\
&= G_\sigma\delta_{\bar\nu}(\Gamma^\sigma_{;\rho})\Gamma^\rho_{\mu\lambda}V^\lambda H^\mu\overline{K^\nu} = \langle\tau\big(\theta(H,V),\overline{K}\big),\iota\rangle.
\end{aligned}$$

The final assertion follows from $\theta(\chi,V)=0$. □

Now we are in condition to give the desired expression for one of the terms of (3.2.1) involving the Hermitian product:

Proposition 3.2.2: *Let $F\colon T^{1,0}M \to \mathbb{R}^+$ be a strongly pseudoconvex Finsler metric on a complex manifold M with constant holomorphic curvature $2c \in \mathbb{R}$. Assume that*

$$\forall H \in \mathcal{H} \qquad \langle\bar\partial_H\theta(H,\chi,\bar\chi),\chi\rangle = 0. \tag{3.2.6}$$

Then

$$\langle\Omega(H,\overline{K})\chi,\chi\rangle + \langle\Omega(\chi,\overline{K})H,\chi\rangle = c\,\{\langle H,\chi\rangle\langle\chi,K\rangle + \langle\chi,\chi\rangle\langle H,K\rangle\}, \tag{3.2.7}$$

for all H, $K \in \mathcal{H}$. In particular, if

$$\forall H,\ K \in \mathcal{H} \qquad \langle\bar\partial_H\theta(H,\chi,\overline{K}),\chi\rangle = 0, \tag{3.2.8}$$

then

$$\langle\Omega(\chi,\overline{K})H,\chi\rangle = \frac{c}{2}\,\{\langle H,\chi\rangle\langle\chi,K\rangle + \langle\chi,\chi\rangle\langle H,K\rangle\} \tag{3.2.9}$$

for all H, $K \in \mathcal{H}$.

Proof: Take V, $W \in \mathcal{V}$ such that $\Theta(V) = H$ and $\Theta(W) = K$ and extend them as sections of $\mathcal{V}$ in any way (and thus extend H and K as sections of $\mathcal{H}$ via Θ). We have

$$V\big(cG\langle\chi,K\rangle\big) = c\big[\langle H,\chi\rangle\langle\chi,K\rangle + G\langle H,K\rangle + G\langle\chi,\nabla_{\overline{V}}K\rangle\big],$$

and, using Lemma 3.2.1,

$$
\begin{aligned}
&V\langle\Omega(\chi,\overline{K})\chi,\chi\rangle\\
&\quad=\langle(\nabla_V\Omega)(\chi,\overline{K})\chi,\chi\rangle+\langle\Omega(H,\overline{K})\chi,\chi\rangle+\langle\Omega(\chi,\overline{\nabla_V K})\chi,\chi\rangle+\langle\Omega(\chi,\overline{K})H,\chi\rangle\,.\\
&\quad=\langle\Omega(H,\overline{K})\chi,\chi\rangle+\langle\Omega(\chi,\overline{\nabla_{\overline{V}}K})\chi,\chi\rangle+\langle\Omega(\chi,\overline{K})H,\chi\rangle,
\end{aligned}
$$

Since (3.2.6) holds, we can use Proposition 3.1.7 (that is, (3.2.5) applied both to K and to $\nabla_{\overline{V}}K$) to get exactly (3.2.7).

Finally, (3.2.9) follows from Lemma 3.1.6. □

The other Hermitian product term of (3.2.1) is obtained in the following:

PROPOSITION 3.2.3: *Let $F\colon T^{1,0}M\to\mathbb{R}^+$ be a strongly pseudoconvex Finsler metric on a complex manifold M with constant holomorphic curvature $2c\in\mathbb{R}$. Assume that* (3.2.6) *holds. Then*

$$\langle\Omega(H,\bar\chi)K,\chi\rangle=c\,\{\langle H,\chi\rangle\langle K,\chi\rangle+\langle\chi,\chi\rangle\langle\!\langle H,K\rangle\!\rangle\}$$

for all H, $K\in\mathcal{H}$.

Proof: First of all, we have

$$\langle\Omega(H,\bar\chi)\chi,\chi\rangle=\overline{\langle\Omega(\chi,\overline{H})\chi,\chi\rangle}=c\,G\langle H,\chi\rangle, \tag{3.2.10}$$

where we used (2.5.2) and (3.2.5). Now let $W\in\mathcal{V}$ be such that $\Theta(W)=K$; then

$$
\begin{aligned}
W(c\,G\langle H,\chi\rangle)&=c\,\{\langle K,\chi\rangle\langle H,\chi\rangle+G\langle\nabla_W H,\chi\rangle\}\,,\\
W\langle\Omega(H,\bar\chi)\chi,\chi\rangle&=\langle(\nabla_W\Omega)(H,\bar\chi)\chi,\chi\rangle+\langle\Omega(\nabla_W H,\bar\chi)\chi,\chi\rangle+\langle\Omega(H,\bar\chi)K,\chi\rangle,
\end{aligned}
$$

so that (3.2.10) yields

$$\langle\Omega(H,\bar\chi)K,\chi\rangle=c\langle K,\chi\rangle\langle H,\chi\rangle-\langle(\nabla_W\Omega)(H,\bar\chi)\chi,\chi\rangle.$$

Now Lemma 3.2.1 gives

$$
\begin{aligned}
\langle(\nabla_W\Omega)(H,\bar\chi)\chi,\chi\rangle&=\langle\tau\big(\theta(H,W),\bar\chi\big),\iota\rangle=\langle\Omega\big(\theta(H,W),\bar\chi\big)\chi,\chi\rangle\\
&=c\,G\langle\theta(H,W),\chi\rangle,
\end{aligned}
$$

again by (3.2.10). But

$$\langle\theta(H,W),\chi\rangle=-G_\alpha\Gamma^\alpha_{\nu\beta}K^\beta H^\nu=-\langle\!\langle H,K\rangle\!\rangle,$$

and we are done. □

Now that we have got the Hermitian product terms, we turn our attention to the symmetric product terms. The first one is almost immediate:

PROPOSITION 3.2.4: *Let $F: T^{1,0}M \to \mathbb{R}^+$ be a strongly pseudoconvex Finsler metric on a complex manifold M with constant holomorphic curvature $2c \in \mathbb{R}$. Assume that (3.2.8) holds. Then*

$$\tau^{\mathcal{H}}(K, \bar{\chi}) = \frac{c}{2}\{\langle K, \chi\rangle\chi + \langle\chi, \chi\rangle K\} \tag{3.2.11}$$

for all $K \in \mathcal{H}$. In particular,

$$\langle\!\langle H, \tau^{\mathcal{H}}(K, \bar{\chi})\rangle\!\rangle = \frac{c}{2}\langle\chi, \chi\rangle\langle\!\langle H, K\rangle\!\rangle \tag{3.2.12}$$

for all H, $K \in \mathcal{H}$.

Proof: As we remarked in Lemma 2.3.8, we have

$$\tau^{\mathcal{H}} = \Omega(\cdot, \cdot)\chi.$$

Thus, using again (2.5.2), we get

$$\langle H, \tau^{\mathcal{H}}(K, \bar{\chi})\rangle = \langle H, \Omega(K, \bar{\chi})\chi\rangle = \langle\Omega(\chi, \overline{K})H, \chi\rangle$$

for all H, $K \in \mathcal{H}$. Then (3.2.9) yields (3.2.11), and (3.2.12) follows immediately. □

For the other symmetric product term we need the weak Kähler condition:

PROPOSITION 3.2.5: *Let $F: T^{1,0}M \to \mathbb{R}^+$ be a weakly Kähler Finsler metric on a complex manifold M such that (3.2.8) holds. Then*

$$\langle\!\langle H, \tau^{\mathcal{H}}(\chi, \overline{K})\rangle\!\rangle = 0$$

for all H, $K \in \mathcal{H}$.

Proof: The weak Kähler condition $\langle\theta(H, \chi), \chi\rangle = 0$ for all $H \in \mathcal{H}$ implies

$$\forall H, K \in \mathcal{H} \qquad \langle(\nabla_{\overline{K}}\theta)(H, \chi), \chi\rangle = 0 \tag{3.2.13}$$

because $\nabla_{\overline{K}}\chi = 0 = \nabla_K\chi$. Now, writing $\theta = \theta^\alpha \otimes \delta_\alpha$, we have $\nabla_{\overline{K}}\theta = (\nabla_{\overline{K}}\theta^\alpha) \otimes \delta_\alpha$ and

$$\nabla_{\overline{K}}\theta^\alpha = \overline{K^\tau}\delta_{\bar\tau}(\Gamma^\alpha_{\nu;\mu})\, dz^\mu \wedge dz^\nu + \overline{K^\tau}\delta_{\bar\tau}(\Gamma^\alpha_{\nu\gamma})\, \psi^\gamma \wedge dz^\nu.$$

Therefore (3.2.13) implies

$$G_\alpha[\delta_{\bar\tau}(\Gamma^\alpha_{\nu;\mu}) - \delta_{\bar\tau}(\Gamma^\alpha_{\mu;\nu})]H^\mu\overline{K^\tau}v^\nu = 0 \tag{3.2.14}$$

for all H, $K \in \mathcal{H}$.

Writing the curvature in local coordinates we find

$$\langle\Omega(\chi, \overline{K})H, \chi\rangle = -G_\alpha[\delta_{\bar\tau}(\Gamma^\alpha_{\mu;\nu}) + \Gamma^\alpha_{\mu\sigma}\delta_{\bar\tau}(\Gamma^\sigma_{;\nu})]H^\mu\overline{K^\tau}v^\nu,$$
$$\langle\Omega(H, \overline{K})\chi, \chi\rangle = -G_\alpha[\delta_{\bar\tau}(\Gamma^\alpha_{\nu;\mu}) + \Gamma^\alpha_{\nu\sigma}\delta_{\bar\tau}(\Gamma^\sigma_{;\mu})]H^\mu\overline{K^\tau}v^\nu.$$

So (3.2.14) yields

$$\begin{aligned}\langle\Omega(\chi,\overline{K})H,\chi\rangle - \langle\Omega(H,\overline{K})\chi,\chi\rangle &= -G_\alpha\Gamma^\alpha_{\mu\sigma}\delta_{\bar\tau}(\Gamma^\sigma_{;\nu})H^\mu\overline{K^\tau}v^\nu\\ &= \langle\!\langle H,\tau^{\mathcal{H}}(\chi,\overline{K})\rangle\!\rangle,\end{aligned}$$

and the assertion follows from (3.2.8). □

We can finally collect all our computations in

THEOREM 3.2.6: *Let $F: T^{1,0}M \to \mathbb{R}^+$ be a weakly Kähler Finsler metric on a complex manifold M. Assume F has constant holomorphic curvature $2c \in \mathbb{R}$ and that* (3.2.8) *holds. Then*

$$\begin{aligned}\mathrm{Re}\Big[\langle\Omega(\chi,\overline{K})H,\chi\rangle - \langle\Omega(H,\bar\chi)K,\chi\rangle + \langle\!\langle H,\tau^{\mathcal{H}}(K,\bar\chi)\rangle\!\rangle - \langle\!\langle H,\tau^{\mathcal{H}}(\chi,\overline{K})\rangle\!\rangle\Big]\\ = \frac{c}{2}\,\mathrm{Re}\Big[G\big\{\langle H,K\rangle - \langle\!\langle H,K\rangle\!\rangle\big\} + \langle H,\chi\rangle\big\{\langle\chi,K\rangle - 2\langle K,\chi\rangle\big\}\Big]\end{aligned}$$

for all H, $K \in \mathcal{H}$.

Proof: It follows from Propositions 3.2.2, 3.2.3, 3.2.4, 3.2.5. □

The computations made above allow us to prove the main result of this subsection:

THEOREM 3.2.7: *Let $F: T^{1,0}M \to \mathbb{R}^+$ be a complete strongly pseudoconvex Finsler metric on a simply connected complex manifold M. Assume that:*

(i) *F is Kähler;*
(ii) *F has nonpositive constant holomorphic curvature $2c \le 0$;*
(iii) *$\langle\bar\partial_H\theta(H,\chi,\overline{K}),\chi\rangle = 0$ for all H, $K \in \mathcal{H}$;*
(iv) *F is strongly convex.*

Then $\exp_p: T^{1,0}_p \to M$ is a Lipschitz E-diffeomorphism at the origin for any $p \in M$. Furthermore, M is foliated by isometric totally geodesic holomorphic embeddings of the unit disk Δ endowed with a suitable multiple of the Poincaré metric if $c < 0$, or by isometric totally geodesic holomorphic embeddings of $\mathbb{C}$ endowed with the euclidean metric if $c = 0$. In particular, if $c = -2$ then F is the Kobayashi metric of M, and if $c = 0$ then the Kobayashi metric of M vanishes identically.

Proof: Fix $p \in M$, and let $\sigma_0: [0,a] \to M$ be a radial normal geodesic in M with $\sigma_0(0) = 0$, and Σ a regular fixed variation of σ_0 such that the transversal vector U of Σ satisfies

$$\mathrm{Re}\langle U^H, T^H\rangle_{\dot\sigma_0} \equiv 0$$

(note that, by Corollary 2.6.5, we single out the same variations of σ_0 by requiring orthogonality with respect to the Riemannian structure and the horizontal lift induced by the real Finsler metric F^o; in other words, the space $\mathcal{X}_0[0,a]$ is the same both working in the complex setting and in the real setting). Then Theorems 2.4.4 and 3.2.6 yield

$$\frac{d^2\ell_\Sigma}{ds^2}(0) = \int_0^a \left\{ \|\nabla_{T^H+\overline{T^H}} U^H\|^2_{\dot\sigma_0} - \frac{c\,G(\dot\sigma_0)}{2} \operatorname{Re}\left[\langle U^H, U^H\rangle_{\dot\sigma_0} - \langle\!\langle U^H, U^H\rangle\!\rangle_{\dot\sigma_0}\right]\right\} dt.$$

Now, the curvature is nonpositive, and Proposition 2.6.1 yields

$$\operatorname{Re}\left[\langle U^H, U^H\rangle - \langle\!\langle U^H, U^H\rangle\!\rangle\right] = \langle (iU^H)^o \mid (iU^H)^o\rangle \geq 0;$$

so

$$\frac{d^2\ell_\Sigma}{ds^2}(0) \geq 0.$$

Furthermore, the second variation may vanish iff $\nabla_{T^H+\overline{T^H}} U^H \equiv 0$, that is iff U is parallel along σ_0. But $U(0) = o_p$; so the second variation may vanish if and only if the transversal vector is identically zero. Recalling (1.7.5), we have just proved that the Morse index form I_0^a is positive definite on $\mathcal{X}_0[0,a]$. But σ_0 was any radial normal geodesic in M; so we can apply the Cartan-Hadamard Theorem 1.7.17, proving that $\exp_p$ is an E-diffeomorphism at the origin. The remaining assertions follows from Theorems 3.1.10, 3.1.15 and (for the case $c = 0$) from standard properties of the Kobayashi metric. □

We recall that a *complex geodesic* in the sense of Vesentini [V] is a holomorphic map $\varphi\colon \Delta \to M$ which is an isometry with respect to the Poincaré distance on Δ and the Kobayashi distance on M. As a consequence of the last theorem we get existence and uniqueness of complex geodesics:

COROLLARY 3.2.8: *Let $F\colon T^{1,0}M \to \mathbb{R}^+$ be a complete strongly pseudoconvex Finsler metric on a simply connected complex manifold M. Assume that:*

(i) *F is Kähler;*
(ii) *F has constant holomorphic curvature -4;*
(iii) *$\langle \bar\partial_H \theta(H, \chi, \overline{K}), \chi\rangle = 0$ for all H, $K \in \mathcal{H}$;*
(iv) *F is strongly convex.*

Then every geodesic complex curve in M is a complex geodesic in the sense of Vesentini. In particular, for any $p \in M$ and $v \in T_p^{1,0}M$ — or for any pair of distinct points p, $q \in M$ — there exists a unique complex geodesic passing through $(p; v)$ — respectively, passing through p and q.

Proof: By Theorem 3.2.7, the exponential at any point $p \in M$ is injective; in particular, then, the geodesics are all globally length-minimizing. Since, by definition, a geodesic complex curve sends geodesics with respect to the Poincaré metric in geodesics with respect to F (which coincides with the Kobayashi metric because $c = -2$ here; Theorem 3.1.15), it follows that every geodesic complex curve is a complex geodesic in the sense of Vesentini.

Finally the last statement follows from the analogous statement for real geodesics, and from the fact that every real geodesic is contained in a geodesic complex curve. □

We explicitly remark that the conditions (i)–(iv) in Theorem 3.2.7 are not necessarily independent. For instance, the results of [P2] seems to suggest that conditions (i)–(iii) directly imply the convexity of the indicatrices of F — but perhaps not the strong convexity.

Finally, an advice: condition (iii) is *not* a consequence of the Kähler condition, unless F comes from a Hermitian metric. In fact, differentiating the weakly Kähler condition one gets

$$\overline{K}\langle\theta(\cdot,\chi),\chi\rangle = \langle\bar{\partial}_H\theta(\cdot,\chi,\overline{K}),\chi\rangle + \langle\!\langle\cdot,\tau^{\mathcal{H}}(\chi,\overline{K})\rangle\!\rangle.$$

3.2.2. The Monge-Ampère exhaustion

Theorem 3.2.7 suggests that in case the Kobayashi metric on a complex manifold M is particularly nice then the geometric picture is quite the same as the one Lempert [L] described for strongly convex domains in $\mathbb{C}^n$. There is still an important property that in this class of domain is available: in terms of the Kobayashi distance it is possible to construct Monge-Ampère potentials with logarithmic singularity at any given point. Our next task is to show that the same can be done in a Finsler setting. An analogous results will be achieved for metrics with holomorphic curvature constantly zero. Of course in this case we are in a parabolic environment rather than in a hyperbolic one, and the model that one should keep in mind is a complex Minkowski space.

Let set up the necessary notations and state our assumptions. Let M be a simply connected complex manifold of dimension n and let us suppose that on M is defined a strongly pseudoconvex Finsler metric F such that

$$F \text{ is complete Kähler;} \tag{3.2.15}$$

$$K_F \equiv 2c \leq 0, \text{ where either } c = -2 \text{ or } c = 0; \tag{3.2.16}$$

$$\langle\bar{\partial}_H\theta(H,\chi,\overline{K}),\chi\rangle = 0 \text{ for all } H,\, K \in \mathcal{H}; \tag{3.2.17}$$

$$F \text{ is strongly convex.} \tag{3.2.18}$$

Of course, in light of Theorem 3.1.15, when $c = -2$ we are considering the Kobayashi metric of M and we are assuming that it has special properties. For simplicity of exposition, for the moment we shall nevertheless treat the negative curved case together with the flat one and we shall underline the differences only later when it will be necessary.

We recall some more notations. Let

$$\tilde{S}_p = \{v \in T_p^{1,0}M \mid F(v) = 1\} \tag{3.2.19}$$

be the unit sphere in the tangent space at p and

$$I_p = I_p(M) = \{v \in T_p^{1,0}M \mid F(v) < 1\} \tag{3.2.20}$$

be the indicatrix of F at p. For $v \in \tilde{S}_p$ let φ_v be the c-geodesic complex curve through $(p; v)$, which exists because of Theorem 3.1.10. We define maps $H_{-2}: \Delta \times \tilde{S}_p \to M$ in case $c = -2$ and $H_0: \mathbb{C} \times \tilde{S}_p \to M$ when $c = 0$ by

$$H_c(\zeta, v) = \varphi_v(\zeta). \tag{3.2.21}$$

Because of Proposition 3.1.11, the maps H_c are C^∞. Finally we define maps $E_{-2}: I_p \to M$ in case $c = -2$ and $E_0: T_p^{1,0}M \to M$ when $c = 0$ by

$$E_c(v) = H_c(F(v), v/F(v)) = \varphi_{v/F(v)}(F(v)). \tag{3.2.22}$$

Again because of Proposition 3.1.11, for $\lambda \in \mathbb{C}$ with $|\lambda| = 1$ and $v \in \tilde{S}_p$

$$\varphi_{\lambda v}(\zeta) = \varphi_v(\lambda\zeta); \tag{3.2.23}$$

therefore for any choice of $\lambda \in \mathbb{C}$ with $|\lambda| = 1$ and $v \in \tilde{S}_p$ we have

$$E_c(\lambda v) = \varphi_v(\lambda F(v)). \tag{3.2.24}$$

Choosing any system of coordinates around p we can identify $\mathbb{C}^n \cong T_p^{1,0}M$ so that E_{-2} will be defined on a strongly convex complete circular domain $I_p \subset \mathbb{C}^n$ and E_0 will be defined on all $\mathbb{C}^n$. Let us collect some immediate properties of the maps E_c. If $\exp_p: T_p^{1,0}M \to M$ is the exponential map of F at p, we have:

PROPOSITION 3.2.9: *Let $F: T^{1,0}M \to \mathbb{R}^+$ be a strongly pseudoconvex Finsler metric on a simply connected complex manifold M satisfying (3.2.15)–(3.2.18). Then for $v \in I_p$ and $c = -2$ we have*

$$E_{-2}(v) = \exp_p\left(\frac{\operatorname{atanh}(F(v))}{F(v)}\, v\right) \tag{3.2.25}$$

while for $v \in \mathbb{C}^n$ and $c = 0$

$$E_0(v) = \exp_p(v) \tag{3.2.26}$$

Furthermore, as a consequence,

(i) E_c is an E-diffeomorphism at the origin;
(ii) if L is any line through the origin in $T_p^{1,0}M$, the restrictions of E_0 to L and of E_{-2} to the disk $L \cap I_p$ are holomorphic;
(iii) if M is a Stein manifold, the map E_c is biholomorphic iff it is of class C^∞ at the origin.

Proof: The equalities (3.2.25) and (3.2.26) are consequence of the definition of E_c in terms of c-geodesic complex curves.

(i) follows immediately from Theorem 3.2.7, Theorem 1.6.2 and the fact that the function

$$f(t) = \frac{\operatorname{atanh} t}{t}$$

is an analytic function.

(ii) is a direct consequence of the definition of E_c. Finally if M is Stein, then it may embedded in $\mathbb{C}^N$ for some suitable N. Then it is well known — for instance it is a consequence of a theorem of Forelli (see [Ru]) — that the components of the composition of E_c with the embedding are holomorphic functions because are smooth and have holomorphic restriction to the complex lines through the origin. □

We shall now introduce on M a nice exhaustion function. For simplicity we set

$$R_{-2} = 1 \qquad R_0 = +\infty;$$

DEFINITION 3.2.1: The *Monge-Ampère exhaustion* $\sigma_c\colon M \to [0, R_c)$ of M is defined as follows. If $q \in M$ then, for some $\zeta \in \mathbb{C}$ and $v \in \tilde{S}_p$, $q = E_c(\zeta v)$ and we set

$$\sigma_c(q) = \sigma_c(E_c(\zeta v)) = |\zeta|^2. \tag{3.2.27}$$

Because of (3.2.23) and (3.2.24), the definition of σ_c is well posed. We gather now the important properties which the function σ_c enjoys.

THEOREM 3.2.10: *Let $F\colon T^{1,0}M \to \mathbb{R}^+$ be a strongly pseudoconvex Finsler metric on a simply connected complex manifold M of dimension n satisfying (3.2.15)–(3.2.18), and let $p \in M$. Then the function σ_c is an exhaustion of M with the following properties:*

(i) *if ρ is the distance from p relative to the metric F, then $\sigma_{-2} = (\tanh\rho)^2$ and $\sigma_0 = \rho^2$;*
(ii) $G = F^2 = \sigma_c \circ E_c$;
(iii) $\sigma_c \in C^0(M) \cap C^\infty(M \setminus \{p\})$;
(iv) *if $\pi\colon \check{M} \to M$ is the blow-up at p, then $\sigma_c \circ \pi \in C^\infty(\check{M})$;*
(v) *$dd^c\sigma_c > 0$ on $M \setminus \{p\}$;*
(vi) *$dd^c \log\sigma_c \geq 0$ on $M \setminus \{p\}$;*
(vii) *$(dd^c \log\sigma_c)^n = 0$ on $M \setminus \{p\}$;*
(viii) *$\log\sigma_c(z) = \log\|z\|^2 + O(1)$ with respect to any coordinate system centered in p.*

In particular M is a Stein manifold.

Proof: Some of the properties are immediate from the definition of σ_c. In fact the statement that σ_c is an exhaustion function, (i), (ii) and (iii) are direct consequence of the definition of σ_c and of (3.2.25), (3.2.26) and (3.2.27). For (iv), as we noticed above the map H_c is of class C^∞ and hence $\sigma_c \circ \pi \in C^\infty(\check{M})$ iff $\rho \circ H_c$ is of class C^∞ and this is obvious from the definition. For (v) we must work more carefully. The proof follows closely the line of Theorems 2.2 and 2.3 of Semmes [S]. In fact our map E_c is essentially what Semmes calls a *Riemann map*. We give the necessary details here.

We start recalling some notations. If $r > 0$, we set

$$B_p(r) = \{q \in M \mid \rho(q) < r\}, \qquad S_p(r) = \{q \in M \mid \rho(q) = r\}$$

and

$$\tilde{B}_p(r) = \{v \in T_p^{1,0}M \mid F(v) < r\}, \qquad \tilde{S}_p(r) = \{v \in T_p^{1,0}M \mid F(v) = r\}.$$

Thanks to (ii), we know that E_c sends $\tilde{S}_p(r)$ onto $S_p(r')$, where $r' = r$ if $c = 0$, and $r' = \operatorname{atanh}(r)$ if $c = -2$. For $v \in \tilde{S}_p(r)$, let $q = E_c(v) \in S_p(r')$; we denote by $H_v\tilde{S}_p(r)$ the holomorphic tangent space to $\tilde{S}_p(r)$ at v, and by $H_qS_p(r')$ the holomorphic tangent space to $S_p(r')$ at q.

Inspired by [L] and [P3], we shall say that a holomorphic embedding φ of the disk Δ (if $c = -2$) or the plane $\mathbb{C}$ (if $c = 0$) in M is *stationary* if the $\bigwedge^{1,0}M$-valued map $\tilde{\varphi}$ given by

$$\tilde{\varphi}(\zeta) = \frac{1}{\bar{\zeta}}\,\partial(\sigma_c)_{\varphi(\zeta)}$$

extends to a holomorphic map defined on Δ (or $\mathbb{C}$). In particular, then, setting $p = \varphi(0)$, the holomorphic tangent spaces to $S_p(r)$ yield a holomorphic fiber bundle along the image of φ.

Then the main point in proving that our E_c is a Riemann map is contained in the following

LEMMA 3.2.11: *Let $F: T^{1,0}M \to \mathbb{R}^+$ be a strongly pseudoconvex Finsler metric on a simply connected complex manifold M satisfying* (3.2.15)–(3.2.18), *and fix $p \in M$. Then:*

(i) *every geodesic complex curve φ_v is stationary;*
(ii) *for every $r > 0$ and $v_0 \in \tilde{S}_p(r)$ the differential dE_c maps $H_{v_0}\tilde{S}_p(r)$ into $H_{q_0}S_p(r')$, where $q_0 = E_c(v_0)$.*

Proof: (i) Recalling Theorem 3.2.10.(i), we immediately get

$$\partial(\sigma_c)_{\varphi_v(\zeta)} = 2\bar{\zeta}\big(1 + (c/2)|\zeta|^2\big)\,\partial\rho_{\varphi_v(\zeta)}.$$

The Gauss Lemma 1.6.10, together with Corollary 2.6.5, says that

$$d\rho = \operatorname{Re}\langle \cdot^H, T^H\rangle_T,$$

where $T(q)$ is the unit tangent vector in q to the unique geodesic joining p to q (see section 1.6.3). Therefore

$$\partial\rho = \frac{1}{2}\langle \cdot^H, T^H\rangle_T, \tag{3.2.28}$$

and so

$$\tilde{\varphi}_v(\zeta) = \big(1 + (c/2)|\zeta|^2\big)\langle \cdot^H, T^H\rangle_T = \big(1 + (c/2)|\zeta|^2\big)^2\langle \cdot^H, \varphi_v'(\zeta)^H\rangle_{\varphi_v'(\zeta)}, \tag{3.2.29}$$

where we used the fact that

$$\varphi_v'(\zeta) = \frac{1}{1 + (c/2)|\zeta|^2}\,T\big(\varphi_v(\zeta)\big), \tag{3.2.30}$$

which follows from Proposition 3.1.3.(iii). In particular, then,

$$\tilde{\varphi}_v(0) = \langle \cdot^H, v^H\rangle_v$$

is well-defined. To prove that $\tilde{\varphi}_v$ is holomorphic, it is enough to prove that the function $\zeta \mapsto \tilde{\varphi}_v(\zeta)(\xi(\zeta))$ is holomorphic for any holomorphic vector field ξ along φ_v. But indeed, since $\partial/\partial\zeta$ is just the tangent vector to the disk φ_v — and thus it is φ'_v — by (3.2.29) we have

$$\begin{aligned}\frac{\partial}{\partial\bar\zeta}[\tilde{\varphi}_v(\zeta)(\xi(\zeta))] &= \frac{\partial}{\partial\bar\zeta}\left[(1+(c/2)|\zeta|^2)^2\langle\xi(\zeta)^H,\varphi'_v(\zeta)^H\rangle_{\varphi'_v(\zeta)}\right]\\ &= (1+(c/2)|\zeta|^2)c\zeta\langle\xi(\zeta)^H,\varphi'_v(\zeta)^H\rangle_{\varphi'_v(\zeta)}\\ &\quad+(1+(c/2)|\zeta|^2)^2\overline{\varphi'_v(\zeta)}\langle\xi(\zeta)^H,\varphi'_v(\zeta)^H\rangle_{\varphi'_v(\zeta)}\\ &= (1+(c/2)|\zeta|^2)^2\Big[\overline{A_c(\zeta)}\langle\xi(\zeta)^H,\varphi'_v(\zeta)^H\rangle_{\varphi'_v(\zeta)}+\langle\nabla_{\overline{\varphi'_v(\zeta)^H}}\xi(\zeta)^H,\varphi'_v(\zeta)^H\rangle_{\varphi'_v(\zeta)}\\ &\qquad+\langle\xi(\zeta)^H,\nabla_{\varphi'_v(\zeta)^H}\varphi'_v(\zeta)^H\rangle_{\varphi'_v(\zeta)}\Big]\\ &= (1+(c/2)|\zeta|^2)^2\Big[\overline{A_c(\zeta)}-\overline{A_c(\zeta)}\Big]\langle\xi(\zeta)^H,\varphi'_v(\zeta)^H\rangle_{\varphi'_v(\zeta)}=0,\end{aligned}$$

where A_c is given by (3.1.4), and we used the holomorphicity of ξ and Corollary 3.1.2.

(ii) Since G is a defining function for $\tilde{B}_p(r)$, it is clear that

$$H_{v_0}\tilde{S}_p(r)=\ker\partial G_{v_0}=\{V\in\mathcal{V}_{v_0}\mid\langle V,\iota\rangle_{v_0}=0\}.$$

On the other hand, (3.2.28) says that

$$H_qS_p(r')=\{v\in T^{1,0}_qM\mid\langle v^H,T^H\rangle_{T(q)}=0\}. \tag{3.2.31}$$

So we have to prove that if $V\in\mathcal{V}_{v_0}$ is such that $\langle V,\iota\rangle_{v_0}=0$ then $d(E_c)_{v_0}(V)^H$ belongs to $H_{q_0}S_p(r')$.

Let $v\colon(-\varepsilon,\varepsilon)\to\tilde{S}_p(r)$ be a smooth curve with $v(0)=v_0$ and $\dot v(0)=V$, and set $\psi_t=\varphi_{v(t)/r}$ for $t\in(-\varepsilon,\varepsilon)$. Since $E_c(v(t))=\psi_t(r)$, it follows that

$$d(E_c)_{v_0}(V)=\left.\frac{\partial}{\partial t}\psi_t(r)\right|_{t=0};$$

so if we define

$$\xi(\zeta)=\left.\frac{\partial}{\partial t}\psi_t(\zeta)\right|_{t=0},$$

it will suffice to prove that $\xi(\zeta)$ belongs to the holomorphic tangent space to $S_p(r')$ at $\varphi_{v_0/r}(\zeta)$ for all ζ, where $r'=|\zeta|$ if $c=0$, and $r'=\operatorname{atanh}(|\zeta|)$ if $c=-2$.

We write

$$\xi(\zeta)=\eta(\zeta)\varphi'_{v_0/r}(\zeta)+g(\zeta), \tag{3.2.32}$$

where $\eta(\zeta)\in\mathbb{C}$, $g(\zeta)\in H_{\varphi_{v_0/r}(\zeta)}S_p(r')$. By definition, ξ is holomorphic; it turns out that both η and g are holomorphic too. Indeed, (3.2.29) yields

$$\begin{aligned}\tilde{\varphi}_{v_0/r}(\zeta)(\xi(\zeta)) &= (1+(c/2)|\zeta|^2)^2\langle\xi(\zeta)^H,\varphi'_{v_0/r}(\zeta)^H\rangle_{\varphi'_{v_0/r}(\zeta)}\\ &= \eta(\zeta)(1+(c/2)|\zeta|^2)^2\langle\varphi'_{v_0/r}(\zeta)^H,\varphi'_{v_0/r}(\zeta)^H\rangle_{\varphi'_{v_0/r}(\zeta)}=\eta(\zeta),\end{aligned}$$

by (3.2.31) and (3.2.30). Since $\varphi_{v_0/r}$ is stationary, it follows that η is holomorphic, and then also g is holomorphic.

Since $\xi(O) = o_p$, we can then write

$$\xi(\zeta) = \zeta\eta_1(\zeta)\varphi'_{v_0/r}(\zeta) + \zeta g_1(\zeta), \tag{3.2.33}$$

with η_1 and g_1 still holomorphic. The rest of the argument now is similar to the one described in Lemma 5.3 of [P3]. Since $\psi_t(\zeta) = E_c(\zeta v(t)/r)$, from Theorem 3.2.10.(ii) it follows that

$$\sigma_c(\psi_t(\zeta)) = G(\zeta v(t)/r) = |\zeta|^2; \tag{3.2.34}$$

differentiating this identity with respect to t and computing for $t = 0$ we get

$$0 = d\sigma_c(\xi(\zeta)) = 2\operatorname{Re}\left[\zeta\eta_1(\zeta)\,\partial\sigma_c(\varphi'_{v_0/r}(\zeta))\right]. \tag{3.2.35}$$

Now setting $t = 0$ in (3.2.34) and differentiating with respect to ζ we get

$$\partial\sigma_c(\varphi'_{v_0/r}(\zeta)) = \bar{\zeta};$$

therefore (3.2.35) yields

$$2|\zeta|^2 \operatorname{Re}\eta_1(\zeta) \equiv 0.$$

Being η_1 holomorphic, it follows that $\eta_1 \equiv ia$ for some $a \in \mathbb{R}$; if we show that $a = 0$ we are done, by (3.2.32). But indeed differentiating $v(t)/r = \psi'_t(0)$ with respect to t (using (3.2.33), of course) and then evaluating at $t = 0$ we get

$$\frac{1}{r}V = \xi'(0) = \frac{ia}{r}\iota(v_0) = \iota_{v_0}(g_1(0)).$$

Now, by assumption $\langle g_1(\zeta)^H, T^H\rangle_T \equiv 0$; for $\zeta = 0$ then (3.2.30) yields

$$0 = \langle g_1(\zeta)^H, T^H\rangle_T\bigg|_{\zeta=0} = \langle g_1(0)^H, \chi\rangle_{v_0/r} = \frac{1}{r}\langle g_1(0)^H, \chi\rangle_{v_0} = \frac{1}{r}\langle \iota_{v_0}(g_1(0)), \iota\rangle_{v_0}.$$

So

$$0 = \frac{1}{r}\langle V, \iota\rangle_{v_0} = \langle \xi'(0), \iota\rangle_{v_0} = \frac{ia}{r}\langle \iota, \iota\rangle_{v_0} = ira,$$

and we are done. □

Now we can resume the proof of Theorem 3.2.10. By Lemma 3.2.11.(ii), dE_c maps at every point the kernel of ∂G onto the kernel of $\partial\sigma_c$; therefore there exists a never zero function g such that $(E_c)_*\partial\sigma_c = g\partial G$. Now, for $v \in \tilde{S}_p$ let j_v denote the inclusion in $T^{1,0}_p M$ of the line (if $c = 0$) or the disk (if $c = -2$) through the origin in the direction of v. By construction, $E_c \circ j_v$ is holomorphic; hence, being $\sigma_c \circ E_c = G$ and so $\sigma_c \circ (E_c \circ j_v) = G \circ j_v$, we have

$$(j_v)_*\partial G = (j_v)_*(E_c)_*\partial\sigma_c = (j_v)_*(g\partial G) = (g \circ j_v)(j_v)_*\partial G$$

and thus $g \circ j_v = 1$. Since this is true for all $v \in \tilde{S}_p$ it follows that

$$(E_c)_*\partial\sigma_c = \partial G.$$

This is a crucial observation since, as a formal consequence, it implies that

$$\begin{aligned} dd^cG &= (i/2\pi)\partial\bar{\partial}G = (1/2\pi i)\bar{\partial}\partial G = (1/2\pi i)d\partial G \\ &= (1/2\pi i)d(E_c)_*\partial\sigma_c = (1/2\pi i)(E_c)_*d\partial\sigma_c = (E_c)_*dd^c\sigma_c. \end{aligned}$$

But then, since dd^cG is a positive definite form, $dd^c\sigma_c$ must be non degenerated. On the other hand from (3.2.25), (3.2.26) and Theorem 1.6.4.(iv) it follows that E_c is differentiable at the origin and $dE_c(0) = \mathrm{id}$. Thus in a neighborhood of p, with respect to a coordinate system centered at p, we have

$$\sigma_c(z) = G(z) + o(\|z\|^2)$$

and hence σ_c is strictly plurisubharmonic nearby p. As M is connected, (v) follows.

Now we shall prove (vi) and (vii). We set $u_c = \log\sigma_c$ and we keep all the notations introduced in the proof of (v). We have

$$(\sigma_c)^2 dd^c u_c = \sigma_c dd^c\sigma_c - d\sigma_c \wedge d^c\sigma_c = \sigma_c \frac{i}{2\pi}(\partial\bar{\partial}\sigma_c - \partial\sigma_c \wedge \bar{\partial}\sigma_c),$$

and hence, arguing as before, it follows that

$$(E_c)_* dd^c u_c = dd^c \log G.$$

As a consequence

$$(E_c)_*(dd^c u_c)^n = (dd^c \log G)^n.$$

Since from the homogeneity properties of G follow at once that $(dd^c \log G)^n \equiv 0$ on $T_p^{1,0}M \setminus \{p\}$, we have also $(dd^c u_c)^n \equiv 0$ on $M \setminus \{p\}$ and hence (vii) holds. Let us prove (vi). Let $q \in M \setminus \{p\}$ be any point and suppose that $q \in S_p(r')$. Then $q = E_c(\zeta v_0) = \varphi_{v_0}(\zeta)$ for some $\zeta \in \mathbb{C}$ and $v_0 \in \tilde{S}_p$. Let $\xi \in T_q^{1,0}M$. Then for some $\lambda \in \mathbb{C}$ and $w \in H_q S_p(r')$ one has

$$\xi = \lambda\varphi'_{v_0}(\zeta) + w = (dE_c)_{\zeta v_0}\big(\lambda\iota_{\zeta v_0}(v_0)\big) + w.$$

On the other hand as dE_c maps at every point the kernel of ∂G onto the kernel of $\partial\sigma_c$ we have $w = (dE_c)_{\zeta v_0}(W)$ for some $W \in H_{\zeta v_0}\tilde{S}_p(r)$ — where $r = r'$ if $c = 0$, and $r = \tanh(r')$ if $c = -2$. Thus, as $\log G$ is plurisubharmonic,

$$\begin{aligned} dd^c u_c(\xi, \bar{\xi}) &= dd^c u_c\Big(\big((dE_c)_{\zeta v_0}(\lambda\iota_{\zeta v_0}(v_0) + W\big), \overline{(dE_c)_{\zeta v_0}\big(\lambda\iota_{\zeta v_0}(v_0) + W\big)}\Big) \\ &= (E_c)_*(dd^c u_c)\big(\lambda\iota_{\zeta v_0}(v_0) + W, \overline{\lambda\iota_{\zeta v_0}(v_0) + W}\big) \\ &= dd^c \log G\big(\lambda\iota_{\zeta v_0}(v_0) + W, \overline{\lambda\iota_{\zeta v_0}(v_0) + W}\big) \geq 0, \end{aligned}$$

and hence also (vi) is proved. Finally, again using the fact that E_c is differentiable at the origin and $dE_c(0) = \mathrm{id}$, we see that in a neighborhood of p with respect to a coordinate system centered at p the function u_c has the same singularity as $\log G$ in a neighborhood of 0, and hence (viii) is immediate. □

Thanks to this result we can introduce the machinery of the complex Monge-Ampère equation in the study of complex Finsler manifolds with constant non positive holomorphic curvature. In fact Theorem 3.2.10 in the language of [Pa2] states that under these assumptions the manifold M is a manifold of circular type, of finite radius if the holomorphic curvature is constant negative, of infinite radius if the holomorphic curvature vanishes identically. It is also interesting to note that, in the hyperbolic case, Theorem 3.2.10 gives precise conditions in terms of complex Finsler geometry for the existence of regular pluricomplex Green functions (cf. [D])

3.3. Characterization of $\mathbb{C}^n$ and of convex circular domains

3.3.1. Geometry of the Monge-Ampère foliation

The geometric theory of the complex homogeneous Monge-Ampère equation gives techniques and criteria useful to characterize and classify the manifold M. In the remaining of this chapter we shall illustrate how this can be done, trying to give an exposition adapted to our setting and as self-contained as possible.

We shall now recall some facts about the geometry of the Monge-Ampère equation. It is worthwhile to do so in the general setting, although we shall limit ourselves to the notions which are relevant for our aims. Details and complete treatments are easily found in the literature. We refer for the basic properties for instance to [BK], [BB], [Bu], [Pa2], [PW], [St], [Wo].

Let M be a complex manifold of dimension n and let $p \in M$. Suppose that there exists an exhaustion function $\sigma\colon M \to [0, R)$ — here R may be a real number or $+\infty$ — with the following properties:

$$\sigma \in C^0(M) \cap C^\infty(M \setminus \{p\}); \tag{3.3.1}$$

if $\pi\colon \tilde{M} \to M$ is the blow-up at p then

$$\sigma \circ \pi \in C^\infty(\tilde{M}); \tag{3.3.2}$$

$$dd^c\sigma > 0 \quad \text{on } M \setminus \{p\}; \tag{3.3.3}$$

$$dd^c \log \sigma \geq 0 \quad \text{on } M \setminus \{p\}; \tag{3.3.4}$$

$$(dd^c \log \sigma)^n = 0 \quad \text{on } M \setminus \{p\}; \tag{3.3.5}$$

$$\log \sigma(z) = \log \|z\|^2 + O(1), \tag{3.3.6}$$

with respect to any coordinate system centered in p.

Because of (3.3.4) and (3.3.5), it follows that associated to σ there is a rank 1 distribution in $T^{1,0}(M \setminus \{p\})$ defined by the eigenspaces relative to the zero eigenvalue of $dd^c \log \sigma$. Formally this distribution is defined by

$$\Sigma = \{v \in T^{1,0}(M \setminus \{p\}) \mid dd^c \log \sigma(v, w) = 0 \text{ for all } w \in T^{1,0}(M \setminus \{p\})\}. \tag{3.3.7}$$

Since $dd^c \log \sigma$ is a closed form, the distribution Σ is integrable. Furthermore, as $dd^c \log \sigma$ is a $(1,1)$-form, the distribution Σ is invariant under the almost complex structure and hence each leaf of the associated foliation is a complex submanifold of $M \setminus \{p\}$ of dimension 1.

Definition 3.3.1: This foliation in Riemann surfaces generated by Σ is called the *Monge-Ampère foliation* associated to σ.

By definition the leaves of the Monge-Ampère foliation are the maximal submanifolds of dimension one along which $\log\sigma$ is harmonic. We shall use later this observation.

It is useful to introduce a further notion. Let h be the Kähler metric — this time is a usual Kähler Hermitian metric — whose fundamental form is $dd^c\sigma$. A vector field f of type $(1,0)$ called the *complex gradient* of σ is defined as the dual with respect to the metric h of the form $\bar\partial\sigma$, i.e., as the unique vector field f such that for all vector field ξ of type $(1,0)$

$$h(f,\xi) = \bar\partial\sigma(\bar\xi).$$

In local coordinates then we have the following expression for f in terms of σ:

$$f = \sigma_{\bar\nu}\,\sigma^{\bar\nu\mu}\frac{\partial}{\partial z^\mu}. \tag{3.3.8}$$

It is worthwhile to list the properties of f. Before doing so we remark that as

$$(\sigma)^2 dd^c\log\sigma = \sigma dd^c\sigma - d\sigma\wedge d^c\sigma, \tag{3.3.9}$$

taking the n-th exterior product and using (3.3.5), we have

$$\sigma(dd^c\sigma)^n = n d\sigma\wedge d^c\sigma\wedge(dd^c\sigma)^{n-1}$$

which in local coordinates is equivalent to

$$\sigma = \sigma_{\bar\nu}\,\sigma^{\bar\nu\mu}\,\sigma_\mu, \tag{3.3.10}$$

or in terms of the complex gradient f

$$\sigma = f(\sigma) = h(f,f). \tag{3.3.11}$$

Using again (3.3.9), we have

$$(\sigma)^2 dd^c\log\sigma(f,f) = \sigma dd^c\sigma(f,f) - d\sigma(f)\wedge d^c\sigma(f) = \sigma^2 - \sigma^2 = 0.$$

Since $dd^c\log\sigma \geq 0$ it follows that f is a global section of Σ. Since from (3.3.11) it follows that f never vanishes on $M\backslash\{p\}$, then Σ is a trivial subbundle of $T^{1,0}(M\backslash\{p\})$ spanned by f. In particular it is possible to express the properties of the distribution Σ and of the associated Monge-Ampère foliation in terms of f.

Definition 3.3.2: The Monge-Ampère foliation is *holomorphic* iff for any point $q\in M\backslash\{p\}$ there exist an open neighborhood U of q, some $r>0$ and a biholomorphic map $\Phi: U\to\Delta_r\times\Delta_r^{n-1}$ such that if L is a leaf of the foliation with $L\cap U\neq\emptyset$, then $\Phi(L\cap U) = \Delta_r\times\{y\}$ for some $y\in\Delta_r^{n-1}$ and conversely if $y\in\Delta_r^{n-1}$ there exists a leaf L of the foliation such that $\Phi(L\cap U) = \Delta_r\times\{y\}$.

This is the case, as it is easy to see, iff the distribution associated to the foliation is a holomorphic subbundle of $T^{1,0}(M \setminus \{p\})$. Thus, because of the considerations made above, we get the following obvious, but important consequence:

PROPOSITION 3.3.1: *The Monge-Ampère foliation associated to σ is holomorphic iff the distribution Σ is a holomorphic subbundle of $T^{1,0}(M \setminus \{p\})$ iff the complex gradient vector field f is a holomorphic section of $T^{1,0}(M \setminus \{p\})$.*

We notice that examples show that very seldom f is a holomorphic section of $T^{1,0}(M\setminus\{p\})$ and hence the Monge-Ampère foliation associated to σ is in general not holomorphic. Nevertheless the vector field f, as a consequence of (3.3.11) is always holomorphic when restricted to a leaf of Monge-Ampère foliation. As we shall not need these considerations, we refer the interested reader to the afore mentioned references. On the other hand, for our applications we shall need some handy geometrical criterion to establish when the Monge-Ampère foliation or — as it is the same — the vector field f is holomorphic. There are in the literature two main results in this regards. We end these general remarks on the geometry arising from the Monge-Ampère equation quoting them.

The first and most general one is related to the Ricci curvature of the leaves of the Monge-Ampère foliation ([BB], [St]):

THEOREM 3.3.2: *Let S be the Ricci form of the metric h associated to the form $dd^c\sigma$. Then*

$$S(f, \bar{f}) = \frac{i}{2\pi} \|\bar{\partial} f\|_h \tag{3.3.12}$$

where $\|\cdot\|_h$ is the metric induced by h on $(T^{0,1})^*(M \setminus \{p\}) \otimes T^{1,0}(M \setminus \{p\})$. In particular the Monge-Ampère foliation is holomorphic iff all the leaves are Ricci flat with respect to the metric h.

As a very clever and deep application of Theorem 3.3.2, Burns [Bu] proved the following

THEOREM 3.3.3: *If all the leaves of the Monge-Ampère foliation associated to σ are parabolic, i.e., are uniformized by $\mathbb{C}$, then the foliation is holomorphic.*

We end this subsection by illustrating how the general theory of Monge-Ampère foliations fits in our context of complex Finsler manifolds of constant nonpositive holomorphic curvature. To start we have the following

PROPOSITION 3.3.4: *Let $F: T^{1,0}M \to \mathbb{R}^+$ be a strongly pseudoconvex Finsler metric on a simply connected complex manifold M of dimension n satisfying (3.2.15)–(3.2.18), and let $p \in M$ so that the function σ_c has all the properties listed in Theorem 3.2.10. Then the leaves of the Monge-Ampère foliation associated to σ_c are exactly the images of the c-geodesic complex curves through p (with the point p removed). Furthermore if H_c is the map defined in (3.2.21) and f is the complex gradient vector field of σ, then*

$$f\big(H_c(\zeta,v)\big) = \zeta H_c'(\zeta,v) = \zeta\varphi_v'(\zeta), \tag{3.3.13}$$

where the prime denotes again differentiation with respect to ζ.

Proof: In this proof we set $\sigma = \sigma_c$. By definition

$$\sigma\big(H_c(\zeta,v)\big) = |\zeta|^2. \tag{3.3.14}$$

Hence $\log\sigma$ is harmonic along the image of φ_v and this is enough to prove the first part of the claim. More formally, from (3.3.14) follows that

$$\sigma\big(H_c(\zeta,v)\big)\sigma_{\mu\bar\nu}\big(H_c(\zeta,v)\big)H_c'^{\mu}(\zeta,v)\overline{H_c'^{\nu}(\zeta,v)} - \sigma_\mu\big(H_c(\zeta,v)\big)H_c'^{\mu}(\zeta,v)\sigma_{\bar\nu}\big(H_c(\zeta,v)\big)\overline{H_c'^{\nu}(\zeta,v)} = 0$$

which is equivalent to

$$dd^c\log\sigma\big(H_c(\zeta,v)\big)\big(H_c'(\zeta,v), H_c'(\zeta,v)\big) = 0.$$

This, together with the fact that $dd^c\log\sigma \geq 0$, implies that

$$dd^c\log\sigma\big(H_c(\zeta,v)\big)\big(H_c'(\zeta,v),\bar\xi\big) = 0 \tag{3.3.15}$$

for all vector field ξ of type $(1,0)$. Thus $H_c'(\zeta,v)$ is tangent to the foliation for every v and hence the claim. For the rest of the statement we proceed as follows. From (3.3.14) it follows that

$$\sigma_\mu\big(H_c(\zeta,v)\big)H'^{\mu}_c(\zeta,v) = \bar\zeta.$$

Since

$$\sigma^2\, dd^c\log\sigma = \sigma\, dd^c\sigma - d\sigma\wedge d^c\sigma,$$

from (3.3.15) we get that

$$\sigma\big(H_c(\zeta,v)\big)\sigma_{\mu\bar\nu}\big(H_c(\zeta,v)\big)H_c'^{\mu}(\zeta,v) = \sigma_\mu\big(H_c(\zeta,v)\big)H_c'^{\mu}(\zeta,v)\sigma_{\bar\nu}\big(H_c(\zeta,v)\big)$$

and hence

$$|\zeta|^2\sigma_{\mu\bar\nu}\big(H_c(\zeta,v)\big)H_c'^{\mu}(\zeta,v) = \bar\zeta\sigma_{\bar\nu}\big(H_c(\zeta,v)\big).$$

We can conclude that

$$\zeta\sigma_{\mu\bar\nu}\big(H_c(\zeta,v)\big)H_c'^{\mu}(\zeta,v) = \sigma_{\bar\nu}\big(H_c(\zeta,v)\big).$$

This, together with the expression in local coordinates (3.3.8) of f, implies

$$f^\mu\big(H_c(\zeta,v)\big) = \sigma_{\bar\nu}\big(H_c(\zeta,v)\big)\sigma^{\bar\nu\mu}\big(H_c(\zeta,v)\big) = \zeta H_c'^{\mu}(\zeta,v)$$

and the proof is complete. □

The last result that we need is the following

THEOREM 3.3.5: *Let $F: T^{1,0}M \to \mathbb{R}^+$ be a strongly pseudoconvex Finsler metric on a simply conncted complex manifold M of dimension n satisfying (3.2.15)–(3.2.18) and let $p \in M$. Then the map E_c relative to the point p is biholomorphic iff the Monge-Ampère foliation associated to σ_c is holomorphic.*

Proof: If E_c is biholomorphic, as the leaves of the Monge-Ampère foliation associated to σ are images via E_c of the intersections of the domain of E_c and the complex lines through the origin, in one direction the result is obvious. Since we have shown in Theorem 3.2.10 that M is a Stein manifold, because of Proposition 3.2.9.(iii) to prove the other implication it is enough to show that E_c is of class C^∞ in a neighborhood of the origin. In fact we shall show that E_c is holomorphic in a neighborhood of the origin. Our assumption is equivalent to saying that the complex gradient f is holomorphic on $U \setminus \{p\}$ for some open set U containing p. Defining $f(p) = 0$ we extend holomorphically f to all of U. As we have seen that $f(H_c(\zeta, v)) = \zeta H_c'(\zeta, v) = \zeta\varphi_v'(\zeta)$ and with respect to any system of holomorphic coordinates $\varphi_v(\zeta) = \zeta v + O(|\zeta|^2)$, it follows that $df_p = \mathrm{id}$. We conclude that in a coordinate neighborhood centered at p, which we denote again by U, we have

$$f = \mathrm{id} + Q. \tag{3.3.16}$$

Let us denote by D a small ball centered at the origin in the domain of definition of E_c, and define for some $r > 0$ a map $A: \Delta_r \times D \to U$ by

$$A(\zeta, v) = E_c(\zeta v) = H_c\left(F(v)\zeta, \frac{v}{F(v)}\right).$$

Thus in particular $f(A(\zeta, v)) = \zeta A'(\zeta, v)$. For every $j \geq 0$ there exists $A_j: D \to U$ so that

$$A(\zeta, v) = \sum_{j\geq 1} A_j(v)\zeta^j.$$

We can conclude that

$$Q(A(\zeta, v)) = \zeta A'(\zeta, v) - A(\zeta, v) = \sum_{j\geq 2}(j-1)A_j(v)\zeta^j. \tag{3.3.17}$$

Define $P(\zeta, v) = \sum_{j\geq 1} A_j(v)\zeta^{j-1}$. If $Q = \sum_{k\geq 2} Q_k$ is the development of Q in homogeneous vector polynomials, then for some vector polynomials Q_{kr} we have

$$Q_k(P(\zeta, v)) = \sum_{r\geq 0} Q_{kr}(A_1(v), \ldots, A_{r+1}(v))\zeta^r$$

so that

$$Q(A(\zeta, v)) = \sum_{k\geq 2} Q_k(P(\zeta, v)) = \sum_{j\geq 2}\left(\sum_{j=k+r} Q_{kr}(A_1(v), \ldots, A_{r+1}(v))\right)\zeta^j.$$

Comparing the two expressions of $Q(A(\zeta,v))$, we get

$$(j-1)A_j(v) = \sum_{j=k+r} Q_{kr}\big(A_1(v),\dots,A_{r+1}(v)\big).$$

Since $k \geq 2$ we have $r+1 \leq j-1$, and there exist vector polynomials R_j so that for all $j \geq 2$

$$(j-1)A_j(v) = R_j\big(A_1(v),\dots,A_{r+1}(v)\big).$$

As $A_1 = df_p = \mathrm{id}$ is holomorphic, by induction, it follows that A_j is holomorphic for all j. For $\varepsilon > 0$ small enough $\sum_{j\geq 1} A_j(v)\zeta^j$ converges uniformly for $\|v\| < 2\varepsilon$ and $|\zeta| < 1/2$. Then we can conclude that, if $\|v\| < 2\varepsilon$,

$$E_c(v) = A(1/2, 2v) = \sum_{j\geq 1} A_j(v)(1/2)^j$$

converges uniformly and hence, as claimed, E_c is holomorphic in a neighborhood of the origin. □

3.3.2. A characterization of $\mathbb{C}^n$

In this and the next subsections we shall apply the results of the previous section to characterize special complex Finsler manifolds. We start with the case of manifolds with vanishing holomorphic curvature. It is easy to provide examples of manifolds satisfying the hypotheses of Theorem 3.2.10 with $c = 0$, just by considering complex Minkowski spaces. It should not be a surprise that the existence of such metrics is a distinct feature of $\mathbb{C}^n$. In fact our result is the following:

THEOREM 3.3.6: *Let M be a simply connected complex manifold of dimension n and let us suppose that on M is defined a strongly pseudoconvex Finsler metric F such that*

$$F \text{ is complete Kähler;} \tag{3.3.18}$$

$$K_F \equiv 0 \tag{3.3.19}$$

$$\langle \bar\partial_H \theta(H,\chi,\overline{K}),\chi\rangle = 0 \text{ for all } H,\ K \in \mathcal{H}; \tag{3.3.20}$$

$$F \text{ is strongly convex.} \tag{3.3.21}$$

Then for every $p \in M$ the map $E_0\colon \mathbb{C}^n \simeq T^{1,0}_pM \to M$ is biholomorphic.

Proof: At this point this result follows immediately from Theorem 3.2.10 and Theorem 3.3.5 since in this case the leaves of the Monge-Ampère foliation associated to σ_0 are, by construction, all biholomorphic to $\mathbb{C}^*$ and hence one may apply Theorem 3.3.3. □

It should be remarked that Theorem 3.3.6 is satisfactory from the point of view of geometric function theory but leaves open a problem in Finsler geometry; in fact it is not yet a characterization of complex Minkowski metrics, as it should be expected.

3.3.3. A characterization of convex circular domains

Let us turn to manifolds of negative constant holomorphic curvature. In this case most of the time the map E_{-2} is not biholomophic, as the the examples of strongly convex domain in $\mathbb{C}^n$ with the Kobayashi metric show. In fact most of them have trivial automorphisms group, whereas a domain biholomorphic to a circular domain has a rich automorphism group. Thus one must expect some additional geometric assumption in order to get the holomorphicity of E_{-2}. Some criteria are available in the literature (cf. [LPW]) and they may be applied in this context. For instance we may prove:

THEOREM 3.3.7: *Let M be a simply connected complex manifold of dimension n and let us suppose that on M is defined a strongly pseudoconvex complex Finsler metric F such that*

$$F \text{ is complete Kähler;} \tag{3.3.22}$$

$$K_F \equiv -4 \tag{3.3.23}$$

$$\langle \bar{\partial}_H \theta(H, \chi, \overline{K}), \chi \rangle = 0 \text{ for all } H,\ K \in \mathcal{H}; \tag{3.3.24}$$

$$F \text{ is strongly convex.} \tag{3.3.25}$$

Then F is the Kobayashi metric of M. Furthermore assume that there exists a point $p \in M$ such that for some $r_0 > r_1 > 0$ there exists a biholomorphism $\Psi: B_p(r_0) \to B_p(r_1)$ fixing p. Then the map E_{-2} is biholomorphic and hence M is biholomorphic to a smooth strongly convex circular domain in $\mathbb{C}^n$. More precisely, M is biholomorphic to the indicatrix $I_p(M)$ of the Kobayashi metric at p.

Proof: According to Theorem 3.3.5 it is enough to show that the Monge-Ampère foliation associated to $\sigma = \sigma_{-2}$ is holomorphic in a neighborhood of p or, that is the same, if f_σ is the complex gradient of σ, then f_σ is holomorphic on $U \setminus \{p\}$ for some open set U containing p. If we denote by S_σ the Ricci form of the metric $dd^c\sigma$, we must show that for some neighborhood U of p we have $S_\sigma(f_\sigma, \overline{f_\sigma}) = 0$ on $U \setminus \{p\}$.

Iterating the map Ψ, we may assume that both $B_p(r_0)$ and $B_p(r_1)$ are contained in any given neighborhood of the point p. At p, as in any other point, the indicatrix of F is strongly convex. If $\mu: \mathbb{C}^n \cong T^{1,0}_p M \to \mathbb{R}$ is defined by

$$\mu(v) = G(p; v),$$

then μ is a strictly convex function which — we recall it here for further reference — satisfies (3.3.1), (3.3.2), (3.3.3), (3.3.4), (3.3.5), (3.3.6) on I_p. Since $d(E_c)_p = \mathrm{id}$, we have that in a small enough coordinate neighborhood centered at p,

$$\sigma = \mu + o(1),$$

and hence in particular it follows that, with respect to the given holomorphic coordinates, σ is strongly convex in a neighborhood of p. As a consequence we may assume that both $B_p(r_0)$ and $B_p(r_1)$ are contained in a coordinate neighborhood U of p and are biholomorphic to a strongly convex domain in $\mathbb{C}^n$. With respect to the given system of coordinates, if μ is as above, using the homogeneity of μ and the fact that $\mu = \sigma \circ E$ we have that for all fixed $z \in U \setminus \{p\}$ and small $|t| \neq 0$, we have

$$\sigma_{\alpha\bar{\beta}}(tz) = \mu_{\alpha\bar{\beta}}(tz) + o(1) = \mu_{\alpha\bar{\beta}}(z) + o(1),$$

and hence $\log\det(\sigma_{\alpha\bar{\beta}}(tz)) = \log\det(\mu_{\alpha\bar{\beta}}(z)) + o(1)$. Differentiating again

$$t^2 \frac{\partial^2}{\partial z^\gamma \partial \bar{z}^\delta}\Big(\log\det(\sigma_{\alpha\bar{\beta}}(tz))\Big) = \frac{\partial^2}{\partial z^\gamma \partial \bar{z}^\delta}\Big(\log\det(\mu_{\alpha\bar{\beta}}(z))\Big) + o(1);$$

so the Ricci forms S_σ and S_μ are related by

$$t^2 S_\sigma(tz) = S_\mu(z) + o(1).$$

Now, the complex gradient f_μ of μ is given by

$$f_\mu = \sum z^\alpha \frac{\partial}{\partial z^\alpha};$$

Therefore, again using the fact that $\mu = \sigma \circ E$

$$f_\sigma(tz) = f_\mu(tz) + o(|t|) = t f_\mu(z) + o(|t|),$$

and thus

$$S_\sigma(f_\sigma, \overline{f_\sigma})(tz) = S_\mu(f_\mu, \overline{f_\mu})(z) + o(1).$$

Since the Monge-Ampère foliation induced on $I_p(M)$ by μ is holomorphic, it follows that

$$S_\mu(f_\mu, \overline{f_\mu})(z) = 0$$

and hence

$$S_\sigma(f_\sigma, \overline{f_\sigma})(z) = o(1). \tag{3.3.26}$$

Now we shall use the remark that $B_p(r_0)$ and $B_p(r_1)$ may be assumed to be biholomorphic to strongly convex domains, and so we can apply the very special properties that the Kobayashi metric and distance enjoy in this case. For $j = 0, 1$ let us denote by ρ_j the Kobayashi distance of $B_p(r_j)$ from the point p. Then, as $B_p(r_j)$ is biholomorphic to a strongly convex domain, the function $(\tanh \rho_j)^2 \colon B_p(r_j) \to [0,1)$ is the unique exhaustion of $B_p(r_j)$ which satisfies (3.3.1), (3.3.2), (3.3.3), (3.3.4), (3.3.5), (3.3.6) on $B_p(r_j)$ (cf. [L], [D]). Since also the exhaustion defined by

$$\frac{1}{(r_j)^2}\sigma \colon B_p(r_j) \to [0,1)$$

satisfies (3.3.1), (3.3.2), (3.3.3), (3.3.4), (3.3.5), (3.3.6) on $B_p(r_j)$, then necessarily it must be

$$\frac{1}{(r_j)^2}\sigma = (\tanh \rho_j)^2$$

and hence, as Ψ being a biholomorphic map is an isometry for the Kobayashi distances,

$$\frac{1}{(r_1)^2}\sigma \circ \Psi = \frac{1}{(r_0)^2}\sigma. \tag{3.3.27}$$

Let $\lambda = r_1/r_0 \in (0,1)$, and define $\tau = \lambda^{-2}\sigma$. Then evidently

$$\tau_{\bar\beta} = \lambda^{-2}\sigma_{\bar\beta}, \qquad \tau_{\alpha\bar\beta} = \lambda^{-2}\sigma_{\alpha\bar\beta}, \qquad \tau^{\alpha\bar\beta} = \lambda^{2}\sigma^{\alpha\bar\beta}. \tag{3.3.28}$$

Because of (3.3.27) it follows that $\sigma \circ \Psi^{-1} = \lambda^{-2}\sigma = \tau$. Then Ψ is an isometry for the Kähler metrics defined by $dd^c\sigma$ and $dd^c\tau$. Furthermore, it is obvious that $\Psi_*(f_\sigma) = f_\tau$. On the other hand, (3.3.28) implies that

$$f_\tau = \sum \tau^{\alpha\bar\beta}\tau_{\bar\beta}\frac{\partial}{\partial z^\alpha} = \sum \sigma^{\alpha\bar\beta}\sigma_{\bar\beta}\frac{\partial}{\partial z^\alpha} = f_\sigma$$

and

$$S_\tau = dd^c \log\det(\tau_{\alpha\bar\beta}) = dd^c \log\det(\sigma_{\alpha\bar\beta}) = S_\sigma.$$

Thus

$$S_\sigma(f_\sigma, \overline{f_\sigma})(z) = S_\tau\big(\rho_*(f_\sigma), \rho_*(\overline{f_\sigma})\big)\big(\rho(z)\big) = S_\tau(f_\tau, \overline{f_\tau})\big(\rho(z)\big) = S_\sigma(f_\sigma, \overline{f_\sigma})\big(\rho(z)\big),$$

and hence for all positive integer k we have

$$S_\sigma(f_\sigma, \overline{f_\sigma})(z) = S_\sigma(f_\sigma, \overline{f_\sigma})\big(\rho^k(z)\big). \tag{3.3.29}$$

Since $\Psi(B_p(r_0))$ is relatively compact in $B_p(r_j)$, a theorem due to Hervé (see [H, p. 83]) implies that the sequence of iterates $\{\Psi^k\}$ is converging, uniformly on compact subsets, to p; therefore (3.3.26) and (3.3.29) yield $S_\sigma(f_\sigma, \overline{f_\sigma})(z) \equiv 0$ on $B_p(r_0) \setminus \{p\}$. This, as observed before, complete our proof. □

We end by characterizing the most special circular domain: the unit ball. Since we do not want to make more restrictive the assumptions on the Finsler metric — it is easy to fall into the case of Hermitian metrics and hence recover well known results! — we shall use the very special properties of the automorphisms group of the ball as it was done for instance in [Pa1]. We have the following

COROLLARY 3.3.8: *Let M be a simply connected complex manifold of dimension n with a strongly pseudoconvex Finsler metric F. Then the following statements are equivalent:*

(i) *M is biholomorphic to the unit ball in $\mathbb{C}^n$;*
(ii) *the assumptions of Theorem 3.3.7 hold for two distinct points $p, q \in M$;*
(iii) *the assumptions of Theorem 3.3.7 hold for a point $p \in M$ which is not kept fixed by $\mathrm{Aut}(M)$.*

Proof: It is clear that (i) $\Longrightarrow$ (ii). Let us prove that (ii) $\Longrightarrow$ (iii). By Theorem 3.3.7, there exist two bounded strongly convex circular domains $D_1, D_2 \subset \mathbb{C}^n$ and biholomorphic maps $\Phi_j: D_j \to M$ such that $\Phi_1(0) = p \neq q = \Phi_2(0)$. In particular, D_1 and D_2 are biholomorphic. By a result of [BKU], there exists a linear isomorphism $L \in GL(n, \mathbb{C})$ such that $L(D_1) = D_2$. Then $\Phi = \Phi_2 \circ L \circ \Phi_1^{-1}$ is an automorphism of M with $\Phi(p) = q$.

Finally we show that (iii) $\Longrightarrow$ (i). By Theorem 3.3.7, M is biholomorphic to a bounded strongly pseudoconvex circular domain $D \subset \mathbb{C}^n$ via a biholomorphic map sending the origin into p. Since the orbit of p under Aut(M) has at least two points, the origin in D is not a fixed point of Aut(D). Then it is known (again by [BKU]) that the orbit of the origin under Aut(D) is the intersection of D with a non-trivial complex subspace of $\mathbb{C}^n$. In particular, Aut(D) is not compact and thus, by B. Wong's theorem [W1], D is biholomorphic to the ball in $\mathbb{C}^n$. $\square$

References

[A] M. Abate: **Iteration theory of holomorphic maps on taut manifolds.** Mediterranean Press, Cosenza, 1989.

[AP1] M. Abate, G. Patrizio: *Uniqueness of complex geodesics and characterization of citcular domains.* Man. Math. **74** (1992), 277–297.

[AP2] M. Abate, G. Patrizio: *Holomorphic curvature of Finsler metrics and complex geodesics.* To appear in J. Geom. Anal. (1993).

[Au1] L. Auslander: *On the use of forms in the variational calculations.* Pac. J. Math. **5** (1955), 853–859.

[Au2] L. Auslander: *On curvature in Finsler geometry.* Trans. Am. Math. Soc. **79** (1955), 378–388.

[BC] D. Bao, S.S. Chern: *On a notable connection in Finsler geometry.* Houston J. Math. **19** (1993), 138–180.

[BB] E. Bedford, D. Burns: *Holomorphic mappings of annuli in* $\mathbb{C}^n$ *and the associated extremal function.* Ann. Sc. Norm. Sup. Pisa **6** (1979), 381–414.

[BK] E. Bedford, M. Kalka: *Foliations and complex Monge-Ampère equations.* Comm. Pure Appl. Math. **30** (1977), 510–538.

[B] A. Bejancu: **Finsler geometry and applications.** Ellis Horwood Limited, Chichester, 1990.

[BKU] R. Braun, W. Kaup, H. Upmeier: *On the automorphisms of circular and Reinhardt domains in complex Banach spaces.* Man. Math. **25** (1978), 97–133.

[Bu] D. Burns: *Curvature of the Monge-Ampère foliations and parabolic manifolds.* Ann. of Math. **115** (1982), 349–373.

[C] E. Cartan: **Les espaces de Finsler.** Hermann, Paris, 1934.

[CE] J. Cheeger, D.G. Ebin: **Comparison theorems in Riemannian geometry.** North-Holland, Amsterdam, 1975.

[Ch1] S.S. Chern: *Local equivalence and euclidean connections in Finsler spaces.* Sci. Rep. Nat. Tsing. Hua Univ. Ser. A **5** (1948), 95–121.

[Ch2] S.S. Chern: *On Finsler geometry.* C.R. Acad. Sc. Paris **314** (1992), 757–761.

[D] J.P. Demailly: *Mesures de Monge-Ampère et mesures pluriharmoniques.* Math. Z. **194** (1987), 519–564.

[DoC] M.P. do Carmo: **Riemannian geometry.** Birkhäuser, Basel, 1992.

[F] J.J. Faran: *Hermitian Finsler metrics and the Kobayashi metric.* J. Diff. Geom. **31** (1990), 601–625.

[Fu] M. Fukui: *Complex Finsler manifolds.* J. Math. Kyoto Univ. **29** (1989), 609–624.

[GR] H. Grauert, H. Reckziegel: *Hermitesche Metriken und normale Familien holomorphen Abbildungen.* Math. Z. **89** (1965), 108–125.

[H] M. Hervé: **Several complex variables. Local theory.** Oxford University Press, London, 1963.

[K] S. Kobayashi: *Negative vector bundles and complex Finsler structures.* Nagoya Math. J. **57** (1975), 153–166.

[L] L. Lempert: *La métrique de Kobayashi et la représentation des domaines sur la boule.* Bull. Soc. Math. France **109** (1981), 427–474.

[LPW] K.-W. Leung, G. Patrizio, P.-M. Wong: *Isometries of intrinsic metrics on strictly convex domains.* Math. Z. **196** (1987), 343–353.

[M] M. Matsumoto: **Foundations of Finsler geometry and special Finsler spaces.** Kaiseisha Press, Ōtsu Japan, 1966.

[P1] M.Y. Pang: *Finsler metrics with the properties of the Kobayashi metric on convex domains.* Publications Mathématiques **36** (1992), 131–155.

[P2] M.Y. Pang: *On the infinitesimal behavior of the Kobayashi distance.* Preprint (1993).

[P3] M.Y. Pang: *Smoothness of the Kobayashi metric of non-convex domains.* Int. J. Math. **4** (1993), 953–987.

[Pa1] G. Patrizio: *Parabolic exhaustions for strictly convex domains.* Man. Math. **47** (1984), 271–309.

[Pa2] G. Patrizio: *A characterization of complex manifolds biholomorphic to a circular domain.* Math. Z. **189** (1986), 343–363.

[PW] G. Patrizio, P.-M. Wong: *Stability of the Monge-Ampère foliation.* Math. Ann. **263** (1983), 13–29.

[R] W. Rinow: **Die innere Geometrie der metrischen Räume.** Springer, Berlin, 1961.

[Ro] H.L. Royden: *Complex Finsler metrics.* In **Contemporary Mathematics. Proceedings of Summer Research Conference,** American Mathematical Society, Providence, 1984, pp. 119–124.

[Ru] W. Rudin: **Function theory in the unit ball of $\mathbb{C}^n$.** Springer, Berlin, 1980.

[Rd1] H. Rund: **The differential geometry of Finsler spaces.** Springer, Berlin, 1959.

[Rd2] H. Rund: *Generalized metrics on complex manifolds.* Math. Nach. **34** (1967), 55–77.

[S] S. Semmes: **A generalization of Riemann mappings and geometric structures on a space of domains in $\mathbb{C}^n$.** Memoirs Am. Math. Soc. 472, American Mathematical Society, Providence, 1992.

[St] W. Stoll: *The characterization of strictly parabolic manifolds.* Ann. Sc. Norm. Sup. Pisa **7** (1980), 81–154.

[Su] M. Suzuki: *The intrinsic metrics on the domains in $\mathbb{C}^n$.* Math. Rep. Toyama Univ. **6** (1983), 143–177.

[V] E. Vesentini: *Complex geodesics.* Comp. Math. **44** (1981), 375–394.

[W1] B. Wong: *Characterization of the ball in* $\mathbb{C}^n$ *by its automorphism group.* Inv. Math. **67** (1977), 253–257.

[W2] B. Wong: *On the holomorphic sectional curvature of some intrinsic metrics.* Proc. Am. Math. Soc. **65** (1977), 57–61.

[Wo] P.-M. Wong: *Geometry of the homogeneous complex Monge-Ampère equation.* Inv. Math. **67** (1982), 261–274.

[Wu] H. Wu: *A remark on holomorphic sectional curvature.* Indiana Math. J. **22** (1973), 1103–1108.

List of symbols

Index

Vol. 1498: R. Lang, Spectral Theory of Random Schrödinger Operators. X, 125 pages. 1991.

Vol. 1499: K. Taira, Boundary Value Problems and Markov Processes. IX, 132 pages. 1991.

Vol. 1500: J.-P. Serre, Lie Algebras and Lie Groups. VII, 168 pages. 1992.

Vol. 1501: A. De Masi, E. Presutti, Mathematical Methods for Hydrodynamic Limits. IX, 196 pages. 1991.

Vol. 1502: C. Simpson, Asymptotic Behavior of Monodromy. V, 139 pages. 1991.

Vol. 1503: S. Shokranian, The Selberg-Arthur Trace Formula (Lectures by J. Arthur). VII, 97 pages. 1991.

Vol. 1504: J. Cheeger, M. Gromov, C. Okonek, P. Pansu, Geometric Topology: Recent Developments. Editors: P. de Bartolomeis, F. Tricerri. VII, 197 pages. 1991.

Vol. 1505: K. Kajitani, T. Nishitani, The Hyperbolic Cauchy Problem. VII, 168 pages. 1991.

Vol. 1506: A. Buium, Differential Algebraic Groups of Finite Dimension. XV, 145 pages. 1992.

Vol. 1507: K. Hulek, T. Peternell, M. Schneider, F.-O. Schreyer (Eds.), Complex Algebraic Varieties. Proceedings, 1990. VII, 179 pages. 1992.

Vol. 1508: M. Vuorinen (Ed.), Quasiconformal Space Mappings. A Collection of Surveys 1960-1990. IX, 148 pages. 1992.

Vol. 1509: J. Aguadé, M. Castellet, F. R. Cohen (Eds.), Algebraic Topology - Homotopy and Group Cohomology. Proceedings, 1990. X, 330 pages. 1992.

Vol. 1510: P. P. Kulish (Ed.), Quantum Groups. Proceedings, 1990. XII, 398 pages. 1992.

Vol. 1511: B. S. Yadav, D. Singh (Eds.), Functional Analysis and Operator Theory. Proceedings, 1990. VIII, 223 pages. 1992.

Vol. 1512: L. M. Adleman, M.-D. A. Huang, Primality Testing and Abelian Varieties Over Finite Fields. VII, 142 pages. 1992.

Vol. 1513: L. S. Block, W. A. Coppel, Dynamics in One Dimension. VIII, 249 pages. 1992.

Vol. 1514: U. Krengel, K. Richter, V. Warstat (Eds.), Ergodic Theory and Related Topics III, Proceedings, 1990. VIII, 236 pages. 1992.

Vol. 1515: E. Ballico, F. Catanese, C. Ciliberto (Eds.), Classification of Irregular Varieties. Proceedings, 1990. VII, 149 pages. 1992.

Vol. 1516: R. A. Lorentz, Multivariate Birkhoff Interpolation. IX, 192 pages. 1992.

Vol. 1517: K. Keimel, W. Roth, Ordered Cones and Approximation. VI, 134 pages. 1992.

Vol. 1518: H. Stichtenoth, M. A. Tsfasman (Eds.), Coding Theory and Algebraic Geometry. Proceedings, 1991. VIII, 223 pages. 1992.

Vol. 1519: M. W. Short, The Primitive Soluble Permutation Groups of Degree less than 256. IX, 145 pages. 1992.

Vol. 1520: Yu. G. Borisovich, Yu. E. Gliklikh (Eds.), Global Analysis – Studies and Applications V. VII, 284 pages. 1992.

Vol. 1521: S. Busenberg, B. Forte, H. K. Kuiken, Mathematical Modelling of Industrial Process. Bari, 1990. Editors: V. Capasso, A. Fasano. VII, 162 pages. 1992.

Vol. 1522: J.-M. Delort, F. B. I. Transformation. VII, 101 pages. 1992.

Vol. 1523: W. Xue, Rings with Morita Duality. X, 168 pages. 1992.

Vol. 1524: M. Coste, L. Mahé, M.-F. Roy (Eds.), Real Algebraic Geometry. Proceedings, 1991. VIII, 418 pages. 1992.

Vol. 1525: C. Casacuberta, M. Castellet (Eds.), Mathematical Research Today and Tomorrow. VII, 112 pages. 1992.

Vol. 1526: J. Azéma, P. A. Meyer, M. Yor (Eds.), Séminaire de Probabilités XXVI. X, 633 pages. 1992.

Vol. 1527: M. I. Freidlin, J.-F. Le Gall, Ecole d'Eté de Probabilités de Saint-Flour XX – 1990. Editor: P. L. Hennequin. VIII, 244 pages. 1992.

Vol. 1528: G. Isac, Complementarity Problems. VI, 297 pages. 1992.

Vol. 1529: J. van Neerven, The Adjoint of a Semigroup of Linear Operators. X, 195 pages. 1992.

Vol. 1530: J. G. Heywood, K. Masuda, R. Rautmann, S. A. Solonnikov (Eds.), The Navier-Stokes Equations II – Theory and Numerical Methods. IX, 322 pages. 1992.

Vol. 1531: M. Stoer, Design of Survivable Networks. IV, 206 pages. 1992.

Vol. 1532: J. F. Colombeau, Multiplication of Distributions. X, 184 pages. 1992.

Vol. 1533: P. Jipsen, H. Rose, Varieties of Lattices. X, 162 pages. 1992.

Vol. 1534: C. Greither, Cyclic Galois Extensions of Commutative Rings. X, 145 pages. 1992.

Vol. 1535: A. B. Evans, Orthomorphism Graphs of Groups. VIII, 114 pages. 1992.

Vol. 1536: M. K. Kwong, A. Zettl, Norm Inequalities for Derivatives and Differences. VII, 150 pages. 1992.

Vol. 1537: P. Fitzpatrick, M. Martelli, J. Mawhin, R. Nussbaum, Topological Methods for Ordinary Differential Equations. Montecatini Terme, 1991. Editors: M. Furi, P. Zecca. VII, 218 pages. 1993.

Vol. 1538: P.-A. Meyer, Quantum Probability for Probabilists. X, 287 pages. 1993.

Vol. 1539: M. Coornaert, A. Papadopoulos, Symbolic Dynamics and Hyperbolic Groups. VIII, 138 pages. 1993.

Vol. 1540: H. Komatsu (Ed.), Functional Analysis and Related Topics, 1991. Proceedings. XXI, 413 pages. 1993.

Vol. 1541: D. A. Dawson, B. Maisonneuve, J. Spencer, Ecole d' Eté de Probabilités de Saint-Flour XXI - 1991. Editor: P. L. Hennequin. VIII, 356 pages. 1993.

Vol. 1542: J.Fröhlich, Th.Kerler, Quantum Groups, Quantum Categories and Quantum Field Theory. VII, 431 pages. 1993.

Vol. 1543: A. L. Dontchev, T. Zolezzi, Well-Posed Optimization Problems. XII, 421 pages. 1993.

Vol. 1544: M.Schürmann, White Noise on Bialgebras. VII, 146 pages. 1993.

Vol. 1545: J. Morgan, K. O'Grady, Differential Topology of Complex Surfaces. VIII, 224 pages. 1993.

Vol. 1546: V. V. Kalashnikov, V. M. Zolotarev (Eds.), Stability Problems for Stochastic Models. Proceedings, 1991. VIII, 229 pages. 1993.

Vol. 1547: P. Harmand, D. Werner, W. Werner, M-ideals in Banach Spaces and Banach Algebras. VIII, 387 pages. 1993.

Vol. 1548: T. Urabe, Dynkin Graphs and Quadrilateral Singularities. VI, 233 pages. 1993.

Vol. 1549: G. Vainikko, Multidimensional Weakly Singular Integral Equations. XI, 159 pages. 1993.

Vol. 1550: A. A. Gonchar, E. B. Saff (Eds.), Methods of Approximation Theory in Complex Analysis and Mathematical Physics IV, 222 pages, 1993.

Vol. 1551: L. Arkeryd, P. L. Lions, P.A. Markowich, S.R. S. Varadhan. Nonequilibrium Problems in Many-Particle Systems. Montecatini, 1992. Editors: C. Cercignani, M. Pulvirenti. VII, 158 pages 1993.

Vol. 1552: J. Hilgert, K.-H. Neeb, Lie Semigroups and their Applications. XII, 315 pages. 1993.

Vol. 1553: J.-L- Colliot-Thélène, J. Kato, P. Vojta. Arithmetic Algebraic Geometry. Trento, 1991. Editor: E. Ballico. VII, 223 pages. 1993.

Vol. 1554: A. K. Lenstra, H. W. Lenstra, Jr. (Eds.), The Development of the Number Field Sieve. VIII, 131 pages. 1993.

Vol. 1555: O. Liess, Conical Refraction and Higher Microlocalization. X, 389 pages. 1993.

Vol. 1556: S. B. Kuksin, Nearly Integrable Infinite-Dimensional Hamiltonian Systems. XXVII, 101 pages. 1993.

Vol. 1557: J. Azéma, P. A. Meyer, M. Yor (Eds.), Séminaire de Probabilités XXVII. VI, 327 pages. 1993.

Vol. 1558: T. J. Bridges, J. E. Furter, Singularity Theory and Equivariant Symplectic Maps. VI, 226 pages. 1993.

Vol. 1559: V. G. Sprindžuk, Classical Diophantine Equations. XII, 228 pages. 1993.

Vol. 1560: T. Bartsch, Topological Methods for Variational Problems with Symmetries. X, 152 pages. 1993.

Vol. 1561: I. S. Molchanov, Limit Theorems for Unions of Random Closed Sets. X, 157 pages. 1993.

Vol. 1562: G. Harder, Eisensteinkohomologie und die Konstruktion gemischter Motive. XX, 184 pages. 1993.

Vol. 1563: E. Fabes, M. Fukushima, L. Gross, C. Kenig, M. Röckner, D. W. Stroock, Dirichlet Forms. Varenna, 1992. Editors: G. Dell'Antonio, U. Mosco. VII, 245 pages. 1993.

Vol. 1564: J. Jorgenson, S. Lang, Basic Analysis of Regularized Series and Products. IX, 122 pages. 1993.

Vol. 1565: L. Boutet de Monvel, C. De Concini, C. Procesi, P. Schapira, M. Vergne. D-modules, Representation Theory, and Quantum Groups. Venezia, 1992. Editors: G. Zampieri, A. D'Agnolo. VII, 217 pages. 1993.

Vol. 1566: B. Edixhoven, J.-H. Evertse (Eds.), Diophantine Approximation and Abelian Varieties. XIII, 127 pages. 1993.

Vol. 1567: R. L. Dobrushin, S. Kusuoka, Statistical Mechanics and Fractals. VII, 98 pages. 1993.

Vol. 1568: F. Weisz, Martingale Hardy Spaces and their Application in Fourier Analysis. VIII, 217 pages. 1994.

Vol. 1569: V. Totik, Weighted Approximation with Varying Weight. VI, 117 pages. 1994.

Vol. 1570: R. deLaubenfels, Existence Families, Functional Calculi and Evolution Equations. XV, 234 pages. 1994.

Vol. 1571: S. Yu. Pilyugin, The Space of Dynamical Systems with the C^0-Topology. X, 188 pages. 1994.

Vol. 1572: L. Göttsche, Hilbert Schemes of Zero-Dimensional Subschemes of Smooth Varieties. IX, 196 pages. 1994.

Vol. 1573: V. P. Havin, N. K. Nikolski (Eds.), Linear and Complex Analysis – Problem Book 3 – Part I. XXII, 489 pages. 1994.

Vol. 1574: V. P. Havin, N. K. Nikolski (Eds.), Linear and Complex Analysis – Problem Book 3 – Part II. XXII, 507 pages. 1994.

Vol. 1575: M. Mitrea, Clifford Wavelets, Singular Integrals, and Hardy Spaces. XI, 116 pages. 1994.

Vol. 1576: K. Kitahara, Spaces of Approximating Functions with Haar-Like Conditions. X, 110 pages. 1994.

Vol. 1577: N. Obata, White Noise Calculus and Fock Space. X, 183 pages. 1994.

Vol. 1578: J. Bernstein, V. Lunts, Equivariant Sheaves and Functors. V, 139 pages. 1994.

Vol. 1579: N. Kazamaki, Continuous Exponential Martingales and *BMO*. VII, 91 pages. 1994.

Vol. 1580: M. Milman, Extrapolation and Optimal Decompositions with Applications to Analysis. XI, 161 pages. 1994.

Vol. 1581: D. Bakry, R. D. Gill, S. A. Molchanov, Lectures on Probability Theory. Editor: P. Bernard. VIII, 420 pages. 1994.

Vol. 1582: W. Balser, From Divergent Power Series to Analytic Functions. X, 108 pages. 1994.

Vol. 1583: J. Azéma, P. A. Meyer, M. Yor (Eds.), Séminaire de Probabilités XXVIII. VI, 334 pages. 1994.

Vol. 1584: M. Brokate, N. Kenmochi, I. Müller, J. F. Rodriguez, C. Verdi, Phase Transitions and Hysteresis. Montecatini Terme, 1993. Editor: A. Visintin. VII. 291 pages. 1994.

Vol. 1585: G. Frey (Ed.), On Artin's Conjecture for Odd 2-dimensional Representations. VIII, 148 pages. 1994.

Vol. 1586: R. Nillsen, Difference Spaces and Invariant Linear Forms. 1994.

Vol. 1587: N. Xi, Representations of Affine Hecke Algebras. VIII, 137 pages. 1994.

Vol. 1588: C. Scheiderer, Real and Étale Cohomology. XXIV, 273 pages. 1994.

Vol. 1589: J. Bellissard, M. Degli Esposti, G. Forni, S. Graffi, S. Isola, J. N. Mather, Transition to Chaos in Classical and Quantum Mechanics. Montecatini, 1991. Editor: S. Graffi. VII, 192 pages. 1994.

Vol. 1590: P. M. Soardi, Potential Theory on Infinite Networks. VIII, 187 pages. 1994.